JOHN FERGUSON

AF873551

BIBLIOGRAPHICAL NOTES

ON

HISTORIES OF INVENTIONS

AND

BOOKS OF SECRETS

Elibron Classics
www.elibron.com

Elibron Classics series.

© 2005 Adamant Media Corporation.

ISBN 1-4212-6773-X (paperback)
ISBN 1-4212-6772-1 (hardcover)

This Elibron Classics Replica Edition is an unabridged facsimile
of the editions published in 1896-1914, Robert Maclehose and Co., Glasgow.

Elibron and Elibron Classics are trademarks of
Adamant Media Corporation. All rights reserved.

This book is an accurate reproduction of the original. Any marks, names, colophons, imprints, logos or other symbols or identifiers that appear on or in this book, except for those of Adamant Media Corporation and BookSurge, LLC, are used only for historical reference and accuracy and are not meant to designate origin or imply any sponsorship by or license from any third party.

BIBLIOGRAPHICAL NOTES

ON

HISTORIES OF INVENTIONS

AND

BOOKS OF SECRETS.

BY

JOHN FERGUSON, M.A.

PART I.

[*Reprinted from the Transactions of the Glasgow Archæological Society.*]

GLASGOW:

Printed at the University Press by

ROBERT MACLEHOSE AND CO.

1896

BIBLIOGRAPHICAL NOTES

ON

HISTORIES OF INVENTIONS

AND

BOOKS OF SECRETS.

Ne sis garrulus, sed pone ori tuo custodiam, et ut filius sapientum, margaritas ante porcos non projicies.

THOMAS AQVINAS, *Thesaurus Alchimiæ secretissimus.*

BY

JOHN FERGUSON, M.A.

[*Reprinted from the Transactions of the Glasgow Archæological Society.*]

GLASGOW:
Printed at the University Press by
ROBERT MACLEHOSE AND CO.
1896

100 copies reprinted in 8vo, 1882.
50 copies reprinted in 4to, 1896.

FROM TRANSACTIONS OF THE ARCHÆOLOGICAL SOCIETY OF GLASGOW.

NOTES ON SOME BOOKS OF TECHNICAL RECEIPTS, OR SO-CALLED "SECRETS,"

BY

JOHN FERGUSON, M.A.,

PROFESSOR OF CHEMISTRY IN THE UNIVERSITY OF GLASGOW.

[*Read at a Meeting of the Society held at Glasgow on 20th April, 1882.*]

THE following notes may serve to some extent as an introduction to a subject wide in itself, and with numerous and important connections. The history of practical invention and of technical progress is one which might well engage the attention of students of anthropology and antiquities, as it throws light on many points connected with the growth of social life and civilization. The desire and the power to turn external objects to his service and convenience are developed to such an extent in man, that, among the many differences between him and other animals, may be reckoned the various arts by which he induces nature to accommodate herself to his wants; among the lower animals one looks in vain for anything parallel to the arts of cookery, medicine, metallurgy—to the systematic use of tools, of clothing, of weapons.

In ancient times the various handicrafts were monopolies of certain families or castes; in the middle ages the handicraftsmen were too glad to pursue their callings in obscurity; it is only in the most recent years that arts and manufactures have acquired such paramount interest, that the special or technical education of those who are to exercise them has come to be thought of national importance. While, in the days of the Greeks and Romans, the artizan was a despicable if not an almost infamous person, and, in the middle ages, was oppressed by the military and ruling classes, against whom, nevertheless, he carried on a ceaseless struggle until he succeeded in asserting his importance, and even his equality with them, it has been reserved for the present day for ignorance of arts and manufactures, and indifference to their progress, to be as discreditable as they were formerly dignified. The history of the growth of the arts themselves, and of the attitude of society towards them, is, therefore, of wider extent, and of greater philosophical interest than at first sight appears. This history has not as yet been written and, as time goes on and material gathers, the more difficult it becomes. The only work in which the attempt has been made is the "*History of Inventions*," of Beckmann, written towards the end of last century.* This work, however, is less a history in the strict sense of the term, than a collection of antiquarian essays upon various objects of manufacture, and upon some technical questions. The essays, besides, are not arranged in any definite order, and have no direct connection with each other, but they are very elaborate, and show a wonderful amount of research and knowledge. Since Beckmann's time, I do not remember any laborious German who has followed up his work. There is certainly nothing on the subject in English.†

---

* Beckmann's work is entitled "*Beyträge zur Geschichte der Erfindungen*," and it was published at Leipzig, in five volumes, between 1786 and 1805. An English version and abstract appeared in 1823, in two volumes, and a new edition was published by Bohn in 1846, in two volumes.

† In writing the above I had forgotten some systematic works on the history of arts and manufactures which are more consecutive and philosophical but less thorough than Beckmann's collections. There is the work of Antoine Yves Goguet—"*Origine des loix, des arts, des sciences, et de leur progrès chez les anciens peuples*," published at Paris in 1758. It was translated into English, and there were two editions, of which that of 1775, in 3 vols., 8vo, is now before me. This book partakes largely of the character of a treatise on antiquities, but it embraces sections on the history of arts and manufactures among all the ancient peoples. It is a curious book, and shows familiarity with the classical writers, but it is of no authority now, after the research that has been expended, not only on almost all the topics that the author includes in his discussions, but also on the authorities to whom he owes his information.

A better and more specialized work is the "*Geschichte der Technologie*" of Johann H. M. Poppe, in three volumes, published 1807-11, and forming part of the Göttingen series of histories of science, arts, and philosophy. In the more recent Munich series of histories there is a "*Geschichte der Technologie*," by Karl Karmarsch, in one volume, 1872.

It is not my intention now to say anything about the progress of arts and manufactures at all, but only to bring under your notice a section of literature which is nearly ignored by bibliographers and antiquaries, and is altogether out of the ken of book-reprinting clubs.

It is hardly to be expected that a practical art can have any literature worth speaking of. The man who is busy practising it can have little time to write about it, and he who wishes to learn it must put to his hand and work at it, and that under the supervision of a master, and not by merely reading books. This is the apprenticeship that every one must serve. No amount of reading will make a sculptor, or a gardener, or a shoemaker, or a surgeon, or a musical executant. The arts must be acquired by practice, and they are extended and improved by practice. Every one who exercises them comes to have special power and certain ways of doing things, which may enable him to surpass others who are similarly engaged. These are his "secrets," which very often he cannot, or will not, reveal to others. Rapid insight into a particular case, power of overcoming physical obstacles, ingenious adaptations of means to ends, exhibition of due care at the right time, enable one man to effect what others cannot.

In earlier times artists were very chary indeed of telling their secrets, and in the great craze of the middle ages—the craze to make the philosopher's stone—the adepts were continually on their guard to conceal their art from the unworthy, while revealing what was thought suitable for the

---

Both of these works give good accounts of the growth of arts and manufactures, and contain numerous references to books and papers on them.

Quite recently I came across a little work entitled "*A pleasant and compendious History of Inventions*," London, 1686, 12mo. In it an attempt is made to give, in briefest outline, the origin of some important inventions. Defective as were both the plan and execution of this booklet, the author had a very clear notion of the importance of his subject, and of its general interest.

So far as I know, no complete and systematic work on the history of arts and manufactures has appeared in this country. A collection of essays by David Bremner on the Industries of Scotland was published at Edinburgh in 1869. They deal chiefly with the then state of the industries, although there are usually short historical narratives prefixed. The work entitled "*Manufacturing Arts in Ancient Times*," by James Napier, London, 1874, 8vo, is occupied almost entirely with the history of metals and of dyeing.

genuine artist to know. The philosopher was warned to admit no one to his laboratory—or to his confidence. Even at the present day, secrets have not wholly died out; there are manufactures which are still undivulged, and any one engaged in the scientific investigation of some phenomenon or law of nature, will not tell his professional brethren unreservedly what he expects to discover, before he has finished his labours.

It would seem, however, that in spite of the precautions of the older artists, their private ways of working, of producing substances, of making colours, and effecting all kinds of material changes, oozed out and became at last public property. But even after their publication, these methods and receipts retained, paradoxically, the name of "Secrets,"* and many collections of them appeared during the last three or four hundred years. It is beyond my power to give a complete list of these; my purpose at present, as I have already said, is to exhibit a few of those to which my attention happens to have been recently directed, and of which some at least are possessed of a certain amount of archæological and bibliographic interest.

This set of books divides into several groups, but perfect classification of them is impossible on account of the way the themes interweave.

I. There are collections of secrets of nature, or treatises on natural history, general science, and cosmogony. Of this set, Pliny's history seems to have been the model.

II. There are treatises on what was called natural magic (as distinguished from black magic or necromancy)—that is, the production of secret effects in optics, acoustics, magnetism, &c., &c., by natural causes. This is the form which natural philosophy originally took.

III. There are treatises which deal chiefly with chemical, pharmaceutical, and medical secrets.

IV. There are treatises on life and generation: physiological secrets.

V. There are treatises on technical or art secrets, strictly so called, and they may be arranged conveniently in two classes: general collections con-

* In the remarks which followed the reading of the paper it was pointed out that not so long ago chemical works were known, everywhere in Scotland at least, as "secret works." Some are still practically such.

taining receipts relating to a variety of arts, and special collections containing receipts of use in one art or handicraft only.

I have examples of each of these classes to exhibit.

The earliest medieval treatise on the subject of the practical arts to which I can refer at the present moment is that of Theophilus. It exists in MS. in several libraries and was first edited with translation, introduction, and notes by Robert Hendrie, in 1847.* The date is not exactly known, but the work seems to have been written in the eleventh century. It is in three books. The first treats of the materials used in painting and illuminating; the second of the making and colouring of glass; the third of metal working, bell making, organ building, lapidary work, colours. It is, therefore, a very important and interesting work for the history both of the sciences and practical arts. It mentions a number of substances and the manner of making them, which involved chemical skill, and it treats of arts, the results of which remain to this day the admiration and the despair of connoisseurs.

One of the biggest, if it be not the biggest, book written in the middle ages and printed in the fifteenth century, was the work of a Dominican monk, called Vincent de Beauvais.† It is a vast compilation or encyclopædia, a mirror of human knowledge as he called it—*Speculum quadruplex*—divided into four main parts, of which science and art form one. Of this huge work I have never encountered a copy, and have only seen a fragment about alchemy reprinted in a collection of such pieces. The author was born about 1190, and died about 1264. His labours, like those of his contemporaries, were chiefly devoted to philosophy—moral, metaphysical, and theological—and he engaged in the conflict then raging

* A French translation had appeared previously in 1843, edited by Count de l'Escalopier.

† For information—not much—about the author, an abstract of the bibliography of his work, and an outline of the contents of it, reference may be made to the article "Vincent de Beauvais" by Daunou, in *Histoire Littéraire de la France*, Paris, 1835, 4to, T. XVIII., pp. 449–519, and the numerous authorities there quoted. Through some oversight, Hain, while giving a cross reference, has omitted Vincent's name in the right alphabetical place in his *Repertorium Bibliographicum*. The Speculum appears to have been printed by Johann Mentelin at Strasburg, and finished by 1476. It is in 9 volumes folio: Speculum naturale, 2; morale, 2; doctrinale, 1; historiale, 4.

between the realists and nominalists, but he was able to turn from discussions on words and definitions, to the contemplation of external objects. The Speculum Naturale, or history of nature, is a commentary in thirty-two books, the text being the narrative of the creation as given in Genesis. This treatise deals less with arts than with cosmogony and natural history.

In this respect it differs, therefore, from the writings of two of his contemporaries, viz., Roger Bacon (1214-1292-4), and Albertus Magnus (1193 (?)-1280).

In his *Opus Majus*, Bacon has given an encyclopædia of human knowledge, and especially of physical science, but it is another work of his which falls to be mentioned in this place, namely, his "*Epistola de secretis operibus naturæ et artis, et de nullitate magiæ.*" This is a short report on the very wonderful effects that can be produced by natural means, and it is in this tract that occurs the often-quoted allusion to the composition and power of gunpowder. This letter was intended as a reply to the persistent attacks of malignant ecclesiastics who accused him of practising "black magic"*—by showing that the apparent mystery was due to the ignorance of his accusers and not to any compact of his with the infernal powers.

Albertus Magnus is credited with a treatise of secrets, but the authorship has been disputed and has been assigned to Henricus de Saxonia, one of his pupils, but whether this be correct or not the work usually, if not invariably, bears the name of Albertus. It deals with physiological secrets only, and that in a very brief manner; so that, both in range and in bulk, it is quite the reverse of an encyclopædia. Notwithstanding, if we may judge by the number of editions, it was one of the most popular treatises from the 15th to the 17th century.† Albertus was the author

---

*The black magician, or necromancer, was believed to have carried on a correspondence with the fiend and entered into a compact by which, on the liberation of the latter from his "prison-house," he would assist the magician in carrying out schemes which, in their expected results, could be called nothing else than supernatural. The correspondence with the magician's name at it, was usually brought up in evidence against him—very much to his annoyance—when he had to implement his share of the bargain.

†It is entitled "*De Secretis Mulierum*," and a list of the early editions is given by Hain (*Repert. Bibliogr.*, Nos. 549-568). In character it closely resembles the "*De phisio-*

likewise of a treatise on the secrets of plants, animals, and stones, in which he described their occult virtues.

In the following century flourished Bartholomew Glanville, a Cornishman, author of a ponderous work on the Properties of Things, divided into nineteen books. As it deals chiefly with natural history it hardly falls within consideration at present, but it gives incidentally technical descriptions. It was first printed in the fifteenth century, the earliest edition being of date 1478, and it was often re-issued, besides being translated into French, Dutch, and Spanish. It was also translated by John Trevisa in 1398 into English, and published by Wynkyn de Worde a century later.*

In the fifteenth century another book which had a considerable share of popularity was printed. It is entitled "*Lucidarius,*" and was written by a monk called Honorius of Strasburg. It first appeared in 1479, but the edition which I have here and which is worth examination for its uncommon type and curious woodcuts, was printed at Strasburg in 1499, by Mathijs Hupfuff. It is very rare, and though Hain mentions it (*Repert. Bibliogr.*, No. 8814), he had no actual copy for collation. He consequently says that the book has twenty-nine leaves, whereas this copy has thirty, the last containing a woodcut of the carrying of the cross. This work is a sort of catechism of natural and supernatural things. The questions are asked by the scholar, and the answers are given by the master, who thus imparts the required instruction in the secrets of creation.

All these works are of a general character, and, except the first, deal with the physical and natural sciences, as these were understood from the twelfth to the fifteenth century—they are examples chiefly of the first class. That some of them were among the books first printed in the fifteenth

---

nomia" of Michael the Scot, and the two tracts were frequently printed together during the 17th century. It was translated into French and German. An edition of the French (Cologne, 1722), entitled "*Les admirables secrets d'Albert le Grand,*" contains, besides the two tracts of Albertus, a third consisting of practical receipts, and a fourth on physiognomy.

*Johnson's *Typographia*, London, 1824, I., p. 354. There is a copy of the Latin edition printed in 1480, in the Euing Collection, Glasgow University Library. It is in folio, in double columns, printed in fine Gothic character.

century and went through several editions and translations, notwithstanding their bulk in certain cases and their frequently absurd contents, shows that even then there were many people anxious to know something about nature and external objects.

The sixteenth century produced no great encyclopædia like some of those I have mentioned. Either the breed of encyclopædists had become extinct, or else knowledge had grown too great to be gathered by one man in his lifetime and put in a book, but we find several less ambitious authors dealing with different kinds of secrets, some of nature, others of arts.

First among those whose works I have here comes Levinus Lemnius, who was born at Zirickzee in Holland, in the year 1505, and after studying at various places became a physician in his native town. Between the years 1559-64 he published a work entitled "*De Miraculis Occultis Naturæ, Libri IV.*" Of this book several editions appeared, of which I have here two, Franckfurt 1604, and 1611, both in 16mo. It was translated into English under the title of "*The Secret Miracles of Nature, in four books,*" London, 1658, in small folio, of which there is a copy here. It was translated into French by the alchemist Gohory and published at Paris [Orleans] in 1568, and it was translated also into Italian.

It is a most heterogeneous collection, heterogeneously piled together, of notions on physiological, physical, medical, religious, and moral topics, with attempts to explain phenomena in nature which subsequent enquiry has shown do not exist at all. The collection is a very curious one, notwithstanding, and furnishes good instances of popular ideas about natural things current three hundred years ago. It would be difficult to bring this collection under any of the classes above mentioned—the only thing tolerably certain is that it contains hardly any practical receipts.

Contemporary with Lemnius, but ten years his junior, was Conrad Gesner, who flourished from 1516 to 1565. One might spend much time over the works of this really distinguished man, who was called the German Pliny, on account of his comprehensive learning. Besides his writings on animals, plants, and minerals, on languages, pharmacy, natural philosophy, and so on, he was one of the first bibliographers and book cataloguers, his "*Bibliotheca Universalis*" being the best and most complete catalogue which appeared in the sixteenth century, and being still a valuable book of reference.

Under the pseudonym of Evonymus Philiatros, he published a book entitled "*Thesaurus de remediis secretis*"—a treasury of secret remedies. It appeared at Zurich in 1554, and in the course of a few years went through numerous editions, and was translated into English, French, and German.

I have here examples of all these:—

1. Latin.—An edition in 16mo, Lyons, 1555, by Balthazar Arnollet, and another without date, printer's name, or place, but as the device on the title page is a tree with five frogs,* doubtless this edition was printed by Froschover at Zurich. Neither of these editions is mentioned by the bibliographers.

2. German.—The German translation appeared at Zurich in 1555, and other editions were published in 1582, 1583, and 1608. Of this last there is a copy here.

3. French.—Like the German, the French translation was executed immediately after the Latin appeared. It was published at Lyons in 1555 in 4to, again in 1557 in 8vo, and in 1559, in 8vo, by Antoine Vincent. I have not observed any reference to the 1559 Lyons edition. As will be seen by the copy here, it is a very prettily printed book with nice woodcuts of herbs and chemical apparatus.

4. English.—It was translated into English by Peter Morwyng, and published by John Day, at London, in 1559. Another edition appeared in 1565, which corresponds page for page with the earlier one, and ten years later, in 1575, a third edition came out. The copy here is of the 1565 edition. It is handsomely printed in bold black letter, and is illustrated by woodcuts of plants and apparatus for distillation. The main purpose indeed of the book is to describe the way of preparing remedies from plants by that method.

In 1569, after Gesner's death, a second part of the treasure was edited

* A more elaborate form of this device will be found on the title page of Gesner's "*Bibliotheca*," ed. Simler, 1582, printed by Froschover. The device is a punning one.

by Caspar Wolff or Wulphius. It became nearly as popular as the previous part, and was translated into French and English.

1. Latin.—It was first printed at Zurich in 1569, but I have no copy to exhibit.

2. The French version was executed by Jean Liebaut, and appeared under his name at Lyons in 1593, in 8vo. Thereafter at Rouen in 1628, and 1643, of which edition a copy is here. I have here also another published at Rouen with the date MVIC, which would appear to mean 1600, and it would therefore be much earlier than the other, but against this date is the fact that it has not the look of a book printed in 1600. Besides it is called on the title page *Derniere Edition*, so that one would suppose it to be subsequent to those above mentioned.

3. The second part was Englished by George Baker, surgeon to Queen Elizabeth, and two editions of it were published, of both of which copies are before us. The first bears date 1576 and is entitled the "*Newe Jewell of Health.*" It was published in London by Henry Denman, in small 4to, and printed in black letter. It is illustrated with numerous woodcuts of apparatus and operations. The second edition appeared in 1599 under the title of "*The Practice of the New and Old Phisicke.*" It was published at London by Peter Short, and is also in black letter.

It will be noticed that this work in particular contains the pharmaceutical and to some extent the chemical knowledge of the time of Shakespeare. The copy of the first part which is here was printed the year after he was born, and it is very likely that the pictures of plants and chemical operations which it contains would not escape his notice even while still a child. The examination of these books gives one some insight into the references which occur in the plays to the physician's art and the works of the apothecary, who was not always then, or for long after, distinguished with sufficient accuracy from the poisoner.

Gesner's work is specially concerned with medical and pharmaceutical secrets, and does not take up either natural history and physical, or trade and technical secrets.

A work more representative of technical art than any of the preceding was the "*Pirotechnia*" of Biringuccio, which was published at Venice in 1540. It deals particularly with the extracting of metals from their ores, their fusion, casting, calcining, and conversion into various compounds, the preparation of salts, the distilling of acids, the founding of cannon, the manufacture of gunpowder and of fireworks. Of the several editions which appeared I have here the first edition of 1540, in small quarto, and the 8vo edition of 1559, also printed at Venice.* The work was translated into French and published at Paris in 1572, of which translation there is a copy here. Copies both of the 1540 Italian and the 1572 French editions are uncommon.

In the compilation of books of secrets the Italians in the sixteenth century showed considerable activity, and the examples of them that are here may be taken together in chronological order.

The earliest of them is ascribed to Don Alessio Ruscelli, a Piedmontese, whose reputed work was first printed at Basel in 1536, in 8vo. It went through a very great number of editions, but although it was one of the most popular of the collections of receipts or secrets, copies of it are now quite unusual in catalogues. The editions were in all likelihood actually worn out by use. The copy here is a comparatively early one, having been printed at Venice in 1568. This collection was translated into English, and the four different parts appeared in succession, the first in 1562, the last in 1578. According to Watt, who gives the titles at length and enumerates different editions,† the secrets appeared in every European language. He adds that an abridgment of it was long a popular book at the foreign fairs, and Nisard mentions a book which consists of extracts from Albertus Parvus, Cornelius Agrippa, and others, but which he thinks is chiefly a rehabilitation of the work of Alexis.‡ This collection, therefore, is still publishing and selling in France by the pedlars, and flying stationers, as they used to be called. The editions Nisard mentions are of 1837 and 1839. It is hardly necessary

---

*I have since got a copy of the second (?) edition, printed at Venice in 1550, 4to.

†An edition of 1595, London, Peter Short, is not mentioned by Watt. It is in small quarto, black letter, and resembles the 1599 edition of the second part of Gesner's work. There is a copy in the Euing Collection, Glasgow University Library.

‡Ch. Nisard, "*Histoire des Livres Populaires,*" Paris, 1854, I., p. 225.

to say that however creditable the Don's compilation may have been to the sixteenth century, it gives one but a poor idea of the progress of true physical and medical knowledge among the mass of the people in the nineteenth, that such books can be sold for actual perusal and reference.

A similar collection to the preceding was made by Gabriello Falloppio, celebrated as an anatomist, who lived between 1523 and 1563. The work is entitled "*Secreti Diversi*," and it appeared after his death in 1566. There is a copy of it here. It contains receipts for preparing different bodies to be used in medicine, for the production of wines, alcoholic extracts of plants, cosmetics and waters. It also explains the chemical treatment of the metals, their alloys, the way of changing their colours, converting them into different kinds of salts and so on. There is no English version of this, so far as I know, but there was a Latin edition, and one in German, Franckfurt, 1641, of which there is a copy here.

Two years later, in 1568, there appeared at Venice another collection of secrets. The author or compiler was Leonardo Fioravanti, a physician of Bologna, who died in 1588. His collection is arranged very much in the same way as Falloppio's and it contains not only secrets of medicine and surgery with the necessary preparation of drugs and remedies, but receipts for several technical purposes as well. Besides the Italian edition of 1571, Venice, 8vo, I have here the German translation of Darmstadt, 1624. Of this work an English translation was made by John Hester, which was first published at London in 1582, in 16mo. It was afterwards reprinted with some other translations by the same hand, and appeared in small quarto in 1652. Of this edition I possess a copy. The translation differs in several details from the Italian, and it embraces only the medical section of the original.

A much better known man than any of these published a collection of curious arts at Naples in 1558. This was the Neapolitan, Giambattista Porta, who lived between 1538 and 1615, made long journeys in search of natural knowledge, and formed an Academy of the Secrets of Nature in his house at Rome, which was suppressed of course.

His work is entitled "*Magia Naturalis*," and it is divided into twenty books, according to subjects. This is a more comprehensive work than some of its predecessors, but I cannot say that in its contents it is much more sensible. It had its share of popularity, however, passed through many Latin editions, and was translated into all the languages. Besides three of the Latin editions (Franckfurt, 1591, Leyden, 1644, and Amsterdam, 1664), I have here a copy of the scarce English translation of 1658, with the still scarcer frontispiece, which contains a portrait of the author, and a representation of the four elements, and of Art and Nature, disposed in compartments. Among the curiosities contained in the first edition of this book, 1558, is an account of the camera obscura as it was known—without the lens—to Leonardo da Vinci. In the 1589 edition it is described with the lens, but there is no proof of what has been stated, that the instrument was either invented or improved by Porta. On the whole, the optical division is one of the best in the Magia Naturalis.

The last of the Italian collections I have to show is that of Leonardo Locatelli, a physician, like most of the older naturalists. The work is entitled "*Theatro d'Arcani,*" and it deals chiefly with chemical and alchemical changes and products. I have here the edition of Milan, 1644, and that of Venice, 1667. So far as I know there was no translation published.

The preceding are almost all mixed collections, but in the sixteenth century there appeared a series of small books, of purely technical character, about which I have failed in getting any information. The only thing to be done therefore is to enumerate and describe the books themselves, which are now before us.

The oldest is entitled "*Kunstbüchlin,*" and it was printed at Augsburg in 1537.* This is a collection of receipts, pure and simple, intended for practical workpeople. The topics are the working of metals, the making of colours, the dyeing and colouring of various objects, the calcining of the metals, and such like.

In 1549 there was published at Amsterdam a small volume, entitled "*Kunst Boeck,*" which I have not seen referred to in the bibliographies. It is said to be compiled and in part translated—*ghecolligeert ende eensdeels*

*Graesse, *Trésor de Livres Rares*, Dresden, 1863, iv., p. 53, quotes this same book, but gives the date, 1538. Other later editions are mentioned.

getraslateert—by Symon Andree. The translated sections are from the "*Künstbüchlin.*" In 1581 the translated part of the preceding was republished, with a second tract containing different receipts from those in the earlier one, and in 1600 there appeared at Amsterdam an edition of all the parts together in Dutch. In 1687 finally a modernized reprint of the "*Künstbüchlin*" appeared at Franckfurt, in 12mo. These collections represent fairly the kind of receipts for practical purposes current in the sixteenth and seventeenth centuries.

Johann Jakob Wecker, a physician at Basle, added another compilation to the preceding. It appeared in 1582, and several editions were published. That which I have here was printed at Basle so late as 1701, and is entitled "*De Secretis Libri xvii.*" It is arranged in a most systematic manner, and in this respect is superior to all the others; but one cannot give much praise to the information which the book contains. An English version was printed in folio, in 1660.

The seventeenth century produced a large number of books of secrets, of which there are one or two here.

A work somewhat similar to Wecker's, but of a more philosophical cast, was written by Heinrich Nollius, and published at Franckfurt in 1619. It is entitled "*Naturæ Sanctuarium: quod est Physica Hermetica . . . methodo perspicua et admirandorum secretorum in Naturæ abysso latentium philosophica explicatione decenter in undecim libris tractata.*" This work contains a discussion on general physical ideas and principles from the standpoint of the alchemical or hermetical philosophy then in vogue, and, as is said on the title, deals with the hid secrets of nature rather than of practical science and art.

Works treating more strictly of natural phenomena and practical physics were compiled by Casper Schott, a Jesuit, and Professor at Würzburg. Reference may be made to two of his works. "*Physica Curiosa, sive Mirabilia Naturæ et Artis Libris xii. Comprehensa,*" printed at Würzburg in 1667, in 4to. This treatise is chiefly on Natural History, and contains a great many drawings, of which those depicting different monsters, both of animals and human beings, show the greatest amount of ingenuity and originality on the part of the artist. The boy with the elephant's head, *puer capite elephantino*; the monstrous cock, with hoggers, and a tail like a cow's, set apparently in a metal socket; the *infans cum promuscide et capitibus animalium*; the *vitulomonachus*, or bull-calf monk; the *equus cute lacerâ*, or horse with the slashed doublet, and numerous others, surpass anything in the way of development devised by recent naturalists.

Ten years later, in 1677, there appeared at Bamberg, Schott's "*Magia Universalis Naturæ et Artis*," in four volumes, 4to. This is a treatise on old Natural Philosophy, and of the arts thereon depending, for instance, under acoustics the author describes organ building. This work is very copiously illustrated, and the general excellence of the drawings of apparatus and experiments is noticeable. The artist has been a much more matter of fact person than the other.

About the same time, but without place or date, there appeared a volume in 4to, entitled "*Joco-Seriorum Naturæ Centuriæ* 4." It bears the name of Caramuelius as the author, but in all probability this is a pseudonym, and the real author was Schott.* It is a collection of curious things to be effected by physical causes, merely for the sake of amusement, and without any practical purpose in view.

In English two technical books were published that had some reputation. Gabriel Plattes' "*Discovery of Subterraneall Treasure, viz., of all manner of mines and mineralls, from the gold to the coal*," appeared at London in 1639. This, as its name indicates, is concerned chiefly with mining and metallurgy, but there is a section at the end devoted to the colours that can be extracted from vegetables and the way of dyeing in fast colours. The other treatise is Sir Hugh Plat's "*The Jewel House of Art and Nature*," London, 1653. This a very mixed collection, as it includes receipts and descriptions from all the sciences. Though not entitled "secrets," these two tracts belong to the category.

In France a certain Mdlle. Marie Meurdrac published a little work

---

* Some ascribe it to Athanasius Kircher, but that may be because a tract by him is added at the end. Internal evidence is in favour of Schott. For example, on page 272, reference is made to "our cryptography in the first book of part four of the Magia," and this reference corresponds with the Bamberg edition, iv., p. 1. As to the date the chronogram with which the preface concludes would seem to point to 1661.

with the quaint title, "*Compassionate and Simple Chemistry written for meritorious ladies.*" Of this there is a German translation by Johann Lange, Franckfurt, 1676. The collection consists chiefly of pharmaceutical and, as was to be expected, cosmetical receipts. The subject is arranged in six books, and the last is devoted specially to the preparation of those substances with which ladies at that time were wont to preserve their beauty.*

During the course of the seventeenth century some arts had so far advanced that treatises dealing with them alone began to appear. This is notably the case with metallurgy, an art which has always occupied a prominent place and has a considerable literature. Glass-making also in the seventeenth century can boast of a special literature. The German chemist Kunckel published a book on the subject at Franckfurt in 1679: a Florentine, Antonio Neri, another, of which I have the late 1686 Amsterdam edition. Of this book Christopher Merrett issued an English translation in 1662; lastly, Blancourt wrote a treatise in French of which an English translation was published in 1699 at London, under the title; "*The Art of Glass. Showing how to make all sorts of Glass Crystal and Enamel. . . . A work containing many Secrets and Curiosities never before discovered.*" †

Another special technical treatise was that of Pietro Maria Canepario—"*De Atramentis,*" which, though dealing professedly with ink and black colours, is much more comprehensive than its name denotes, as it includes the technology of pyrites, of vitriol, of the oil of vitriol, and of different colours. I have here two editions, the first published at Venice in 1619, and the second at London in 1660.

It must not be forgotten, however, that parallel with these empirical

---

* In books of receipts cosmetics occupy sometimes a considerable space, but there are, besides, works specially devoted to the subject. Amongst these I may mention an exceedingly scarce work by Jean Liebaut, to whom as translator of Gesner's *Thesaurus* reference has been made above. It is entitled "*Trois Livres de l'Embellissement et Ornement du corps humain.*" Paris, 1582. 8vo. pp. 16—464—16.

† The most recent book on the subject which I have seen is by the French chemist Peligot. "*Le Verre. Son Historie, sa Fabrication.*" Paris, 1877. 8vo. A popular work, also from the French, was published at London in 1870, "*Marvels of Glass-making in all ages,*" by A. Sauzay. 8vo.

books, progress was making, though slowly, in those sciences on which the successful and sound development of their applications to the arts depends. In chemistry, and especially in pharmaceutical chemistry, that is in the preparation of natural substances for use in medicine, which occupies so large a place in some of these receipt books, there was considerable advance; and, in addition, a very great impulse to technical and applied chemistry was given towards the end of the seventeenth century by Glauber and by Becher.

But, notwithstanding, the books of secrets in the eighteenth century in many cases showed a retrogression towards credulity and absurdity, for this century has a broad shadow of ignorance and superstition athwart its vaunted enlightenment. It was this ultra-rational age which encouraged such books as the "*Secrets Merveilleux de la Magie Naturelle et Cabalistique du Petit Albert,*" and a host of similar would-be magical and cabalistical books. They are of interest only as studies in the aberration of the human mind. But even in the works of men who ought to have known better we encounter the most ridiculous and gross ideas. The French chemist Lemery made a collection of receipts which was translated into English and published at London in 1711. Its title is "*New Curiosities in Art and Nature: or a Collection of the most valuable secrets in all Arts and Sciences.*" If any one is desirous of seeing the puerile credulity which, in the same age and country, may run parallel with an equally contemptible scepticism, let him turn to the pages of this precious compilation. Some of the collections, however, are a little more practical and more rational than the preceding, but in all of them there is a conspicuous empiricism.

A few years later, in 1723, Dr. William Salmon, a great compiler of books, issued a work containing "*Choice experiments and observations on Building, Husbandry, Gardening, Mechanics, Chemistry, Painting, &c.*" London, 8vo. Of this I have no copy, but I have here another similar and earlier work by the same author. It is entitled "*Polygraphice, or the Arts of Drawing, Engraving, Etching, Limning, Painting, Varnishing, Japanning, Gilding, &c. . . . Eighth edition.*" London, 1701. Though professing to deal with the arts of design it is far more comprehensive and includes the arts of dyeing and staining, alchemy, chemistry, chiromancy, arts of beautifying and perfuming, and so on.

I possess also two small treatises; one is without date, but belongs obviously to the last century. It is entitled "*Arts Treasury of Rarities: and choice Inventions. . . . The fifth edition. London . . . G. Conyers. . . . Price* 1*s*." This is really a book of practical receipts, including such subjects as the dyeing of cloth, silk, hair, bone, leather; gilding, lacquering; removing spots of tar, grease, oil; preserving from moths; cementing broken glass, and so on. The other is entitled "*La Magie des Artistes . . . ou collettion complette des secrets utiles. . . . Harlem. . . .* 1783." 12mo. This work is similar in contents to the preceding, and both are free from the absurdities already spoken of.

With the progress of the sciences and with insight into the causes of chemical and physical change, books of secrets like the preceding can no longer show any reason for existence. If, therefore, they circulate at all it is as chap books, the hawkers' reprints of Aristotle, Albertus Magnus, &c., already referred to, which have little interest from a bibliographical, and none from a scientific point of view, or else as collections of trivial receipts which are of no practical use. Any one wishing sound information on a practical matter will now go to such a work as Cooley's Cyclopædia of Practical Receipts, or Ure's Dictionary of the Arts, or to some of the numerous works on applied science and technology in which the rationale of the processes, or at least rational processess are given.* Arts and manufactures are passing more and more from the empirical to the scientific state, and to appreciate the immeasurable improvement that takes place when a manufacture is based on a rational foundation and not on mere empiricism and what is called familiarly "rule of thumb," a comparison may be made, say of Canepario's account of sulphuric acid making with the treatise by Lunge on the same subject, or the tiny chapters on iron

* One still more recent is now to be had. Its title is "*Spon's Encyclopædia of Industrial Arts, Manufactures, and Raw Commercial Products*," and it has just been published in London in two volumes. The author, editor, publisher, or whoever is responsible for the advertisement of it, makes a special merit of its containing a minimum of scientific and historical details.

smelting in Agricola or Ercker, with Lowthian-Bell's Studies of Blast Furnace Phenomena or Percy's Treatise on Iron.

On another occasion I hope to communicate to the Society extracts from some of the books now exhibited, and to supplement the list with notes on other works of a similar kind, which I have not in the meantime at hand.

Postscript.—I have just become aware of the existence of a "*History of Inventions*," by F. S. White; but of the book I, as yet, know nothing.

BIBLIOGRAPHICAL NOTES

ON

HISTORIES OF INVENTIONS

AND

BOOKS OF SECRETS

SUPPLEMENTS I.-VI.

READ TO THE ARCHÆOLOGICAL SOCIETY OF GLASGOW
MARCH 1894—FEBRUARY 1908

BY

JOHN FERGUSON, M.A., LL.D.,

F.S.A. LOND. AND SCOT., ETC.

MEMBER OF THE GERMAN SOCIETY FOR THE HISTORY OF MEDICINE
AND THE NATURAL SCIENCES

REGIUS PROFESSOR OF CHEMISTRY IN THE UNIVERSITY OF GLASGOW

GLASGOW

Printed at the University Press by

ROBERT MACLEHOSE & CO. LTD.

1910

BIBLIOGRAPHICAL NOTES

ON

HISTORIES OF INVENTIONS

AND

BOOKS OF SECRETS

FIRST SUPPLEMENT

BY

JOHN FERGUSON, LL.D., F.S.A. LOND. AND SCOT.
REGIUS PROFESSOR OF CHEMISTRY IN THE UNIVERSITY OF GLASGOW
PRESIDENT OF THE PHILOSOPHICAL SOCIETY
AND OF THE ARCHÆOLOGICAL SOCIETY OF GLASGOW

GLASGOW
STRATHERN & FREEMAN, 145 WEST NILE STREET
MDCCCXCIV

Read to the Archæological Society of Glasgow, 15th March, 1894.

100 Copies Reprinted.

BIBLIOGRAPHICAL NOTES ON HISTORIES OF INVENTIONS AND BOOKS OF SECRETS.

FIRST SUPPLEMENT.

BY

JOHN FERGUSON, LL.D., F.S.A. LOND. AND SCOT.,

REGIUS PROFESSOR OF CHEMISTRY IN THE UNIVERSITY OF GLASGOW,

President of the Society.

1. When to the title of the sixth part of my "Bibliographical Notes on Books of Secrets," which was read to the Society in January, 1888, and was printed in 1890,[1] I appended the word "Conclusion," I knew, and stated categorically, that it was an arbitrary stop which I was making. Strictly speaking, I had come only to the end of the books which I had been able to inspect, not to the end of those which existed; still less had I exhausted all that there was to say even about those which had been enumerated. My opinion, however, was that these papers contained enough to indicate generally the character and the extent of the early literature of the arts and of technology, and I was content to let my consideration of it end there; but, during the last four or five years, I have found that the subject would not rest, but kept itself in evidence whether I wished it or not. Books, which I had not heard of previously, have come unexpectedly into my hands; editions, which I knew of by repute but had not seen, have thrust themselves upon my attention; copies of books mentioned in my earliest papers, a dozen years ago, have emerged from their hiding places after I had given up all thought of them; of certain books which I was fain to revere, as nearly, if not altogether unique, other copies have risen up to

[1] *Transactions*, N.S., Vol. II., p. 1.

spoil my descriptions and chill my exultation. One result hitherto of my experience in Bibliography is to make me deny the existence of a unique book. There may be many so considered, but their single-blessedness cannot be demonstrated, and I take leave to doubt it, as a cardinal principle of the science. The duty of every one, whenever such a statement is made, is to produce evidence—in the form of at least one other copy of the unique book—that it is incorrect.

2. When I have contemplated the amount of material which has thus accumulated, I have found myself unable to ignore it. To do so would be to leave the first research not so complete as it is in my power to make it, and I have thought, therefore, that an account of the books which I have acquired within the last half-dozen years is a suitable, if it be not a necessary supplement to the original series. To obtain a proper survey of this material, however, I have been forced to deal with it in two ways. So much has been amassed relative to certain authors and to certain books, that the only suitable method of exhausting it all seems to be to devote a special monograph to such of them as deserve it. The majority of the books, however, which I have recently collected, do not require this elaborate treatment, and I have thought that these might be disposed of now in the manner pursued in my previous papers. This is all the more appropriate, as the larger proportion of the works to be noticed consists of different editions of treatises already referred to.

3. The collection of as many editions of a book as possible, even when there is no variation in the text or contents, will not be regarded with much, if any, favour by those who think that books have lost all their value as soon as a new edition appears, and that old editions should make way for new as rapidly as possible to supply the wants of "The Student," "The General Reader," and "The Reading Public." The holders of such views may not be aware that these phrases, wide as they are, do not include all those who make use of books. There are antiquaries, bibliographers, printers, students, who require to consult early editions, and those who are not "general" but "special" readers, who, from such a collection, may gather information on a variety of topics, not only connected with their own subjects, or with the author, but with the history of the book itself, and with the history of the

science or of the stage of mental development of the time when it was in use, or with other subjects which need not be particuiarised at present. Only this may be said, that the librarian who gets rid of edition *n* as of no further use, because he has got edition $n+1$, knows neither his business nor his duty.

Whatever view, however, be taken of this matter as to its intrinsic importance, the fact remains that different editions of certain books do exist, and in the present case I have no alternative but to record as many of them as it has been possible for me to become acquainted with. My aim in these papers is not to ignore the wide circulation and active life of a treatise on the arts, or to suppress the fact of its numerous editions. It is just the reverse: to give the fullest account of it that is possible for me to do, to enumerate all the editions of as many works as I can inspect. The intention is to demonstrate the extent and amount of the early literature on the arts and sciences coming under a certain appellation, and thereby the wide diffusion of interest in these topics themselves.

4. The absence from the list of any notable number of quite new books of "Secrets" under any category, and the presence of editions of so many works already enumerated, indicates, I think, that my previous lists are fairly full in the matter of separate treatises, however far they may be from comprehending all editions. I have no doubt, however, that there are still collections of "Secrets" of medicine and the arts which have escaped me, but there cannot be so many as when I first took the subject up. Of some of the books, too, there are editions which I have not seen, even in large libraries, and which must, therefore, be of very rare occurrence. Some, even, of the books to be described now are of very great rarity indeed. Specially so is the first edition of the second part of the *Treasure* of Evonymus, edited by Wulphius, for the copy described in this supplement is the only one I have come across in the last dozen years. Some of the editions of Polydore Vergil's *History of Inventors*, of Alexis' *Secrets*, some of the works on the Powder of Sympathy, the little tract of André le Fournier, and several others are extremely scarce and very fascinating on that account. They display the knowledge of the time; they give a clue to what were the topics of interest; the number of editions shows what authors were most appreciated, and were thought to have

dealt best with their subjects. To enumerate all the rare books, however, would amount almost to a recapitulation of the whole paper; there is hardly a book mentioned in the following to which the epithet cannot be attached.

As in the original papers, the subject in this supplement is arranged in two divisions—(1) Histories of Inventions; (2) Books of Secrets. The books are arranged chronologically by the dates of their authors, so far as I have been able to ascertain them. This has the advantage of displaying the books on various subjects which were contemporaneous, and thereby the state of knowledge.[1]

I.—Histories of Inventions.

5. First in the list of histories comes, of course, Polydore Vergil's book *De Inventoribus Rerum.* The investigations which I was led to undertake with reference to this book became so extensive that I was forced to devote a monograph to it. The first draft of this was communicated to the Society of Antiquaries of London on the 18th of June, 1891. The paper, however, was so much altered subsequently by additions that it has been necessary to treat it as a distinct work. The complete bibliography does away now with the necessity even of a summary, nevertheless I may mention the dates of the different editions which have come into my possession since my last reference to the book was printed in 1889. In the following, the second column and all those marked with an asterisk in the first column are in my own collection.

---

[1] To save constant repetition, references will be made to the six parts of the original series of Notes, and not to the volumes of the Transactions of the Society. The following shows the volumes in which the different parts appear:—

Part I.—	*Transactions*,	. . .	Vol. II.	pp. 180-197.
II.—	,,	. . .	Vol. II.	pp. 229 272.
III.—	,,	*New Series*,	Vol. I.	pp. 188-227.
IV.—	,,	. . .	,,	pp. 301-336.
V.—	,,	. . .	,,	pp. 419-460.
VI.—	,,	. . .	Vol. II.	pp. 1-33.

List of Editions down to 1889.		Editions acquired since 1889.
* 1499.		
1503.		
		1507.
* 1509.		
Senant's Edition.		
* 1513.		
		1516.
		1521.
		1525.
* 1528-29.		
		1532.
		1540.
		1543.
1544.		
* 1546 April.	London.	1546 Gryphius.
1546 April—January.	R. Grafton.	Han's Edition.
* 1546 January.		
* 1546 Basil.		
* 1550 Venice, Italian.		1550 Basil.
* 1551.		
		1558.
		1560.
* 1563.		
* [1570.]		
		1575.
		1576 Lyons, French.
		1576 Rome.
		1582.
* 1585.		
		1587.
		1590.
		1599.
1604.		
* 1606.		
1626.		
* 1644.		
* 1659.		
* 1663.		
* 1671.		
* 1680.		
1868.		

The copies in the second list are rare, and two or three may be called unique, in so far as they are the only copies known at present, and are referred to nowhere else. The majority are omitted by the bibliographers, and are not enumerated by Beckmann in his valuable catalogue.[1] In the second list it is curious that the date of every copy is prior to 1600. In the first list the majority of dates are also before 1600. This accords with my observation that the later editions are less common, and that those of last century are the hardest to get. Of certain of them I have never been able to see a copy in any collection. This may be accounted for in various ways. One may suppose that of the earlier editions large numbers were printed, while of the later there were only limited issues. On the other hand, the book may not have been latterly in request, and the copies were destroyed wholesale; if so, then of certain editions not one appears to have escaped.

My own is the largest collection I know of in the meantime, containing, as it does, 37 numbers, besides duplicates of the editions of 1499, 1529, 1546, (1570), 1644, 1663, and some others. The British Museum contains 32; the Bibliothèque Nationale 20. Other libraries contain two or three, or may get the length of half a dozen, or even of a dozen. All these together, however, represent but a part of the subject, for there is on record upwards of a hundred different editions: of which some 75 have been described from actual copies in my monograph referred to. The remaining fourth I have not seen. A full account of all these editions, including an analysis of the places where and dates when they were printed, and a consideration of a number of questions relating to the book's history, presents a considerable amount of interest to the bibliographer. Everything conspires to show that Polydore Vergil's book was a popular and prominent one.

6. The following little tract, which is in Sir William Hamilton's collection in the University Library, was overlooked by me in my former notices, though it is mentioned by Beckmann and described in considerable detail.

Iohannis Matthæi | Lvnensis | Libellvs | De Rervm In- | ventoribvs | Ex recognitione Aug. Iuſtiniani | Epiſcopi Nebienſis. | M. Antonii Sabellici | De Rervm Et Artivm | inventoribus Poëma. | * | Hambvrgi, | In Bibliopolio Michaelis Heringii. | Anno cIↃ IↃc xiii.

Small 8vo. Signatures A to E in eights; pagination [2] 76 (which is an error for 78, because 61-62 are repeated).

---

[1] *Transactions of the Archæological Society of Glasgow*, 1883, II. p. 233.

Collation:

(A 1) Title. verso blank.

(A 2) Epistola Nuncupatoria of Justinianus to Robertus Geduinus, "Regis Francorum à ſecretis" ends on A 3 *recto.* (p. 3) dated Parisiis, 16 Cal. Maias. MDXX.

A 3 *verso.* Text of Matthæus' tract begins and ends E 3 *verso*, pp. 4-66 *(sic)*

E 4 *recto.* Sabellicus poem, ends E 8 *verso*, pp. 67-76 *(sic)*

In his collation, Beckmann (*Geschichte*, 1792, iii. p. 559), simply gives the pagination as numbered incorrectly, 76, and omits the title leaf.

What little is known about the author is gathered from Justinianus' preface. Joannes Matthæus belonged to Luna, a town in the extreme north-west of Tuscany, a famous place, but at the beginning of the sixteenth century in ruins. The harbour still remained one of the best according to sailors, but the name was changed from Luna to Portus Veneris. About the close of the fifteenth or beginning of the sixteenth century Matthæus was born, and was distinguished as a skilful rhetorician and first-rate antiquary. He wrote a book, *De mulieribus claris*, which, says Justinianus, "we will make accessible to all students very soon," This book, edited by Justinianus, was printed in 1520 at Paris,[1] and it is commended for its uncommon learning and wide interest.

Matthæus began a work, *De rerum inventoribus*, which, although unfinished, Justinianus resolved to print, so that it might not be lost. Then it occurred to him to dedicate it even in its inchoate state to Geduinus, and after the customary compliments, he concludes: *Parvum verò mole libellum eo animo accipe, quo accipere soles cuncta seu parva, seu magna, quæ ab amicis proficiscuntur. Vale. Parisiis. 16 Cal. Maias, MDXX.*

The *Peplus Italiæ* of Matthæus, Paris, 1578, was re-printed by Fabricius, *Conspectus Thesauri Litterarii Italiæ*, Hamburgi, 1730, p. 369, and Jöcher (*Allgemeines Gelehrten-Lexicon, Leipzig*, 1751, III. col. 287-8) quotes Latin poems by him. They are the Psalms, sacred poems, hymns and poems, published at Paris by Joannes Auratus, in 1576.

As for the present tract it would seem as if it had been printed about 1520 from the unfinished MS. Beckmann never saw a copy, and the Hamburg

---

[1] *Bibliotheca Instituta et Collecta . . . a Gesnero . . . amplificata per Johannem Jacobum Frisium . . .* Tiguri, 1583, p. 394.

edition of 1613, he says, is so rare that the only copy he could meet with was in the ducal library at Wolfenbüttel.

Lessing (*Kollektaneen*, Berlin, 1790, Vol. II., p. 142) mentions Matthæus, and expresses his indebtedness to him for notes of certain inventions and discoveries. What he says about the author himself is taken from Justinianus' preface and from Jöcher, and he seems to consider him the earliest modern author on the subject. Supposing the work had been printed in 1520, involving its being written say even twenty years earlier, it would still be subsequent to Polydore Vergil, though anterior to Sardi, who was only born in 1520. It is not easy, therefore, to explain how Lessing has placed him prior to Polydore Vergil.

7. The history of Alessandro Sardi, which was intended as a supplement to that of Vergil, is usually printed along with the *De Inventoribus Rerum* of the latter. It will be found, for example, in the editions of 1600, 1604,[1] 1606, 1613, 1626, 1726. In my first account of it (Part II., p. 242) I had only these 1604 and 1626 conjoined editions before me, and Chappuys' French translation of 1584, an edition to which I have observed no allusion anywhere, but, as I then stated, the first edition was printed at Mayence in 1577 as a supplement to a work *De ritibus gentium*. This work first appeared under the following title:

Alexandri Sardi | Ferrariensis, | De Moribvs Ac Ritibvs | Gentivm | Libri III. | Nunc primum in lucem editi. | Quod opus quanta rerum uarietate refertum, | quamq; omnibus ingeniis utile ac neceſſarium | ſit, ſequens elenchus, & auctoris procemium | indicabunt. | Cum Priuilegiis. [Device.]

Venetiis, | Ex officina Stellæ Iordani Zilleti. | 1557.

Small 8vo. *, A-Q in eights, R in six; or pp. [16] 265. [1 with register and colophon, 1 blank, 1 with device.] The dedication is to Pope Iulius Cananus, not dated.

It is a very prettily printed book.

The second edition is entitled:

De Moribus ac Ritibus Gentium Lib. III. Moguntiæ, Per Franciscum Behem, Anno MDLXXVII.

It is in small 8°, *, A to N in eights, or pp. [16] 207 [1 blank], and is printed in italics.

---

[1] The edition of Sardus, 1604, mentioned in the British Museum Catalogue under *Plinius* [721.b.23(2)], is merely the appendix to Stoer's edition of Polydore Vergil of that date, bound separately.

Then follows, with separate title page, signatures, and pagination, the tract *De Inventoribus* as follows:

> Alexandri Sardi Ferrariensis, De Rervm Inventoribvs, Libri Dvo. Iis Maxime, Qvorvm Nvlla mentio eſt apud Polydorum. In Qvibvs Omnivm ſcientiarum, omniumq; ferè rerum principium quoddam quàm breuiſſimè continetur.
> Mogvntiæ Per Franciscvm Behem. Anno M.D.LXXVII.
> Small 8°. *a* to *d* in eights, *e* in four, or pp. 64 [5, 1 blank, 2]. Printed in italics.
> *a* 1. Title, *verso* is blank.
> *a* 2 to *d* 8. Text, pp. 1-64.
> *e* 1-3 *recto*. Index. At the foot of *e* 3 is the imprint: Coloniae Agrippinae, Typis Godefridi Kempenſis.
> *e* 3 *verso* is blank, and *e* 4 (blank ?) is wanting.

This book is in the British Museum [802 . b . 9 (1)], the second number in the same volume being, by the way, the rare work of Pastregicus, which I formerly described in detail. I have a copy of the book *De rerum inventoribus* with that date, and so far as appearances go it might pass of course for a separate publication.

If this be the first edition, it is certainly curious that it should have appeared at Mayence. That the work, *De Ritibus Gentium*, which was first printed at Venice in 1557, should be reprinted at Mayence is intelligible, but how an unpublished work by the author should be appended to it is not quite so intelligible. One would have expected the first edition of a book by an Italian, which was meant to supplement a work by another Italian, to have appeared somewhere in Italy, in Ferrara most likely, as being the place where the author spent his life; but, if not there, in Venice, where the previous works of both the authors concerned had been printed.

There seems, however, to be no good ground for doubting it, if any reliance can be placed on the following.

The tract, *De Ritibus*, is dedicated to Pope Julius, and the dedication is dated 1557. In the 1577 edition there is another dedication to the same pope, in which it is said that after twenty years the book returns to him once again, only more correct in this German edition than in the former one at Venice, and in order that this iteration may not be distasteful, there is now added by way of novelty a tract on those inventions which were omitted by Polydore. This, therefore, would seem to be actually the first issue of the tract, unless the whole thing, preface and all, has been reprinted bodily from an Italian edition of the same or an earlier date, only with the necessary change involved in the adjective "German." I have seen no notice of an Italian edition of or prior to 1577, and therefore the Mayence edition may be the first, with Sardi's actual supervision. This, too, is possible, though remarkable in the case of a man who seems never to have gone from the town where he lived and died. But with regard to these dedications there is something of a hitch. Pope Julius III. bore rule from 1550 to 1555, in which year he died. Although not printed till 1557, the author may have written the dedication before 1555, and may not have seen any need to change it. But how he could re-address it to the same Pope, twenty-two years after his death, not only taking no notice of that fundamental change, but actually saying that to keep him from being bored by the sight of this old book, he had added something new, is a problem for which I have no solution handy. I have found a nearly parallel case in Vigani's dedication. Is it a way Italians had to dedicate their books to patrons who had been dead for a quarter of a century or so? This question, I think, has not been discussed by Mr. Wheatley in his book about Dedications to Patron and Friend.

8. Although the *History of Inventions*, by Guido Panciroli,[1] was originally written in Italian, it first appeared in the Latin version of it, which was made by Heinrich Salmuth.[2] While, therefore, on the ground of chronology that translation ought to be taken first, it seems more correct to describe the Italian original from which the Latin was presumably translated.

It is as follows from the copy in the British Museum, 7955. c. 32:—

> Raccolta | Breve | D'Alcvne Cose Piv | fegnalate c'hebbero gli antichi, e d'alcune | altre trouate da moderni. | Opera Dell' Eccell. S. Dottore | Gvido Panciroli | Da Reggio. | Con l'aggiunta d'alcune confiderationi curiofe, & utili di | Flavio

[1] Formerly (Part II. p. 243), I called him Pancirollo, but the true form seems to be Panciroli. There are also two incompatible statements on the same page—the first is that I knew no Italian edition of the work; and then a few lines further on the Italian edition of 1612 is mentioned. What was really meant, though it was so inaccurately expressed, was that I knew no Italian edition prior to 1599, and concluded that Salmuth therefore must have translated from MS. Besides, I had not examined the 1612 edition, and did not know whether it was the original or a retranslation from the Latin, which was not impossible.

[2] [See Note on Panciroli's book at the end of this paper.]

Gvaltieri da Tolentino, Dottor Teologo. | Dedicata al Serenissimo | D. Carlo Emmanvele | Duca di Sauoia &c. | Con Privilegio. | [Device] |
In Venetia, M D C XII. | Presso Bernardo Giunti, Gio. Battista Ciotti, & Compagni. |
4°, or large 8°. *, †, ††, in fours; A to Z, Aa to Ee in eights, or pp. [24], 443. [1].
Collation:
*1 Title.
*2 *recto*: Gualtieri's preface to Carlo Emmanuele, from which one gathers that this is the first edition in Italian.
*3 *recto*: A'lettori.
*4 *recto*: Panciroli's address to Carlo Emmanuele.
†1 *recto*: Ordine de Capi.
†2 *verso*: Indici delle cose, ends ††4 *verso*.
A1 *recto*, p. 1: Text begins, and ends Ee 8 *recto*, p. 443.
Ee 8 *verso*, not paged: Correttione.

From the MS. of this work Salmuth must have made his version, which was published at Amberg in 1599-1602. This book is uniform in size and style with the later 8vo editions, only it is more neatly and elegantly executed. Not that it is of any great beauty, but the later editions degenerated both in paper and printing. There is a copy of the 1599 volume in the British Museum.

On the present occasion I have the second edition of both parts to show. The first volume was printed at Amberg in 1607, by Michael Forster, and the second similarly in 1608, in small 8vo. The volumes are in very commonplace style. The third edition, Amberg, 1612, in 8vo, was formerly described. The fourth (?) edition, Franckfurt, 1622, in 8vo, is also got up in an inferior manner. It is in the Museum (7942. aa. 31). I am not sure if this was the last 8vo edition, but those of 1631, 1646, 1660, are in 4°.

Besides the English translation already described, there is a brief abstract of both parts in French made by Pierre de la Noue, not from the Latin, but from the Italian as well, two editions of which were formerly mentioned, Lyons, 1608, 1617. I have the later edition:

Livre Premier Des Antiqvitez Perdves, Et Si Av Vif Representees par la plume de l'Illustre Iurisconsulte G. Pancirol qu'on en peut tirer grand profit de la perte; Accompagné D'Vn Second, Des choses nouuellement inuentées & auparauant incogneües. En faueur des Curieux Traduits tant de l'Italien que du Latin en François. Par Pierre De La Nove.
A Lyon, Pour Iaqves Gavdion, M.DCXVII. Avec Priuilege du Roy.
18mo. Signatures: † in six, A to L in twelves, of which L 12 is blank; or pp. [12 261 [3 blank].

It is rather a poor little book, but it is a scarce one. The above is taken from my own copy. There is another in the British Museum, 1137. b. 4 (1).

There is a book by Michael Watson with the following title:

> **Theatrum Variarum Rerum exhibens Excerpta & Annotata in Libb. de Rebus Memorabilibus Pancirolli & Salmuth.**
> **Bremæ, clɔ lɔc LXIII. in 8vo.**

There is a copy in the Museum, 1137, b. 6. This is partly a description of Panciroli's history and partly a sort of running commentary on it.

On this book one or two remarks may be made. It is in two parts—lost inventions of the Ancients and newly discovered inventions of the Moderns. From the first edition being dated 1599-1602, I conclude that it was in two distinct volumes, certainly like the second and third editions, and probably like the fourth. My observation is that copies of every edition of the second part are much scarcer than of the first. That may be accounted for in two ways: 1st, fewer copies of it may have been printed in the respective editions; 2nd, the second volume may have been used much more, and so the copies of it may have been consumed. The fact that the second volume referred to new discoveries may have been quite enough to attract attention, while the discoveries of the ancients would be passed by without regard by the majority of readers. They might be interesting to scholars, or antiquaries, or historians, but not to "practical people" who wished to know the latest improvements or inventions.

In some respects the second part of Panciroli's history is even now the more interesting. It is always curious to know when a thing was an absolute novelty, however old it may be now, and it is curious also to see what inventions and discoveries impressed themselves upon a man leading a quiet life in a comparatively small and obscure town. No apology, therefore, is required for selecting the following titles from the contents of the book:

The New World,	**Porcelain,**	**Alchemy,**
The Bezoar Stone,	**Sugar, Manna,**	**Rhubarb,**
Distillations,	**Clocks,**	**The Mariner's Compass,**
Printing,	**Paper,**	**Cypher,**
Saddles,	**Squaring the Circle,**	**Cannon,**
Greek Fire,	**Jousts,**	**Water Mills,**
Bird-catching,	**Silk,**	**Caviar,**
Etc.	Etc.	Etc.

Of course, at a later date, some of these subjects got much fuller treatment at the hands of Beckmann. Still, Panciroli tried conscientiously to make his discourse on these and similar themes thorough and attractive. Panciroli's book was, however, handled in somewhat rude fashion by Goguet. In the preface to his work upon the *Origin of Laws, Arts, and Sciences*,[1] he refers to it in the following terms:

> We have a very bad book of Pancirolus's, intituled, *Rerum memorabilium sive deperditarum*, &c. This is, in general, a very crude indigested compilation, in which he sticks at nothing. The falsest facts and most ridiculous tales are adopted for truths. This work is an example of the greatest negligence, joined to the strongest itch of making a book. In what Pancirolus says of certain arts, which, according to him, were known to the ancients, and have been since lost, there are almost as many mistakes and puerilities as words. The arts which he speaks of, either never existed, or they exist to this day, and in a more perfect state than ever. It would be easy to demonstrate this, if the work was worth the trouble.

Goguet's superlatives require a considerable amount of pruning, before they can be accepted as positively accurate.

9. Some account has already been given of Paschius, who wrote a book on the history of inventions and science. It was the second edition in 4to which was then mentioned, but to complete the account a notice of the first edition was desiderated. This I can now give:

> Georgii Paschii Philof. Mor. in Academia Kiloniensi Prof. Ord. Schediasma De Curiosis Hujus Seculi Inventis, Quorum Accuratiori Cultui Facem Prætulit Antiquitas.
>
> Kiloni, Sumptibus Joh. Sebaft. Riechelii, Literis Joach. Reumanni. Anno MDCXCV. Small 8vo. Signatures:)(, A to Y in eights; or pp. [16] 342.

This is a most attractive book, for it tells us what were considered new discoveries just two hundred years ago. Among these, not the least interesting for my present audience is the account of the diving-bell of Professor George Sinclar.[2]

10. One of the most important works for elucidating the arts of the eleventh century is the well-known treatise of Theophilus the Monk. In 1882, when describing the English edition by Hendrie, I had not seen, but could only

---

[1] Edinburgh, 1775, I., p. vii.

[2] [His writings occupy a not inconspicuous place in the "Old Glasgow" Exhibition, which has been opened since the above was read. August 6th, 1894.]

refer to the French edition by Count De L'Escalopier. A copy of that edition has now come into my hands, and one cannot but be satisfied with the very handsome way in which it has been got up. There are two titles, in Latin and French respectively. They are as follows:

Theophili | Presbyteri et Monachi | Libri III. | Seu | Diversarum Artium Schedula. | Opera et Studio | Caroli De L'Escalopier. | Lutetiæ Parisiorum. | Excudebant Firmin Didot Fratres, | Via Jacob, 56. | MDCCCXLIII.

Théophile | Prêtre et Moine. | Essai Sur Divers Arts, | Publié | Par Le Cte Charles De L'Escalopier, | . . . | Et Précédé | D'Une Introduction, | Par J. Marie Guichard. |

Paris, | J. A. Toulouse, . . . 1843.

It is a handsome finely printed 4to, pp. [4] LXXII; 1 leaf, facsimile of MS.; 314 [1, 1 blank].

This edition contains some preliminary remarks by L'Escalopier on the time at which Theophilus lived, on the nature of his book, and on the scope of the present edition. This is followed by Guichard's elaborate introduction upon the history and contents of the book, and its bearing upon the state of the fine arts in the 12th or 13th century. Then follows the Latin text, with various readings from the MSS., and a parallel French translation. The notes with which the volume closes are arranged alphabetically, and form a sort of brief glossary, though, as the editor remarks, there is much in Theophilus which still requires elucidation. The edition is indispensable for the history of the practical art secrets of the Middle Ages, but even its bibliography would require a dissertation to itself.

11. The following is the most valuable addition I have been able to make to the list of histories of inventions:—

Original Treatises, dating from the XIIth to XVIIIth centuries, on the Arts of Painting in oil, miniature, mosaic, and on glass; of gilding, dyeing, and the preparation of colours and artificial gems; preceded by a general introduction; with translations, prefaces, and notes. By Mrs. Merrifield. . . . In two volumes.

London: John Murray, Albemarle Street. 1849.

8vo. Vol. I. pp. cccxii. 1-321 [1]
Vol. II. pp. v. [i] [323—5] 326-920.

This work ought to have been included among the histories long ago, but it has become rare, and it is only recently that I have been able either to see or get a copy. During all these investigations I have come across no book of more importance than this for the history of technical receipts, and for elucidating the literature of secrets down to the present time. One can see the sources from which the old compilers gathered their information, not always intelligently, for they were sometimes mere compilers without any practical knowledge either of the materials or processes. One can also see how the results obtained by the old artists and craftsmen by patient trial and practice, and by the desire to make their work as good as possible, have incited their successors to equal them if possible, and how, therefore, their methods and substances have been handed on traditionally from generation to generation. The book is a necessary companion to that of Theophilus, which deals with similar topics.

12. As a set off to the historians who have done their best to display human skill and perseverance in the inventions of byegone times, there have never been wanting the pessimists whose task it has been to depreciate these qualities and to moderate that vanity about mankind which his successes appear to foster. Some of these, Faustinus, Cornelius Agrippa, Thomas Baker, have been already quoted.

To what has been already said[1] about the *De Triumpho Stultitiæ* of Perisaulus Faustinus, printed in 1524, it seems superfluous to add anything. The little volume, however, is so very rare, so little known, or rather so quite unknown and so confusedly described, that anything that would elucidate its history might be considered of value. In the hope of finding a variation from my previous description, I acquired the copy which occurred last year (1893) in the Manzoni sale (No. 4168). On examining it I was both satisfied and disappointed. Satisfied, because it agreed exactly with my other copy, and therefore confirmed all my arguments; disappointed, because in so agreeing it gave no help towards solving the difficulties in the book's history. The Manzoni copy, however, is an addition to the list of those already known, for, so far as one can judge, it cannot be identified with any of the others. This copy is entire, having the final blank leaf. The only difference between it and my other copy is in the relative position of the red and black lines of the title page. In the Manzoni copy the red lines are less accurately adjusted, and almost touch the black lines immediately above them. In every other respect

[1] *The Library*, February, 1890.

the two copies tally, and so once more we meet with Hieronymus Soncinus on the title page, but in the colophon "Jo. Franc. & Jo. Antonius de Rusconibus, Fratres." Notwithstanding the appearance of this probably unknown copy, the book remains as rare as ever. Doubtless copies of it exist in Italian libraries and collections, but meantime, if there be such copies, they are unknown, and we can enumerate only the half-dozen which I have described.

13. With the rarity of Faustinus' pæan of folly, copies of which are to be counted by units, may be contrasted the abundance of Cornelius Agrippa's fantasia on the same motivo: "The Vanity of Sciences," editions of which are to be counted almost by the score. The great circulation which it had shows that it responded to some necessity of the time, but it is too much to suppose that its scepticism produced any effect on the progress of science and of the arts, or prevented men from pursuing any path—whether of labour, study, amusement, or dissipation—which allured them. The book itself, however, is particularly attractive to the bibliographer and chronicler. It presents many fine points of interest; the number of editions, the rarity of some of them, the excisions in the expurgated and licensed copies, the translations; these and others lend themselves agreeably to investigation. To treat these details in full would take too much space here; besides I considered some of them in a communication to the Society of Antiquaries in London. Since then, however, additional matter has come into my hands which requires new treatment. Here, however, I may briefly state certain points of importance. The book was printed in 1530, in 4to, of which rare edition I got a duplicate in 1888. There followed some seven or eight issues, all dated 1531, and all at different places. Edition after edition appeared, and then, about 1536, the book was expurgated, and thereafter the emasculated editions went widely into circulation. The excisions and variations were tabulated by Clément, and they are useful for comparison to ascertain whether an edition is complete or not. Only the earlier editions are complete, and on that account desirable. The book was translated into various languages—French, English, Dutch, German. There are two English versions—the earlier, 1569, and 1575, both in 4to, and the later in 8vo, dated 1676, 1684, 1694. It is not probable that the book is ever looked at at the

present day—even the English version of it—but it repays examination, not merely by its style and vein of thought, but still more by its serving the purposes of a history. It is not a systematic work of that kind, but it shows such erudition and such a knowledge of the habits, customs and events of the time, that, in its unconscious simplicity and general accuracy, it is preferable to a set history in certain dramatic effects. It is rather a criticism with a historical basis than a mere narrative.

14. A small volume entitled *The Vanity of Human Inventions* was published at London in 1666, anonymously, though the author is said to have been a certain John Wilson. In case those who come across this little work should be led by the title into thinking that the author resembled Perisaulus Faustinus, Agrippa, or Baker, in his treatment of the subject, I may say that he does not do so, but deals only with religious rites and ceremonies. It is in fact merely a pamphlet in the anti-papal discussions of the time. I have included it here only to have the opportunity of saying that it does not properly belong to the series of books now considering, as its title might suggest.

15. When bringing forward formerly Thomas Baker's *Reflections upon Learning* as a piece of shrewd scolding, I could show only the third edition in English, printed for A. Bosvile, in 1700, and was unable to say anything about previous editions. I have since ascertained that in the same year there were other two editions, one without any number and therefore presumably the first, and one with identical title page, but having in addition the words: "The second edition corrected." These two exhibit variations in the text, but I do not know how far the third edition agrees with or differs from them. This would be conclusive as to the publication of the book, but in the Library of Emmanuel College, Cambridge, there is a copy dated 1699. Whether this points to a cancelling of this title page, and the substitution of one dated 1700, or to two distinct editions, is a point I have not yet settled. Obviously, however, the bibliography of Baker's book is worth looking into.

II.—Books of Receipts and Secrets. (Italian.)

16. Coming to this second division, the amount of material at my disposal is so great that for convenience it must be divided. As it happens that books of secrets in Italian and by Italian writers make up nearly one third of those which I have recently seen, I shall take them by themselves in this first division, leaving the remainder for subsequent consideration. This arrangement brings into prominence the number and the activity of the Italian writers down to the eighteenth century, the topics which were most in vogue, and the steady demand which must have existed for information, however defective or empirical, on practical matters. The first in the list is the following remarkable work on medical secrets.

Early in the fourteenth century Gulielmus Varignana, a native of Bologna so far as I have ascertained, wrote a remarkable treatise which he entitled, *Secreta Medicinæ.* In my account of Kiranus I shall have repeatedly to make reference to this collection, but it is of sufficient importance to warrant separate notice of it here. I have consulted and compared three editions of this work in the British Museum—Lyons, 1539; Basil, no date, printed by Henric-Petri; and one printed by Sebastian Henric-Petri at Basil, in 1597. Of these, by far the most attractive is that of 1539. It is a thin 8vo, printed in double columns in black letter, with numerous contractions. The next is a little square 8vo. It is in Roman character, and is rather pleasing to look at, but the text varies considerably from the previous edition, and is corrupt. The last, edited by Caspar Bauhinus, may have been the most useful, but it is excessively plain and commonplace as a book. The editor, however, has restored the text in most places, so that it agrees very closely, if not entirely, with the edition of 1539. The plan on which the book is constructed is simple and convenient for rapid consultation. The author begins at the head, and goes through all the diseases in detail right down to the feet, describing their treatment and the remedies to be employed. His information, however, is taken from the old writers, and the result is but an "aggregation" like the more massive volume which was produced by his contemporary, Jacobus de Dondis, Aggregator, of Padua.

17. In the British Museum (1038. d. 35) there is a volume containing over a dozen small receipt books to which reference will have to be repeatedly made. They are chiefly in Italian of the sixteenth and seventeenth centuries, and several are mere chap books. The earliest of them to be mentioned belongs to this category. It is No. 12 in the volume.

Recettario Nouo probatiffimo a molte infirmita, & etiãdio di molte gentilezze vtile a chi levora prouare. Cofa noua non piu ftampata. M. D. XXXII.
Small 8vo. Signature A in eight; or, pp. [16].
Colophon:
¶ Stampato in Vinegia ad inftantia di Zuan maria Lirico Venitiano. Nelli anni del Signore. M. D. XXXII.

This is a book of medical receipts of the usual kind. The title is enclosed in a border similar to that on the title-page of the Italian version of Michael Scotus' *Physionomia*, Venice, 1532, from which it may be inferred that the ornament was a common one with the Venice printers of the time.

18. The first treatise on any art can hardly fail to be an interesting one. Even if it be defective, it will still furnish a summary of the knowledge of the time, which is invaluable. Moreover, there will be a flavour of originality about it which subsequent books must necessarily lack, and it will be the work of an expert who has practised the art and knows the details. This is the interest, for instance, of Neri's work on glass-making, and it is equally so of the work on dyeing by Gioanventura Rosetti. So far as I know this is the first separate work on the subject, and it is of very great rarity. Receipts about dyeing are given in a good many of the secret books, more particularly in the *Kunstbüchlein*, *Ettliche Künste*, by Alexis, etc., etc. The present work, however, deals with the whole subject in a systematic way and not only so, but gives woodcuts of the plant required for the various operations.

The technical details are hardly suitable for exposition in this paper, and are besides so lengthy that they had better be dealt with separately, but as it is distinctly a book of technical receipts, and the only one of the kind which I have met with in the whole course of my researches, it may be described. Singularly enough, of this book there was only one edition, so far as I know.

Plictho De Larte
De Tentori Che Insegna
Tenger Pāni Telle Banbasi
Et Sede Si Per Larthe Magiore
Come Per La Comvne.

4to. Signatures: A to L in fours. No pagination.

The title-page is decorated with an elaborate border and a curious coat of arms, if it can be called such, and it is entirely woodcut, even to the title itself.

At the end of the text, L2 *verso*:

Compofto per Gioannuentura Rofetti prouifionato ne lo Arfena dallo Illuftrifsimo Senato Venetiano.

In Venetia per Francefco Rampazetto.
M.D.XL.

L3 and 4 contain the table of contents.

There is a copy of this book in the British Museum, 1044.i.19 (1). It is mentioned by Brunet, *Manuel*, 1863, vol. iv., col. 1396; Graesse, *Trésor*, 1865, vi. i. 165; but what is of more importance as indicating its value, it is in the *Catalogue of the Choicer Portion* of the Libri Library, 1859, p. 315, No. 2362. It is there said to be extremely scarce. Brunet gives the prices which this volume has fetched: 55 fr., Libri in 1847; 25 fr., Libri in 1855; and 19 sh. in 1859, so that the Libri copy or copies seem the only ones which have been in the market for fifty years.

19. The *Pirotechnia* of Vanoccio Biringuccio, treating in ten books of metallurgy and related technical and chemical arts, differs from a large number of works enumerated in these lists in two respects:—1st, its bibliography is quite simple and the book presents no difficulties as to its own history and meaning; and 2nd, its contents are more interesting historically, than the bibliography. On this occasion, however, I must confine my remarks to the book and leave its contents to be reviewed hereafter.

The first edition, Venice, 1540, is an annoying book. When I became acquainted with it, which was long before 1882, it was considered very rare. Whether this was correct or not at that time, I do not know, but, as a matter of fact, it is no longer rare, but has become comparatively common; at all events I have repeatedly encountered copies in sale catalogues, and I should think there would be no difficulty in getting a copy of it at any time. Some of the later editions are rarer, and this is evidenced by the fact that after mention made (Part I., 1882), of the editions of 1540, 1550, 1559, and the

French translation, 1572, there has been no opportunity to refer to the book again in any of my subsequent papers.

Recently, however, some other copies have come into my hands. The rarest of them, probably, is the first edition of the French translation, Paris, 1556, in a handsome quarto, superior in its style to the edition of 1572. There is also another edition in Italian (Venice, 1558-59), in 4°, and an octavo edition published at Bologna, without a date, but about the middle of the sixteenth century. By far the most interesting and rarest item, however, is the sole fragment that remains of a proposed and partially executed translation into English. It consists of the introduction and the sections which deal with gold and silver, occurring in the signature *4 to 8 and ff. 1 to 10 *verso* of the 1540 edition. It was executed by Richard Eden, and is contained in his edition of Peter Martyr's *Decades of the New World*, printed at London in 1555, ff. 326 *verso* to 342 *recto.* There are few books rarer than this last. Besides the copy I now exhibit I know only of the copy in the British Museum, and another in the University Library, Cambridge. The section from Biringuccio would merit reprinting as an appendix to a separate dissertation on the knowledge of mines and metals, possessed by the author in the early part of the sixteenth century.

20. As the modern scientist deems his duty to long-suffering humanity but hardly discharged if he has not published a text book on what he designates "our science," the older representatives of the class acted as if an obligation were laid on them to discuss the whole system of the universe. They did not conceal their aim, and they certainly showed considerable power in either collecting and repeating, or in inventing explanations of the wonders and mysteries which the physical, spiritual and moral worlds contain.

Among these the work of Jerome Cardan, of Milan, *De Subtilitate*, is conspicuous. It was first published in 1550; it passed through many editions, and was translated into various languages.

There are three copies of this book to refer to.

> **Hieronymi Cardani Mediolanensis Medici, De Svbtilitate Libri xxi. Nvnc Demvm ab ipſo autore recogniti atque perfecti.**
>
> **Lvgdvni, Apud Guliel. Rouillium. 1554.**
>
> **8vo, signatures: α, β, γ, δ, in eights; ε in four; a to z, A to Z, Aa to Ee, all in eights, of which Ee 8 is blank; or pp. [72] 813 [3 blank].**

The present copy is a fine example of what could be done in the middle of the sixteenth century to produce a book for ordinary circulation. The paper is good and the printing clear and well arranged; the title page is curiously illustrated, other ornaments are tasteful, and the woodcuts, which are numerous, excellent. It is strongly bound in wooden boards covered with white pigskin, nicely stamped, and dated 1557, so that the binding is of the same time as the book. One clasp is wanting, but that is the only imperfection. Nothing better could be desired as an example of a book of three hundred and fifty years ago. It contains, besides p. 428, the passage which caused its author some trouble. Clément (*Bibliotheque Curieuse*, 1756, T. vi., p. 273) speaks of the rarity of this edition, which apparently he had not seen, and says: "Elle est presque inconnue."

The next copy is a French translation.

Les Livres De Hierome Cardanvs Medecin Milannois, intitulez de la Subtilité, & fubtiles inuentions, enfemble les caufes occultes, & raifons d'icelles. Traduits de Latin en Françoys, par Richard le Blanc. Nouuellement reueuz, corrigez, & augmentez fur le dernier exemplaire Latin de l'Auteur, & enrichy de plufieurs figures neceffaires. [Device].

A Paris. Chez Pierre Cauellat, ruë S. Iaques, à l'enfeigne de la Fleur de Lys. 1578.

8vo. *, ā in fours, ē, ī, ō in eights, ū in four; a to z, A to Z, Aa to Oo, in eights, Oo 8 wanting; or ff. [36] 478 [1 with scroll] and 1 blank (?) wanting.

This also is in excellent state, though the binding is at least a century and a half later than the book. It is more closely printed than the Latin, but it is nevertheless nicely executed. Cavellat was the publisher also of Levinus Lemnius' Secrets, translated by Gohory, and of some of the editions of Alexis.

The third copy is that contained in Cardan's works, which were collected in ten volumes folio, and printed at Lyons in 1663. The present treatise is in the third volume.

The treatise, "De Subtilitate," is a summary of philosophy, science, and the arts, and it is only as such that it is entitled to be embraced in this list. Yet, as it does contain a description of secrets both of nature and art, it cannot reasonably be excluded, even though its name does not specify the secrets which it aims at elucidating. The French title, however, shows its claim more clearly than the original Latin, since it says that it treats of "subtile inventions."

This book was long a standard one, it passed through many editions, and its bibliography was studied by Clément, *Bibliotheque Curieuse*, 1756, T. vi., p. 268, sqq.

The following editions are enumerated:

1.	Norimbergae, apud J. Petreium,	1550, fol.
2.	Lugduni, apud Rouillium,	1551, 8°.
3.	Paris, ap. Mich. Fezandat et Rob. Granjon,	1551, 8°.
4.	Basileæ, ap. Lod. Lucium,	1554, fol.
5.	Lugduni, ap. Rouillium,	1554, 8°.
6.	Lugduni, ap. Guil. Rouillium,	1559, 8°.
7.	Basileæ,	s.a. fol.
8.	Basileæ, ap. Henric Petri,	1560, fol.
9.	" "	1560, 8°.
10.	Lugduni, ap. Bart. Honoratum,	1580, 8°.
11.	Basileæ, ap. H. Petri,	1581, 8°.
12.	Basileæ, ap. Seb. Henricpetri,	1582, 8°.
13.	" "	1611, 8°.
14.	" ap. Emanuel König,	1664, 4°.

In French, by Richard le Blanc:

15.	Paris, chez Charles l'Angelier,	1556, 4°.
16.	" "	1578, 8°.
17.	" "	1584, 8°.
18.	a Rouen,	1642, 8°.

An abstract in German was appended to the translation of Cardan's *de rerum varietate* by Hulderich Fröhlich von Plawen under the title, *Offenbarung der Natur und natürlicher Dinge*, Basel, 1559, 1591, fol.

Of these various editions I have managed to see only No. 2, 1551, which in good condition would be a fine book; No. 4, 1554, which is printed in large type, and is almost too big; No. 5, 1554, described above; No. 12, 1582, printed in small italics, yet containing pp. [80] 1148 [4], and got up in the ugliest possible way; No. 14, 1664, printed at Basel, "Impensis Henric-Petrinorum," who, not to be false to the firm's traditions, have printed it on soft paper that has foxed, and have made it as repellant as a working quarto of the date could be; and No. 16, 1578, described above.

21. Another of the Italian receipt books mentioned above [B.M. 1038.d.35 (2)] has the following title:

> Notandissimi Secreti De L'Arte Profvmatoria: A Fare Ogli, acque, paſte, balle, moſcardini, uccelletti, paternostri, e tutta l'arte intiera, come ſi ricerca coſi ne la citta di Napoli del Reame, come in Roma, e quivi in la citta di Vinegia nuoua

mente impresſi. [Device: Fortune (?) with a sail, standing on a ball, which is floating on the sea, than which there could hardly be a better symbolizing of instability and uncertainty.]

In Vinegia.

Small 8vo. Signatures: A to K in 8; or ff. 74, and table [6].

After the table, the colophon:

In Vinegia per Franceſco Rampazetto. Ad instantia di Marchio Seſſa, Ne l'anno del Signore. M . D . LV.

Printed in italics.

At the foot of f. 73 *verso* is the note: E tutte queſte coſe ſon compoſte per Giouanni Ventura Roſeto Veneto. Was he also the author of *Plictho*, § 18, above?

From its contents this is to be compared with Le Fournier's tract, and with he corresponding sections in the Secrets of Alexis which followed close upon it.

22. Among the books which first drew my attention to the early literature of technology, the foremost place is due to the *Secrets* of Alexis. It was mentioned by me in 1882, and it has recurred again and again in subsequent communications. It is undoubtedly one of the most interesting books of its class as regards both its own history and its contents. The historical detail, however, from its extent, must be considered separately, and the contents must be similarly dealt with, but the editions which I have recently acquired may be mentioned. From the book having been published in four parts, and the translations also in parts, it is hardly possible to arrange them in strictly chronological order, for the reissues overlap, and a tabular arrangement is almost a necessity. This method I have preferred for the separate paper on the subject for which my material has been collected.

The so-called first edition of 1555-6 still remains a desideratum, but I have an unknown and undescribed edition of the first part, printed in 1557 at Pesaro. It is entitled, *Secreti del Reverendo Donno Alessio Piemontese, Nuouamente posti in luce.* It contains Alexis' preface to the reader narrating the origin of the book, the contents, and the text of the six books, and it forms a small octavo volume, printed in italics, and containing 12 unnumbered and 143 numbered leaves. The colophon is on the recto of f. 143, and the verso is blank, so that the last leaf of the signature, which is wanting, is probably blank also. This edition is entirely new.

The Lyons reprint, 1558, in 16°, of the second edition was noticed in Part II. of the "Notes." I havé now got another copy which is identical with it in all respects. It is uniform also with several other books of secrets printed at Lyons about the same time, such as the *Treasure* of Evonymus, the *Prodigies* of Obsequens, and the *Natural Magic* of Baptista Porta. These books have an unmistakable character, and the Lyons press of the sixteenth century was a vigorous and active, if not always a scrupulous one.

The next copy I have bears date Antwerp, 1560, and it forms a small narrow 24°. It is a reprint of Wecker's translation into Latin of the six books of Part I, which was printed at Basel in 1559, and a copy of which I exhibited to the society in 1888. The Antwerp reprint contains Wecker's dedication to Abbot Caspar Muller, as well as Alexis' usual preface to the reader. This reprint is equally rare with the Pesaro edition, for, so far as I have observed, it is not mentioned by anyone in connection either with Alexis or with Wecker.

In 1561 a French translation was published also at Antwerp, by Christopher Plantin, of which I have a copy. This is an enlarged edition, consisting of three parts. The first contains the first part of Alexis' secrets in six books. The second part, also in six books, translated from the Italian, is ascribed to Alexis, but it differs entirely from the second part which is found in the English translation. At the moment I am unable to compare it with any other version. The third part consists of secrets from other authors, and it agrees on the whole with the third part as translated into English. Portions of this third part belong to the *Kunstbüchlein* series which I have described elsewhere.[1] At Lyons, in 1564-65, Rouille printed a French translation, in thick 12mo. It does not coincide exactly with Plantin's.

The English translation from the way it appeared is somewhat difficult to manage, and each part has to be dealt with individually. Two copies which I have acquired are collected in the following way:—

London, John Wight: Part I. 1568; Part II. not dated, but 1568, both printed by Bynneman; Part III. 1566; Part IV. 1569; both printed by Denham.

London, Nycolas England: Part I. 1562; Part II. 1563; Part III. 1562; all printed by Roulande Hall.

Another volume contains Part I. 1562; Part II. 1563; and Part III. 1566.

A fourth, but very imperfect copy, is composed of Part I. 1559; Part II. 1563; and Part III. 1562. It is quite obvious, therefore, from these as well

[1] *Proceedings of the Philosophical Society of Glasgow*, for session 1893-94.

as from the copies in the British Museum, that sets were made up as wanted from different editions of the various parts. These sets were made up prior to 1569 so that the fourth part does not appear in them. All these editions are well known. Apparently the least common of all is the 1560 edition of Part II. The latest edition in English with which I am acquainted was printed in London in 1615. It is in black letter and forms a substantial volume in small quarto. It contains the whole four parts, the last three of which are provided each with a separate title page, though the folio numbers and signatures are continuous. This is a convenient edition, but it labours under the disadvantage of being rather rare.

Amongst the absolute novelties must be placed two editions of a translation into Dutch. The translator's name is not given. One appeared at Amsterdam in 1658, in small 12°, printed in black letter, as seems to have been the fashion at that time in Holland as well as at home. It contains the first part in six books with Alexis' preface, and also a second part. This, though stated to be translated from the French, does not agree with the second part in Plantin's edition or any other French reprint I have seen, but tallies with the second part as contained in the English. This would indicate apparently two distinct French versions, but I am not aware that there are such.[1] The other edition, which is of the same size and in the same style, bears the imprint: Amsterdam, 1670.

23. When referring formerly to the *Magia Naturalis* of Baptista Porta, I described copies of the second and enlarged form of the work in twenty books, as published in the Elzevir editions of 1664 and the English translations of 1658. There was, however, an earlier edition in four books, of which I have two different copies to show. One is the earliest which came from Plantin's press, Antwerp, 1560, and is a thin small octavo. The other is a stout little 16mo, printed at Lyons by Rouillius in 1561. The Plantin edition is in italics, with an ornamental title-page, and is rather pretty. The other has been, and is still, a nice copy in the original boards and white stamped hog-skin binding. Of the later editions also, I have one printed at Rouen, 1650, which, like many other contemporary works which issued from the same

[1] Plantin printed Alexis in Dutch in 1558, 1561, 1564, 1571. I have not seen these editions, and cannot say if the 1658 book be a reprint or not.

centre, is as ugly and unattractive as there was any need for making it. This edition contains the twenty books.

There was a large number of editions of this famous collection, both in four books and in twenty. In the former, besides those of 1560 and 1561 just mentioned, there are Naples, 1558, folio, which is the first edition; Antwerp, 1561, 8°, the second Plantin issue; and again, 1564 in 18°; and in Italian: Venice, 1560, 8°; and 1665, 12°. Of the enlarged form: Naples, 1589, folio; Frankfurt, 1607, 8°; Hanau, 1619, 8°; Leyden, 1651, 12°; Amsterdam, 1664, 12°; London, 1658, 4°; and in Italian, Naples, 1611, 4°. These are in the British Museum, but Plantin besides printed a Latin edition of the four books in 1567,[1] and a Dutch version of the same in 1566.[2]

24. In the case of books printed as late even as the seventeenth century, one can never be certain that copies having the same date, and even place of publication, are necessarily of the same issue or edition; they may or may not be so, but anyhow, I have found a sufficient number of instances of their not being duplicates, to make me invariably investigate any instance of identity of date, so as to settle whether or not the books are identical altogether. In a subsequent supplement I shall give one instance at least of apparent duplicates proving to be entirely different, and, with that in my mind, I was not disposed to take the following on trust, though as it happens the present examples are really duplicates. Isabella Cortese wrote a book, *I Secreti*, printed at Venice in 1565, in small 8vo. Later editions, 1625 and 1677, appeared at the same place, and have been mentioned already (Part II. 258; III. 225).

In the Museum volume of pamphlets [1038.c.4 (2)] I found what proves to be the first edition. The account is as follows:

> I Secreti De La Signora Isabella Cortese. Ne' Qvali Si Contengono cose minerali, medicinali, arteficiose, & Alchimiche, & molte de l'arte profumatoria, appartenenti a ogni gran Signora. Con Privilegio. [Large Device: Female figure, standing, with a mirror.]
>
> In Venetia, appresso Giouanni Bariletto. MDLXI.
>
> Small 8vo. Signatures: *, A-L in eights, or, ff. [8], containing title, Isabella's dedication, and contents. Text, ff. 88.
>
> Printed in italics. It contains three books.

The copy of this book which I have recently acquired corresponds with

---

[1] Léon Degeorge, *La Maison Plantin A Anvers*, Paris, 1886, p. 150. [2] *Ib.* p. 148.

this throughout, so that it is another copy of the same edition. As the contents in no wise differ from the later editions, it is not necessary to say more about it at present, but, as may be seen from the contents in the title, it must be classed along with other collections of cosmetic secrets which have been referred to.

25. Pietro Bairo[1] was born at Turin in 1468, studied medicine there and became so distinguished as a physician, that he was consulted by persons of the highest rank. He ultimately obtained the title of physician to Charles II. duke of Savoy. He died on the 1st of April, 1558, at the advanced age of ninety. He wrote a treatise on the plague, which went through a number of editions, a treatise on medicine, and some minor disquisitions on medical topics. There was also a work on medical secrets, the first edition of which, as I think, is in the British Museum (1038. c. 4 (1).). The title is as follows:

> Secreti Medicinali Di M. Pietro Bairo Da Tvrino, Gia Medico Di Carlo Secondo Dvca Di Savoia. Ne Qvali Si Contengono i rimedi che ſi poſſono uſar in tutte l'infermità che uengono all' huomo, cominciando da capelli fino alle piante de piedi. Et queſto libro per l'utilità ſua ſi chiama Vieni Meco. Con priuilegio per anni xv. [Device]. In Venetia.
>
> Small 8vo. Signatures: *, A–Z, AA–KK in eights. Text ends on KK 6. Register on KK 7. KK 8 blank; or ff. [8] containing title, address to the reader and contents. Text ff. 262 [but ff. 119-120 *bis* and 167-168 dropped]. f. 263, the register; 264 blank. After the register comes this colophon: In Venetia, Appresso F. Sansovino. M.D.LXI. Printed in Italics.

According to Panckoucke's notice, there was an edition dated Venice, 1585, 8vo. This I have not seen, but I have a third dated 1602.

> Secreti | Medicinali | Di M. Pietro Bairo | Da Tvrino, | Già Medico di Carlo II. Duca di Sauoia. | Nei Qvali Si Contengono | i rimedij, che ſi poſſono vſar in tutte l'infermità, | che vengono all' huomo, cominciando da capelli | fino alla pianta de piedi. | Con nuoua giunta poſta nel fine: | Et queſto Libro per l'utilità ſua ſi chiama, | Vieni Meco. | [Vignette, St. George and the Dragon.]
>
> In Venetia, MDCII. | Appreſſo Nicolò Tebaldini. |
>
> Small 8vo. †, A to Z, Aa to Kk in eights, but Kk 7 and 8 possibly blank, or containing the device or colophon, wanting; or ff. [8] 262. Printed in italics.

This is a general treatise on the art of healing, and is topographically arranged like the works of Jacobus Paduanus and Varignana. Impressed by the vastness of the subject, he endeavoured to bring it into reasonable compass,

---

[1] Panckoucke, *Biographie Médicale*, I. p. 509. He is responsible for the dates.

and as it saved the practitioner the trouble of constantly referring to the great authorities, he, like Valescus de Taranta, called his book *Vieni Meco*, Vade Mecum. He cannot therefore claim that title as original.

The medicines employed are sometimes of a startling character—such as suffumigation by a dead man's tooth, which must have been used for the occult virtues of the tooth or of its owner, rather than for any other reason. Extraordinary remedies, however, formed the rule, so long as physicians knew neither the physiology nor the pathology of the human economy.

26. Notwithstanding the way in which Fioravanti was denounced as a quack, his books maintained their circulation. In addition to those already shown and described, I have here some fresh copies to exhibit. One is a copy in Italian of his *Secreti Rationali* in five parts, printed at Venice in 1630. There is nothing noteworthy about the volume, except that it is in the chap-book style of the period.

The other is the first edition of the French translation of the *Miroir Vniversel des Arts et Sciences en General, . . . diuisé en trois liures*, by Gabriel Chappuys. It was printed in 1584 at Paris by Cavellat, who was also the printer of Cardan's *Livres de la Subtilité*, mentioned above. It is an 8vo, and this copy has been much read and thumbed. Chappuys' editions must have been popular, for I have never succeeded in meeting with any of them in good condition. The second edition, 1586, was mentioned in Part II., p. 251.

Fioravanti's Discourse upon Chirurgery went through several editions. The first edition of the English translation was printed at London by Thomas East in 1580, in small 4to, and in black letter (Part V., p. 452). In addition to a copy of this first edition, I have got also a later one, edited by Richard Booth, Gent., and printed at London by Edward Allde in 1626. This is also a small 4to, 2 leaves, sigs. B to Q in fours, R in two, or pp. [4] 117 (misprint for 118) [6], and to this, as in the other, are added "most notable "secrets found out by the said Authour."

The later edition differs from the earlier one by omitting Hester's dedication to the Earl of Oxford and the coat of arms, Hester's address to the reader, and Fioravanti's address to the reader, and by inserting an address to the reader by Richard Booth. At the end is also added, in the second edition, "the making of certaine precious waters taken out of John Vigo, Chirurgion."

27. A book devoted entirely to medical secrets, without the admixture of anything artistic or technical, was composed by another Italian:

> **Marauigliosi Secreti Di Medicina E Chirvrgia, Dell' Eccellente Medico Il S. Gio. Battista Zapata Nvovamente Ritrovati. Con l'aggiunta d'altri ſecreti, raccolti dalli ſuoi Diſcepoli. Con Licenza De' Svperiori, & Priuilegio per anni XX.**
> **In Roma, Per Tito, E Paolo Diani. M. D. LXXXVI.**
> **8vo. Signatures: a, b in eights, c in four, A to R in eights, or pp. [40] 272.**

This volume presents a rather peculiar appearance by the pages being surrounded by lines, by the catchwords, the marginal rubrics, and the running title being divided off by lines, and by the chapters and their titles being separated by three lines. Though this book deals professedly with the treatment of a great number of ailments, and explains the preparation of the necessary remedies, it contains incidentally a good deal of chemistry as applied to pharmacy. This makes it valuable for the history of the science, for here we get the processes employed three hundred years ago for preparing certain well-known compounds, uncomplicated by any theoretical views about elements or transmutation. Amongst these, Chapter XV. is of special importance, for, according to it, Zapata had invented a method of making oil of vitriol:[1] *Oglio di vetriolo, secondo la nostra inuetione*, p. 127. He had also various compounds of antimony, a "litharge water" or solution of litharge in acetic acid, and other preparations, which may be omitted at present.

There are other editions: Venice, 1641, 1677; in Latin, Ulm, 1696; in German, Franckfurt-a-M., 1605; all of which are in the British Museum.

Zapata seem to be very little, if at all, known, as his name does not occur in the biographical dictionaries.

In the Bodleian Catalogue (Oxon. 1843, III., p. 411) there is this entry:

> **Scientia (Gioseppe)**
> **Li maravigliosi secreti di medicina e chirurgia, raccolti dalla prattica di Gio. Battista Zapata.**
> **8°. Venet. 1595.**

I have not seen this copy.

---

1 **This process is quoted by Caneparius under the title, "Oleum Vitrioli potabile ex Zapata," in his book *De Atramentis*, Roterodami, 1718, Lib. VI., cap. xl., p 479 for which see § 36 below.**

28. It so happens, by the accident of date, that most of the items contained in the volume of pamphlets already quoted (B.M. 1038. d. 35), have come all together. They are small, require no comment, are mostly anonymous, and several have no dates. In their case I have risked 1620 as about the probable time of their publication. Those which have dates present, of course, no difficulty. They begin with 1600, 1604, 1606, then five belong to about 1620? and one is dated 1626. They are as follows:—

Tesoro Di Varii Secreti Natvrali. Tratta da diuerſi Auttori famoſiſſimi, & più volte ſprementati in molte perſone, Opera molto gioueuole alla ſanità de' corpi humani, Con la dichiaratione di molti dubij. Et il modo di far & conſeruar il Vino. Con Licenza de i Superiori.
In Venetia, MDC. Appreſſo Gio. Battiſta Bonfadino.
Small 12mo. Sig. A in 12.

There is nothing special about this except the preparation of wine. The first part contains some remedies for common disorders. (1038. d. 35 (4)).

This was followed by a book with a similar title. (1038. d. 35 (11)). It is, in fact, a reprint of a portion of the above, in chap book form.

Tesori Di Varii Secreti Natvrali. Raccolti da diuerſi Eccellentiſs. Auttori. [Device.]
Stampata in Venetia, e riſtampata in Verona, Per Franceſco dalle Donne. 1604.
Small 8vo. Title within a narrow border. Sig. A 4.

29. In the same volume [1038 . d . 35 (8)], under Antonio Venetiano, there is the following:

Opera Nella qualle trouerai molti bellisſimi Secreti. Vtili, Et Necessarii li quali ſono ſtati eſperimentati. Nouamente dati in luce, per beneficio d'ogni corpo humano. Ad' inſtantia d'Antonio Viniciano. [Vignette.]
Stãpata in Papoa con licentia ſuperiori.
Small 8vo. A in 4. Chap-book, no date, but about 1620.

The vignette is a man with a bagpipe sitting, one with a staff standing, a dog, and some sheep, two trees, and a house or two in the background. It is a very large display for such meagre results, for there are only a few medical receipts for burns, worms, deafness, white hands, and for sleep.

30.

B.M., 1038 . d . 35 (9):

Secreti Mirabili Per diuerſe infirmità ſperimentati Da molti Eccellentiſſimi Medici, cauati da diuerſi Authori. Et io Oratio di Gouerri hauendone fatto l'esperienza, mi è parso di metterli in luce, a beneficio, & sanità de gli corpi humani.

Small 8vo. Title enclosed in a broad border. No date, place, or printer, but about 1620.

Sig. A in 8. Printed in italics. Last leaf contains the table.

These secrets are entirely medical, and, though similar to the other collections, are different from them.

31.

B.M., 1038. d. 35 (10):

Secreti Vtilissimi e nuoui. Doue s'intendono molti particolari vtili per salute delli corpi humani. Nuouamente posti in luce, ad instanza del Pesarino gran Giocator di mano. [Device, the full sun.]

Stampati in Brescia, con licenza de' Superiori. E ristampato in Verona.

Small 8vo. A in 4. Italics. No date [1620].

Miscellaneous receipts—to make hair grow, to remove hair, to make strong vinegar, against intestinal worms, to send a secret letter *(by a carrier pigeon)*, etc., etc.

One cannot expect very much of a chap-book of four small leaves.

32.

B.M., 1038. d. 35 (13):

Giardino Et Fioretto De Secretti. Tradotto da molti Autori, si antichi, come Moderni, à beneficio di ogni persona. Con Gratie, & Priuilegij. [Device.]

Stampato in Venetia, ad instantia di Fiorauanti Cortese.

Small 8vo. Sig. A in 4. The title is within a narrow border. The device is a winged lion holding a book, with a tower in the background.

This is a collection of receipts like the others. It is a chap-book about 1620. The author, or printer, or publisher seemed to know what he was about when, with a touch of humour, he put the names of Fioravanti and Cortese together as concerned in its publication.

33.

B.M., 1038. d. 35 (14):

Il Medicinal Tesoro. Di Vespasiano Angelico Cittadino Veronese. Opera a chi brama la Sanità vtilissima, Con alcuni secreti Filosofici. Nouamente stampato, con licenza del collegio di Medici di Padoa, Et di Signori Superiori.

Small 8vo. Ornamental border round the title. ff 4. No signature, no date. An early 17th century chap book. Same kind of collection as the others.

Vespasiano Angelico seems to be a pseudonym, but I have not found the name noted as such by Placcius or by Melzi.

34.
B.M., 1038. d. 35 (7):
Centvria Di Secreti Politici, Cimichi, e Naturali; Dispensati da mè Francesco Scarioni da Parma. A' requisitione de' Spiriti curiosi, & amatori di virtù.
In Venetia, Verona, Parma, Lodi, Milano, Bologna, & in Padoua per Pietro Paolo Tozzi. M.DC.XXVI. Con Licentia de' Superiori.
Small 8vo. Sig. A in 8.

Contains 100 brief receipts, and an index of them. Scarioni may be the author. I have not been able to check him, but there is just a suspicion of quackery in putting half-a dozen places on the title-page of a pamphlet of eight leaves.

35. Glassmaking, of all the arts, is one of the oldest and most interesting, and the literature relating to it, though quite modern, is also interesting. The oldest work on the subject, that of Neri, is a rather fascinating production. The interest of it lies in the light it throws on this branch of technical chemistry, for the chronology of the book is quite simple. A discussion, however, of the contents must be undertaken elsewhere. Besides another copy of the rare first edition of 1612, and of the English translation of 1662, I can show a copy of a reprint, in cheap form, Firenze, 1661. The book is apparently hardly out of date even yet in Italy, for I have also a copy dated Milan, 1817, which would indicate that the demand for it continued two hundred years. There is also here the reprint of the English translation, which appeared in a preposterous folio at Middlehill, in 1826. There are other editions in the British Museum, but the account of them must be reserved in the meantime. Immediately connected with Neri's art of glass is the German translation of it amplified with annotations by Johann Kunkel. The edition of this work of 1679 has been already mentioned by me. I have now got a reprint of it dated 1689. There is no essential difference between these issues. Geissler's German translation I formerly said (Part II., p. 268) contained nine plates, but in a copy which I have since acquired there are ten plates. The French work on glass by Blancourt was previously noticed. At the time I had the English translation in one volume, dated 1699—indicative of an earlier French version. Such I have not seen, but I have one in two volumes, 12°, printed at Paris in 1718. No allusion is made to Neri, but it is essentially based on Neri's book, and the illustrations are practically the

same as his. At all events in any review of Neri's book and art, it would be necessary to take into account both these translations as illustrating the original work.

36. In a handsome quarto volume the monograph of Caneparius *De Atramentis* was reprinted at Rotterdam in 1718. The earliest edition I know of and have already mentioned was printed at Venice in 1619,[1] and an edition appeared at London in 1660. Both of these are inferior in style to the edition of 1718. The London edition differs from that of Venice by the omission of the long table of contents of the six sections into which the book is divided. The Rotterdam edition also omits this important table, but in it the table of authorities is re-arranged in a more strictly alphabetical order. The text, so far as I have compared the present with the earliest edition, is a literal reprint without modification or addition. It is well worth observing that a treatise on such a practical subject as is discussed in this volume could literally stand still for a whole century, and be then reprinted without change, presumably for the use of those interested in it. One can only suppose that the subject itself stood still for a hundred years. This, however, would not be correct, for in the interval a very great deal had been done to elucidate the nature of salts and their reactions, the making of sulphuric acid by the bell had been introduced, and there had been a great deal of discussion about the nature of the chemical elements or principles, and on the cause of combustion. The author, of course, was not alive to take cognisance of these changes, but it is instructive to consider that nothwithstanding its being behind all these advances, it was still thought worth reprinting for practical purposes. It would hardly be possible to reprint a book on a practical technical art published at the end of last century for use at the present moment.

At the present time, however, Caneparius's book is full of interest to the student of the history of chemistry. The first part treats of pyrites and analogous minerals, and of various theoretical questions which arise out of it as to the generation of metals, the nature of fire, etc.; the second part of metallic *atraments;* part three is devoted specially to green vitriol; part four to different kinds of writing and printing inks; part five to inks of different

---

[1] There is said to be an edition of Venice, 1618; Clément, *Bibliothèque Curieuse*, 1756, T. vi., p. 195.

colours, and incidentally to pigments, including cinnabar, red lead, lapis lazuli, etc., etc. The last book is taken up entirely with oil of vitriol.

37. The fertility of Italian writers in producing these books of secrets is evinced by another, which is quite new to me. It is entitled:

> **De' Secreti Vniversali Di D. Timoteo Rosselli. Parte Prima, Si per huomini, & donne di alto ingegno, come ancora per Medici, & ogni ſorte di Artefici induſtrioſi, & ad ogni perſona virtuoſa accomodati. Con Privilegio.**
>
> **In Venetia MDCXXXXIV. Appreſſo il Barezzi.**
>
> **8vo. Signatures: a, A to R, in eights, or ff. [8] 136.**
>
> **Parte Secondo; *Nella quale ſi contengono innumerabili rimedi vtili a diuerſe infermità, & altre coſe belliſſime, ingegnoſe & artificioſe.* a, A to Q in eights, R 4; or ff. [8] 132. The second part has a distinct title.**

It is in two parts, and the contents fully confirm its title, for it gives receipts for every conceivable contingency in medicine and practical arts. How far the author is a compiler it is not easy to determine, but I should suppose that he has not been loth to avail himself of the collections of the pioneers along the same path, such as Alexis and the author of the *Kunstbüchlein*. The determination, however, of that point would lead me into a discussion of the contents, for which this is hardly a fitting time and place.

38. Auda's *Breve Compendio di Maravigliosi Secreti*, in four divisions, dedicated to Francesco Maria Febei, was printed at Rome in 1663. It was already in its fifth edition, to which were added other "bellissimi Secreti" by the author. It is an 8vo volume, printed roomily with nice legible type, and it contains †, ††, A to V, all in eights; X in three, or pp. [32] 326. X 4, probably blank, is wanting. There is a copy in the British Museum, 1038. d. 17.

An edition of later date purports to be the same. The title is as usual, the most important part being this addition:

> **In queſta quinta impreſſione ricorretto, & ampliato di belliſſimi Secreti dall' iſteſſ' Autore.**
>
> **It was printed: In Torino, Per Gio. Sinibaldo. M.DC.LXV, and forms a small square 8vo, sigs. +, A to Y in eights, or pp. [16] 352.**

The additional secrets, if we may trust the bastard title which ushers them in, appeared for the first time "in queſta vltima impreſſione di Torino," enlarged by the author to more than double, "with most beautiful and useful secrets." To check this statement it would be necessary to compare this fifth edition with the fifth edition of Rome, which I have not been able to do.

Then there appeared in 1673, both at Venice (B.M. 1038. a. 16), and at Bologna (B.M. 1038. c. 23), an edition in 12mo. These books are much less pleasant to deal with than that of Rome, 1663. They are larger, however, containing a few extra receipts at the end. Otherwise the three copies are identical. There came another dated "Venetia, 1676. A San Giuliano. Appresso Giacomo Zini." It is a mere reprint, and the added secrets are put in without any reference to their source. It agrees with the Venice edition of 1692, formerly described, being in 12mo, and containing: ✠ in ten, A to M in twelves, N in fourteen, or pp. [20] 316.

Whether any editions appeared under Auda's name subsequent to 1692, I do not know. If there were, the success of the book, which in various editions carried it through an interval of not less than fifty years, was sufficiently assured to make Auda indisposed to withdraw his name from it. It forms, therefore, a bit of a bibliographical puzzle to know why, in 1711, the book appeared not only without Auda's name, but with the entirely new name of Quinti.

39. Allusion has already been made (Part IV., p. 322) to a small volume by a Dr. Quinti in French, which purports to be a translation from the Italian, but which is neither more nor less than a version of Auda's Secrets. My notion then was that Quinti had translated the book, and had claimed the authorship, for he speaks of the original Italian having been often printed; or else that Quinti was merely a fictitious name for Auda. Auda's book goes back to 1663 at least; Quinti's translation "printed at Venice and sold at Liege"—a rather queer sort of mixture—is dated 1711, so that the two names may have been used by one person.

The following, however, which I found in the British Museum (1035.a.35) seems to have been the original of the French translation:

> Maravigliosi | Secreti | Medicinali | Chimici | Sre. Giuseppe Quinti, | Doctore di Venetia. | Ricolte per molti cure & lavori, che | sono stati da egli stesso isperimenta- | ti, & provati più volte in parecchie | infermite. |
>
> Venetia, | Et si vendono | Leodii, Dal J. F. Broncart, | in Supremo-Ponte | M.DCCXI.
> Small 12mo, *, A to O in twelves, pp. [24] 336.

*1 is an engraved frontispiece. The rest of the signature is occupied by the title, dedication to "Molt Illustre Sign," by Quinti, and is dated: Milano li 23 Decembre, 1702 (not dated in the French version), followed by the preface, Sonnets and Tavola as in Auda's edition. Then follows the text. Some of the receipts are differently arranged in Auda's and Quinti's editions, but practically the two books are identical. The French version runs parallel with this even to the place and date, but the two differ in the last book. The variation, however, does not affect the fact of their simultaneous—or seeming simultaneous—publication, and it makes it still more difficult to understand how Quinti, if different from Auda, could reprint the book of the latter with his own name, and then in a French version refer to the Italian original which, he says, had gone through several editions. And if Quinti was a name assumed by Auda late in life, what did he mean or gain by it?

In his dedication to some illustrious person, not named, Quinti says:

> Ces Secrets de Medecine que je fais paroître au jour, ont été si bien reçûs du Public, qu'encore qu'ils ayent été imprimez plusieurs fois à Venise; on n'a pas laissé de les redemander avec empressement aussi-tôt que les Exemplaires en ont été distribuez: C'est pourquoi afin d'obliger tout le monde & principalement ceux de ma profession, & leur donner une plus grande intelligence & plus de facilité en cet Art, je fais imprimer de nouveau ces Secrets, avec l'augmentation d'un quatriéme Livre. J'espere qu'il sera aussi bien reçû que les trois autres qui avoient paru ci-devant:

and then adds the usual flourish about the protection of the Tres-Illustre Seigneur, which it is unnecessary to quote.

This is ingeniously put, for while it might and does leave the impression that Quinti is speaking of a book of his own, he does not say it in so many words, although on the title-page, both of the Italian and French, the book is ascribed to him. At the moment I have no explanation to suggest as to this republication of Auda's collection under another author's name.

40. Auda was the author of a treatise on the theory and practice of pharmacy. It is entitled:

> Practica | De' | Spetiali | Che per modo di Dialogo con- | tiene gran parte anco di | Theorica. | Opera | Vtile, e neceſſaria per queſti, che deſiderano | ben comporre li Medicamenti. | . . . | Con vn Trattato delle Confettioni noſtrane | per vſo di Caſa. | Et vna nuoua aggiunta de' Segreti vtiliſſimi. | . . . | In Venetia, M.DC.

LXXIV. | Preſſo gli IIH: di Gio: Battiſta Ceſtari | A ſpeſe di Zaccaria Conzatti. | . . .

Small 12mo. Sigs. A to O in twelves; or pp. [1-12] 13—329 [5]. O12, probably blank, is wanting.

The Trattato delle Confettioni and the Nvova Aggivnta Di Secreti have title-pages both dated M.DC. LXXIII., both included in the signatures and pagination.

The first or theoretical part gives explanation about operations and general descriptions of certain drugs. Then follow electuaries, pills, syrups, lozenges, plasters, oils, waters, etc., etc. The second tract deals with a few preserves, olives, citron peel, etc. The additional admirable and useful secrets are receipts for alleviating certain troubles and maladies, both external and internal.

From the fact that these are said to be added, I presume this is not the first edition of the book, but I have no means of ascertaining, as it is not mentioned by any of the authorities I have consulted, and there is no copy of the book in the British Museum.

41. Among the little medical secret books which swarmed in Italy in the seventeenth century, was this:

Secreti Medicinali Del Sig: Alessandro Ventvrini Ne' quali ſi contengono i più ſcelti rimedi, che ſi cauano da gli Animali per ſalute dell' Huomo. Nuouamente accreſciuti d' importanti Secreti dal Sig. Francesco Pignocatti Al Molt' Illuſtre Sig Il Sig Gasparo Bassi.

In Bologna, M.DC.LXXII. Per Gio: Recaldini. Con licenza de' Super.

Small 12mo. Signatures: 4 leaves, A to E in twelves, F in eight. Pagination: [8] 136.

This is apparently an Italian chap book; anyhow it is shabby enough to have been an English one. It has, however, the merit not merely of being an entirely new book to me, but of dealing with a new subject in a way to which I remember no strict parallel, and only a partial resemblance in the works of Lovell and of Schröder. It gives a very brief summary of all the drugs obtainable from man and other animals. It is, therefore, strictly a book of medicines, rather than a book of medical secrets, for it would require a counterpart, or at least an index of diseases, with reference to all the drugs useful for these.

These are the various Italian collections which I have either acquired or examined during the last few years. Considering their number, and that a proportion of them are quite new, I should infer not that this is an end of them, but that there must be a considerable number still, especially of chap-books. Accident may bring some more of these into daylight, as it has already done with those described.

In the next part of this supplement, I shall consider another portion of the books which I have collected.

Note to § 8. October 10th, 1894.

In the Bibliothèque Royale at Brussels I have found, since my paper was in type, some additional information as to Panciroli's book. A copy of it was in Van Hulthem's collection, and appended to the entry in the catalogue (*Bibliotheca Hulthemiana, ou Catalogue Méthodique de la riche et precieuse Collection de Livres et des Manuscrits délaissés par M. Ch. Van Hulthem.* Gand, 1836. 5 Tomes, 8vo) there is the following note:

> Traité curieux, dont la traduction par *Salmuth* parut en 1599 avant l'original, composé pour le duc de Savoie. Comme on ne peut retrouver la copie du texte primitif, Gualterio, à la prière des neveux de Panciroli, suit l'ouvrage en italien, d'après le texte latin. Voy. *Biogr. univ.* vol. 32, p. 479, et la note.

This is an unexpected confirmation of the suggestion made in my note to § 8, that the Italian edition might have been re-translated from the Latin.

In the same library I found the only copy I have met with of the first edition of the second part, accompanying the first edition of the first part: 1599 and 1602. The 1599 volume is also in the British Museum, as I have already said, but not that of 1602. No apology, therefore, is required for the following account of the first issue of this book:

> Rervm | Memorabilium | Iam Olim | Deperditarum: | & contrà | Recens Atqve Ingeniose | Inventarum: | Libri duo, | à | Gvidone Pancirollo: | Ic. Clariss. | Italicè primum conſcripti, nec unquam | hactenus editi: | Nunc verò & Latinitate donati, & Notis quam- | plurimis ex ICtis, Hiſtoricis, Poëtis & Phi- | lologis illuſtrati | Per | Henricum Salmuth. |
>
> Ambergae, | Typis Forſterianis, | Cum privilegio. | M. D. IC. |
>
> Small 8vo. Signatures:)(, A to Z, Aa to Zz, Aaa to Ccc, in eights; or pp. [16] 752, 31 [1 blank]; last leaf, blank (?), is wanting.

Signature)(contains the Title, Salmuth's Præfatio to Frederick Count Palatine, dated: Amberg, 5th March 1599, Joachim Camerarius to Salmuth, commendatory verses, titles of the chapters, and a quotation from Scaliger. The text occupies pp. 752. Then follow a list of laws discussed in the book, Index Rerum et Verborum, and Errata.

On Ccc 7 *recto* is Forster's device, and beneath:

> Ambergæ, | Ex Officina Typographica Mi- | chaëlis Forsteri. | M. D. XCIX. |

The verso is blank, and Ccc 8 is wanting.

The second volume has the following title page:

> Nova | Reperta, | Sive | Rerum Memorabilium, | Recens Inventarum, | et | Veteribus Plane | Incognitarum | Guidonis Pancirol' | li I C. | Liber Secundus. | Jam primùm ex Italico Latinè reddi | tus, & Commentariis illuſtratus | Ab | Henrico Salmuth. |
>
> Ambergae. | Typis Forſterianis. | M.D.CII. |
>
> Small 8vo. Signatures: (:) in eight,):(in two, A to Z, Aa to Zz, in eights, Aaa in four; or, pp. [20], 719, [21, 2 blank].

The 20 preliminary pages contain the Title, Salmuth's preface to Christian, Prince of Anhalt, dated: Ambergæ, Kal. Martii, 1602, dedicatory and commendatory verses, and heads of chapters. The text occupies pp. 719. The unnumbered pages at the end contain the laws discussed in the volume, Index Verborum, and table of Errata. The last leaf is blank.

The earliest editions, therefore, seems to have been as follows:

Part I.			Part II.		
First edition,	Amberg,	1559.	First edition,	Amberg,	1602.
Second edition,	Amberg,	1607.	Second edition,	Amberg,	1608.
Third edition;	Amberg,	1612.	Third edition,	Amberg,	1612.
Fourth (?) edition,	Frankfurt,	1622.	(See Part II., p. 243.)		

The original Italian had been circulated in manuscript, and a copy had reached Salmuth, who had been a student under Panciroli at Padua, and who translated it into Latin. The translator, however, was a Lutheran; the book acquired a heretical strain, and it was ultimately condemned. Carlo Emanuele, Duke of Savoy, however, and the nephews of Panciroli, thinking this was a severe punishment of the innocent author, at last succeeded in getting Flavio Gaultieri to make an orthodox version in Italian, excluding the objectionable matter which Salmuth had introduced, and this was published at Venice by Barnardo Giunti in 1612, 4°, as described above. This Italian version, therefore, is not the original, and Salmuth's translation, apparently, is not an accurate rendering of the original, so that what passes under Panciroli's name is not his book at all, but only an approximation to it. In this respect it is possibly unique among all the works I have had to specify. See Zeno's note in *Biblioteca Dell' Eloquenza Italiana di Monsignore Giusto Fontanini*, Parma, 1803-04, 4°, Tomo II., p. 276, note (*a*). Reusch, *Der Index der verbotenen Bücher*, Bonn, 1883-85, II., p. 161. *Index Librorum Prohibitorum et Expurgatorum*, Genevæ, 1619, 4°, p. 431, where a list of the condemned passages will be found.

BIBLIOGRAPHICAL NOTES

ON

HISTORIES OF INVENTIONS

AND

BOOKS OF SECRETS.

SECOND SUPPLEMENT.

BY

JOHN FERGUSON, LL.D., F.S.A. LOND. AND SCOT.

REGIUS PROFESSOR OF CHEMISTRY IN THE UNIVERSITY OF GLASGOW.

GLASGOW:

Printed at the University Press by

ROBERT MACLEHOSE & CO.

1897.

Read to the Archæological Society of Glasgow, 15th March, 1894.

100 *Copies Reprinted.*

No. VI.

BIBLIOGRAPHICAL NOTES ON HISTORIES OF INVENTIONS AND BOOKS OF SECRETS.

(*SECOND SUPPLEMENT.*)

BY

JOHN FERGUSON, LL.D., F.S.A. Lond. & Scot.,
REGIUS PROFESSOR OF CHEMISTRY IN THE UNIVERSITY OF GLASGOW.

(*Read at a Meeting of the Society, held on 15th March, 1894.*)

1. The "Notes" which were communicated to the Society on March 15, 1894, formed a supplement embracing the histories and receipt books which I had examined during the previous six years. In printing them, however, it has been found necessary, in consequence of their bulk, to take them in sections. The first of these, called the First Supplement, was printed in the *Transactions* for 1895, Vol. II., p. 364. It contained a list of histories of inventions and of collections of receipts written in Italian or by Italians. I had intended to pursue this method of division by countries with the remaining examples, but trial has shown me that while an arrangement based on language or nationality may be valuable for some purposes, it is not suitable historically for giving one a survey of the general progress of the arts, and still less bibliographically, since it separates translations from the originals, and thus fails to afford a complete view of the distribution and importance of a book. Keeping the bibliographical as the more accurate method of description before me, I have reverted to my original plan of treatment. The authors, therefore, in this Second Supplement, follow in chronological order, so far as I can

ascertain it, and the editions of their books are taken together, irrespective of the language in which they were written, or into which they have been translated.

Histories of Inventions.

2. In the previous supplement I had to enumerate the histories of inventions which I had acquired. Since then I have seen only two or three items which could be ranged under this head.

3. The first is a copy of Polydore Vergil's three books, without place, date, and printer's name, but supposed to have been printed at Strasburg by Schurerius about 1509. It is a small 4to, with an ornamental border round the title, and it contains the poem of Sabellicus at the end. There are copies in the British Museum and Cambridge University libraries. This is probably the first edition of the book which appeared in Germany, and it is different from that of Schurerius dated 1509, quoted in the list given in the first supplement.

In the same list there was included Han's undated edition in 8vo of the German translation, which contains a number of woodcut illustrations. Before it, however, the same translation had appeared in small folio, printed by Heinrich Stayner at Augsburg—the first edition in 1537, the second in 1544. Both are rare; for, during all the time I have searched for these books, it is only quite recently that I have acquired a copy of the 1544 edition. These editions also are enriched with woodcuts, boldly executed, and important as depicting the costumes, weapons, implements, buildings, actions, and ideas of people three and a half centuries ago. The 1544 edition contains the eight books, and has 10 preliminary leaves, and clxxi of text. There is a copy of the first edition in the University library at Cambridge, and of the second in the British Museum. Both are mentioned by Beckmann as being in the Göttingen library.*

There is also a translation of the first three books into Dutch, which appeared at Amsterdam in three volumes in 16mo. Besides my own copy, the only one I have seen is in the University library at Ghent. It has no date, but it seems to have been issued late last century.

---

* *Beyträge zur Geschichte der Erfindungen*, Leipzig, 1792, III., pp. 574, 575.

4. In the second part of my "Notes" (p. 244) reference was made to the work in which Dutens endeavours to show anticipations by the ancients of modern inventions. It seems to have acquired a certain amount of reputation, for at long intervals editions of it appeared. Those originally quoted were the English of London, 1769, and the French of London, 1796—the latter entitled the third edition. I have now got a copy of the so-called fourth edition, in French, published at Paris in 1812, in two 8vo volumes, containing respectively: I., pp. [4] 281; II. [4] 420. Whether the numbering of these editions is consecutive or not, I am unable to say.

5. The only other work belonging to this section of the subject I have come across, is a little volume prepared for the instruction of youth, entitled *Petit Dictionnaire des Inventions*. The third edition was published at Paris in 1825, and contains pp. [2] 11–286, with a symbolical frontispiece. The author gives a succinct account of ancient and modern discoveries in a manner as likely as possible to attract the readers for whom the book was prepared. It may be viewed as a modern counterpart to the *Pleasant and Compendious History of Inventions*, which was published at London in 1686, and was an alphabetical abstract of Polydore Vergil. From what is said on p. 284, this summary was written in the year 1812, and it is possible that this is not really a new edition, but a portion of the first issue with a new title page. The fact that pp. 1–10 are wanting seems to point to the cancelling of some preliminary matter which may have been thought not appropriate to the new title page. The book is interesting now as indicating what was considered worth knowing about discoveries and inventions at the beginning of the present century.

Books of Secrets.

6. The books which have come together for consideration on this occasion, refer as it happens, with one or two exceptions only, to medical secrets. They are all prior to 1600, at all events in composition, and for the most part in publication.

7. Under the name of Aristotle there has passed current for centuries

a treatise on the secrets of generation. So far as I am aware, it does not correspond to any of his original works, though it may be a later compendium from some of them. It is not to be confounded with the *Secret of Secrets*, another work ascribed to him, to which I shall refer presently. That upon generation was in great vogue, and cheap editions, accompanied by other tracts of similar tenor, used to be abundant, and are possibly still to be had. The treatise, however, is accounted spurious, and it is put among the pseudonymous Aristotelian works in the British Museum catalogue; but the authorship is a question of secondary importance at the present moment; it is sufficient that the book exists. The edition I have is not common, and it may, therefore, be described with some minuteness.

> Aristotle's | Manual | of Choice | Secrets, | Showing | the Whole Myſtery of | Generation. | With | Receipts to prevent | Barrenness, | and Cauſe | Conception. | Very Neceſſary to be Known | and Practiced by all *Midwives*, | *Nurſes*, & *Young Married Women*. | Tranſlated out of *Latin* by *J. P.* | *London*, Printed for *John* | *Back*, at the *Black-Boy* on | *London-Bridge*, 1699. |

It is a small narrow 24mo, signatures A to F in twelves, misnumbered pp. 135 for 143, and 1 blank. This is a volume of varied interest: it is a London Bridge book; it seems to be the first edition, but this I should not like to affirm too strongly; it has a frontispiece of a birth, in which the attendant gossips are all—like so many hospital nurses—attired in a sort of uniform, with queer head-dresses; I cannot find it in the British Museum catalogue, although there are other editions there which resemble it; and I conclude that it is rare. The contents of this tiny manual—all things considered—are sober and sensible; though, of course, in some matters the author could not deliver himself from certain misconceptions of his time. As for the Latin original, I have not seen it, and I have failed to identify J. P. the translator.

Lowndes takes no notice of the book at all, but one finds instead this note: "Many other works have been published to which the name of Aristotle has been falsely affixed, but they are of no value or estimation." The comment upon which is that one wants descriptions of the books, not opinions as to their demerits; "I do not desire you to please me; I do desire you to sing!" If would-be bibliographers acted on the principle of mentioning only books of merit, what need would there be for bibliographies at all?

8. The other work ascribed to Aristotle—a letter on the regulation of life, addressed to Alexander the Great—was mentioned in Part IV., p. 325, in connection with the English abstract of 1702. I have now got a copy of the complete work. The title is as follows:

SECRE
TA SE-
CRETO-
RVM.
ARISTO
TELIS:
Cum priuilegio.

It is a small square 8vo; A to O in eights (P dropped?), Q in 2; folios numbered i-cxiii, the last leaf not being numbered. Black letter, 31 lines to the page. On the *verso* of the last leaf is the colophon:

> Explicit feptifegmentatum opus ab Alexandro Achillino ambas ordinarias & philofophie & medi cine theorice publice docente: vt non amplius in te- nebris latitaret editus Et impreffus Parifiis An no domino. 1520.

The title page has a quaint appearance. The title, arranged irregularly in large Roman capitals, is in red, and is enclosed in a border of separate pieces in black. The words *Cum priuilegio*, in gothic letter, are also in black.

No copy of this book is mentioned by Panzer, Brunet, or other bibliographer. It is a coincidence that, of the various editions of the book before and after 1501, it should be that of 1520 which has come to be described now, after the lapse of eleven years from my former notice of the work. For it was this very edition of 1520 which the editor of the abstract of 1702 says was so rare, that "it was not to be bought or seen (as far as he could find) anywhere but in the Bodleian Library at Oxford, and with one Gentleman in London."* In the British Museum there is a copy of the

* See Part IV., p. 326.

1520 edition (520.a.12), but I have not as yet examined it. Can it be the copy which belonged to the "Gentleman" just named?

In the Bodleian catalogue there is no copy of 1520, but I find instead the following entries under Aristoteles:

> De secretis secretorum ex Arab. MS. in Lat. per Philippum quondam clericum. 4° s. l. et a.*
>
> ... acc. de signis tempestatum, ventorum et aquarum et de mineralibus, Lat. ed. Alex. Achillino. fol. Bon. imp. Bened. Hectoris. 1516.

I have not seen either of these books, and am unable therefore to speak with certainty; but I cannot help thinking that the former, *De secretis secretorum*, is identical with the first section (or part of it) of the 1520 edition, entitled "Secretum Secretorum," and that the latter corresponds with the section immediately succeeding that in the 1520 edition: "Maximi phylofophi Aristotelis de fignis Aquarum: ventorum: & tempeftatum," and "Aristotelis de mineralibus." I infer, therefore, that these two editions were practically reprinted in 1520. The later edition contains besides: "Alexander (*sic*) Aphrodifei de intellectu"; "Averrois de beatitudine anime"; "Alexandri Achillini Bonienfis de Univerfalibus"; and "Alexandri Macedonis ad Ariftotelem de mirabilibus Indie."

The treatise "Secretum Secretorum" has been known for the last eight or nine hundred years. It seems to have been originally an Arabic compilation which was made from the Greek, amongst other sources, and was turned by clerk Philip, whoever he was, into Latin. The part relating to physiognomy formed the groundwork of the tract on the same topic, which was edited by Michael Scotus, so that what there is to say about it had better be reserved for my bibliography of the works of that writer.

The English version of 1702 includes only a part of the treatise as it appeared in 1520, and of what is given portions are condensed, so that the editor quite correctly announces that it is "contracted." From what he further says, the version published in 1528 was also shortened, but I am not in a position to confirm the statement. The work, however, as I said formerly, falls within the scope of these notes rather on account of its name than of its contents.

---

* This seems to be Hain's No. *1779.

9. Of the famous works ascribed to Albertus Magnus, *De Secretis mulierum* and *De Secretis naturæ* or *Liber aggregationis*, several copies have recently come into my hands. Notwithstanding the multitude of editions through which they passed, copies are not common. As for those printed previous to 1501, they seldom appear in catalogues at all.

The earliest example which I have met with lately has no date or place, but was printed at Paris in the early years of the sixteenth century. The device on the title page is that of André Bocard,* who began to print at Paris a few years prior to the sixteenth century. The present work, I have no doubt, is subsequent to 1501. It is not in Hain, and I have failed to identify it with any of the editions in the Museum catalogue, or in Brunet or Graesse.

Secreta mulierū et virorum | ab Alberto magno compoſita. | [Device.]
No place, date, or printer.
Small 4to. Signatures: a8, b and c in six, d8. No pagination. 42 lines. Black letter.
Collation: [ai] *recto*: Title page. The device consists of two tree trunks, with flowers. On the dexter trunk is suspended a shield with the arms of the University of Paris; on the sinister trunk a shield with the arms of the city of Paris; above all are two angels carrying a shield with three fleurs-de-lys, the arms of France, and above the shield is a crown. Round the device are the words: HONNEVR · AV · ROY · ET ⁘ ALACOVRT ◇ SALVT · ALVNIVERSITE ◇ DONT · NR̄E · BIEN · PROCEDE · ET · SOVRT ⁘ DIEV · GART · DE · PARIS · LA · CYTE. ⁘ The *verso* is blank.
aij *recto*: ſCribit philoſophus p*h*orum priceps quarto ethicoꝝ | Homo eſt optimū eoꝝ que ſunt in mūdo. Et mūdus | etc.
Ends dviij *verso*: Ca. duodecimum *e* vltimum de gn̄atione ſpermatis in viro. |

Since the preceding was written, I have acquired another Paris edition:

Secreta mulierum | Et virorum nuperrime corre- | pta et emendata. 58 | [Device.]
Uenundātur Pariſiis ab Joan | ne Paruo cōmorāte ī vico ſancti | Jacobi ſub ſigno Lilii aurei. |

Small square 8vo; a to e in eights, f in four. In small black letter, 33 lines to the page. The device on the title page is Petit's initials on a shield,

* Louis-Catherine Silvestre, *Marques Typographiques*, Paris, 1853, No. 5. He says Bocard printed from 1496 to 1531, but according to the following authority a book seems to have been printed by him in 1494; *Gravures sur bois tirées des Livres Français du XVe Siècle*, Paris, Adolphe Labitte, 1868. 4to. Bocard's device is Fig. 324, and is on the title page of this volume; a suitable vignette for a Paris book.

suspended on a tree, with cupids and birds in the branches, the shield supported by a lion and a leopard.*

An edition of the parallel work of Michael Scotus was also published by Jean Petit, but the two are not quite uniform.

An edition similar to Bocard's, undated, but printed early in the sixteenth century at Paris, I found recently (April, 1896) in the Mazarine Library. I made the following note upon it:

Secreta mulierū | et virorum ab alberto magno compſita (*sic*).

[Device. In a double border of flowers and grotesque figures is a tree with a shield: 3 roses above a crescent supported by two winged stags rampant, with crowns round their necks (gorged). Tied to the trunk of the tree by a ribbon are the initials, R.G., and below all: ROBERT GOVRMONT.] †

4to or large 8vo, a to d in sixes. Small black letter, double columns.

ai Title.

aii *recto*: Scribit philoſophus philo | ſophorū princeps quarto | ethicorū. ... the text ends dvi *recto*: on the *verso* Gourmont's device is repeated.

The copy which I got at Klemm's sale in 1889, at first sight seemed to be merely a duplicate of that in the British Museum, described in my *Notes*, Part III., p. 210. More minute examination, however, and direct comparison of the two copies, showed that though both were printed in uniform style at Antwerp in 1538, by the same person, they are different throughout. As the present is another case of two different issues with the same date, it may be worth while to indicate some of the variations by which one issue can be discriminated from the other. The following will suffice, though similar instances will be found on every other page.

The transcript of the title page of the British Museum copy already given, suits the present one except in the following minute points:

	Museum copy.	Present copy.
Title, l. 10:	| philoſophiæ, quàm Me- |	| philoſophiæ quàm Me- |
— l. 15:	olfaciēs.	olfaciens.
—	In the shield in the centre of the printer's ornament there is a cross.	The shield is blank.

* Silvestre, *Marques Typographiques*, Paris, 1853, No. 25. Petit printed between 1493 and 1541. The present volume belongs to the early years of the sixteenth century.

† Gourmont's device is given by Silvestre, *Ibid.*, No. 81. He printed between 1498 and 1518.

The following are other differences:

Museum copy.	Present copy.
Title leaf or Ai *verso*:	
Prohemivm. | Scribit Philoſophus Philoſophorum prin- |	Prohemivm. | Scribit Philoſophus philoſophorũ princeps : |
Ai *verso*, last line ends:	
Genimi (*sic*), Leo. |	Gemini, Leo. Illis correſpon- |
Aii *recto*, last line:	Last lines:
medicinalis. |	partim medi- | cinalis.
Aii *verso*. The small print of the note runs on continuously.	Divided into two paragraphs at l. 23, beginning: *Sapientia eſt cognitio*, etc.
Aiij *recto*, last lines:	
... appetitu intellectua- | li, quia vidit vtilitatem cognitionis earum. |	... appetitu intellectuali, quia vidit | vtilitatem cognitionis earum. |
Aiij *verso*, line 1: Secundo	ſecundo
last line: ...Vnde natu | ra |	... Vnde nam | nihil |
Diiij *recto*, l. 1. V is an ornamented capital.	plain.
Fviij *recto*, l. 3. A is ornamented.	plain.

In the Museum copy the numbers ij, iij, iiij in the signatures are written separate, in my own copy ij, etc., are written together. Contractions occur in the Museum copy which are wanting in mine, and *vice versa*. In the fourth line of the colophon (Gvij *verso*) the misspelling *recongnitis* as it stands in the Museum copy is correctly given *recognitis* in the other. The two issues differ typographically throughout, and the whole book must have been set up twice. It is difficult to believe that this happened in one year, and the fact that both issues have precisely the same date, seems rather to indicate that one is a reprint of the other, including the date, so that there may have been an interval of time between them. It could not have been long, however, for they have certainly issued from the same office. If this supposition be correct, I should think that the Museum copy belonged to the earlier edition, and that the other, in which the errata *Genimi* and *recongnitis* are wanting, was the later. Anyhow it is an event in the history of the book that there were two similar issues of the same year, just as happened in the editions of Lyons, 1566, as described in Part III., p. 210.

Another copy was acquired by me subsequent to this supplement being read to the Society. It is dated Lyons, 1584, and is a small square 16mo, pp. 381, [11]. It contains also the analogous tract of Michael Scotus, and appears to be the second edition in this form. There was another at Lyons in 1580, but I have not seen it. This is one of the early examples of the pocket editions of Albertus, of which so many appeared in the following century.

Corresponding to this, and undoubtedly one of the least known of all these books is the French version of Michael Scotus' *Physionomia*, made by Nicole Volkyr de Serouille, and published at Paris in 1540, in a small square 16mo. This is a book to which there is no reference by Brunet, and of which I have never seen any copy but one. It is not in the British Museum. An account of it, however, belongs properly to the bibliography of the wizard, and must be omitted here.

The following German version I found in the British Museum (07581. df. 19):

Albertus Magnus | Von haimlichait der frawen· | Auch ist dauon schreijben Arestoteles · Galienus · Boe | cius · Constantinus · Ypocras · Auicenna &c. Als dann | hernach volget jnn disem büchlein. Du vinndest auch | mancherlaij frag menschlicher vñ thierlicher natur. | [Vignette.]

Small 4to. a, b, in eights, c in six, no foliation. Black letter.

The vignette is an elderly man, presumably Albertus, expounding secrets to a woman.

cvj *recto* has this colophon:

¶ Hije ennden sich mancherlaij fragen mennschlicher | vnd thierlicher natur vnnd geschicklichait. Als die na | türlichen maijster dauon schreijben sind. Als Albertus | magnus · Arestoteles · Boecius · Cõstantinus · Egidius | Galienus · Auicenna · vnnd Ypocras &c · Anno &c · jm | zehenden jare. |

verso is blank.

The B. M. catalogue assigns the date 1510 to the book.

This is not a translation of Albertus' book, but is a collection from the authorities mentioned. The subject is discussed in catechism form: "Why?" is this, that, or the other thing—and the answer is: Aristotle, or Albertus, or Galienus says so and so. It is therefore a popular summary of physiological teaching, specially interesting to women. It contains, of course, a number of curious notions, but valuable as illustrating the beliefs of the sixteenth century.

10. Next come the copies of the *Liber Aggregationis* as it is called, of

which quite as many editions appeared separately and conjointly as of the other. It deals with the Secrets of Plants, Stones, and Animals, and in some of the editions there is included another tract on the Wonders of the World. Several editions in Latin have been described in previous parts of these "Notes," those which are now to be mentioned are translations into German and English.

Drei Bü- | cher Alberti Mag- | ni, Von wunder natürlichen | wirckungenn der Kreutter, | Edelgesteyn vnnd Thier, | mit leblicher Conter- | feytung derselbigen. | ∴ | 1536.

Small 4to. Signatures A to E in fours (E four is blank) or, ff. [1–2] 3–19. The title is enclosed in an elaborate border, and there are illustrations through the text.

Eiij *verso:* ₡ Gedruckt zu Franckfurt am Meyn | Bei Christian Egenolph. | 1536.

Eiiij is blank.

At the top of the title in a compartment is a battle of nude warriors. At one side of the title are the three Graces; at the other a female holding a child on her right arm and a flaming heart in her left hand, with another child at her feet. In a large compartment at the bottom is represented a walled place with trees and a fountain. In it are nine female figures, some nude, others draped, and a male figure with a laurel wreath. The females are occupied in different ways, and the whole may be meant for Apollo and the Muses. On the *verso* is a woodcut to illustrate Exod. 17. The woodcuts of the plants are lifelike and spirited if rough, and those of beasts and birds are humorous. The text in this edition is curtailed as much as possible, every word and phrase that could be dispensed with having been omitted. It contains only the briefest account of plants, minerals, and animals.

Albertus Magnus. | Das Bůch der haymligkayten Magni | Alberti, von Artzney vñ tugendẽ der Kreüt- | ter, Edel geſtayn, vnd von etlichen wol- | bekañten Thieren. | [large woodcut.] M.D.XXXX.

Small 4to. Signatures A to K in fours, no leaf or page numbers.

Kiij *verso.* Gedruckt zů Augspurg durch Haynrich Stainer.

Kiiij *recto* a woodcut.

Two-thirds of the title page are filled by a woodcut of a man in a study, sitting at a desk writing. It gives one an excellent idea of the furniture of such a place in the middle of the sixteenth century. The woodcut at the end represents a doctor, with gown and biretta, sitting at a reading desk consulting a book. In front of him is another holding up a glass *matula* and looking at the contents.

This is a larger book than the preceding, for besides a fuller version of the secrets of plants, minerals, and animals, it contains an introductory chapter on "Signs of Birth and Fruitfulness," and at the end a number of receipts against ailments, the preparation of certain waters and powders, and a series of directions against the plague. The last may be compared with those given by Vicary (see § 17).

The translation is different from that which appeared in 1725 (Part III., p. 214).

The *Liber Aggregationis* and the tract *De Mirabilibus Mundi* were translated into English, and I gave an account of the black-letter edition of 1617, in Part III., p. 214. A copy is in the British Museum. There are other two editions now to specify. One is that printed by William Copland in the sixteenth century; the other appeared early last century. A copy of the former exists in the University Library, Cambridge. The following is a brief account of it. I am not certain whether this is the first edition of the English translation or not.

Ai Title—wanting.		
Aii *recto.*	¶ The firſt booke of the vertues of certayne Herbes.	ends Bviii *recto.*
Bviii *recto.*	¶ Here beginneth the ſecond booke of the vertues of certayne ſtones.	ends Ei *recto.*
Ei *recto.*	¶ The thyrde booke of Albertus Magnus, of the vertues of certayne beaſtes.	ends Fii *recto.*
Fii *recto.*	¶ Here are ended ſome ſecretes of Albartus (*sic*) Magnus of Colone vpon Natures Vertues, and effects of certain Herbes, Stones, and Beaſtes. And here foloweth in what houre, euery Planet hath his dominion. (∴)	ends Fvii *verso.*

Fviii *recto*. ¶ Heere beginneth thee booke of the marueyles of the worlde, ſet forth by Albertus Magnus. ends Liii *verso* as follows:

¶ An ende of the ſecreetes of Nature, ſet foorth by Albertus Magnus in Latine, newlye tranſlated into Engliſh.

¶ Imprynted at London by me Wyllyam Copland.

(.:.)

Small 8vo. Signatures A to K in eights, L in four, no pagination or foliation. Biiii is misprinted Ciiii. Black letter throughout.

22 lines to the page, excluding the head line and catchword line.

This copy unfortunately wants the title (Ai) and Liiii. There is no date, but Copland printed from 1548 to 1568-9.* His edition is quoted by Lowndes and by Graesse (*Trésor*, I., p. 56). It is not in the British Museum, and it is not mentioned by Herbert or Dibdin, so I take it to be most rare. The only copy I know of is that just described.

Cooper (*Catalogue of Chymical Books*, London 1675) speaks of an edition of 1637, which keeps entirely out of sight.

The other copy which has just come into my hands is comparatively modern: it is in its original appropriate sheep cover, and is clean and whole. So far as English books go, this is one of the greatest novelties, and is certainly another rarity. During all my observation and collecting of these books I have never seen or heard of another representative of this edition. As of similar books, the copies were probably worn out by use.

The | Secrets | of | *Albertus Magnus:* | Of the Vertues of Herbs, | Stones, and certain Beasts. | Whereunto is newly added, | A ſhort Diſcourſe of the Seven Pla- | nets, governing the Nativities of | Children. | Alſo a Book of the ſame Author, | Of the Marvellous Things of the World, | and of certain things, cauſed of | certain Beaſts. |

London, | Printed by M. H. and J. M. and are to | be Sold by J. Wright, J. Clarke, | W. Thackeray, and T. Paſſinger. |

---

* *Dictionary of National Biography*, XII., p. 174. The present work is not mentioned there.

16mo. Signatures A to H in eights. No pagination. Roman letter. The title is in a narrow border. No date, but about the end of the 17th or beginning of the 18th century. A chap-book.

A1 *r.* The title. *verso:* to the Reader.

A2. Blank.

A3 *r.* Text of the Secrets begins, and ends E1 *r.*

E1 *v.* The order of the Planets, ends E8 *v.*

F1 *r.* Book of the Marvels of the World, ends H8 *r.* The *verso* is blank.

Passenger in 1691 published *The Complete Servant Maid.* He was then on London Bridge. This book is of about the same date.

11. Since 1883, when I considered certain copies, both in print and in manuscript, of Bartholomew Glanville's work *De proprietatibus rerum*, and 1884, when I commented on the charge of plagiarism which had been brought against him, and showed a 15th century MS. of the same work, I have met with one or two items which had not been already alluded to. Besides these, however, I am able to include another and very fine copy of the Strasburg 1485 Latin edition, credited to Wynken de Worde, and a copy of Berthelet's English edition, 1535. The latter, which is in the Hunterian Library, I had already described, but could not exhibit. The present copy can be exhibited, without the necessity of describing it. A good deal of what seems to me now rather misplaced laudation has been expended on this volume. Lowndes says it is probably the finest specimen of Berthelet's printing. I have too limited a knowledge of his printing to either challenge or confirm that statement, but I have seen certain small tracts which to me are more attractive, both from the type used and the appearance of the page. In Glanville's book the letters seem commonplace, and the page is loaded and heavy for its size; and while it is undoubtedly a respectable solid folio, I could not say that it is a "magnificent book." It is certainly not to be compared with the 1480 edition in Latin. Still, if it be not absolutely beautiful, it is sufficiently rare, and it is sufficiently in demand on account of its contents and the English in which they are couched, to make it a prize when it does occur.

One of the novelties which I have to bring before you is the reprint that appeared fifty years later, edited by Stephen Batman. This is an important book for several reasons, and a description of it may be given.

Batman | vppon Bartholome, | His Booke | *De Proprietatibus Rerum*, | Newly corrected, enlarged and amended: | with ſuch Additions as are requi- | ſite, vnto euery ſeuerall | Booke: | Taken foorth of the moſt approued Authors, the like here- | tofore not tranſlated in Engliſh. | *Profitable for all Eſtates, as well for the benefite of* | *the Mind as the Bodie.* | 1582. | London | Imprinted by Thomas Eaſt, dwel- | *ling by Paules wharfe.* |

Small folio. Signatures: 2 leaves; ¶, ¶¶, B to Z, Aa to Zz, Aaa to Zzz, Aaaa to Cccc in sixes; or, 14 leaves not numbered and 426 numbered. Black letter. Double columns of 53 lines each.

Collation: Title, surrounded by a broad border of scrolls. *Verso*, arms of Batman (?) with an interpretation of them in 14 lines of verse.

Leaf 2, *recto.* Batman's dedication to Lord Henry Cary.

— *verso.* Batman to the Reader.

¶ 1 *recto,* the Prologue of the Translator. *verso,* a collection of the Authors. Ends ¶ 6 *recto,* the *verso* of which is blank.

¶¶ 1 *recto.* Generall Table of the chapters, ends ¶¶ vj *verso.*

Bi *recto.* The Text begins, and ends Cccc vj *verso.* At the end of the Text is the printer's coat of arms, with the crest of the black horse having a crescent argent on its left shoulder, with the motto

MIEVLX · VAVLT · MOVRIR · EN · VERTV · |QVE · VIVRE · EN · HONCTE.

¶ Imprinted at London by Thomas East.

Part, at least, of the value of this book depends on its being not a mere reprint of Glanville's encyclopedia, but a veritable new edition; for Batman has made considerable "Additions," which are always distinguished from the original by that word as a heading or on the margin, and he has been at very considerable pains to make the book as useful as possible. The table of authors is an addition of his, and it is not a mere list, but it gives biographical notices of various writers quoted throughout the work. These are necessarily brief, but they are both useful and curious, even for their inaccuracies, as when Cornelius Agrippa is called a "Spaniard of excellent learning." As Batman was therefore fully alive to the advance of the times, and did his best to accompany it, his revision is a necessary adjunct to the two earlier editions of the English translation.

Batman's own account "To the Reader" of the occasion of his editing the work, is quite worth quoting in full:

This booke, beeing firſt ſet forth, in the yere of our Lord God. 1360. The Author whereof is reported to be a Franciſcan Frier, of the noble familie of the Earles of Suffolke (named Barthelmew Glantuile) who ſtudeouſly gathered this ſingular worke, for the moſt part, of the properties of thoſe things the which he had found written in the Bible, to the onelye benefit of his Countrie: which booke, was had in great eſtimation among the learned, as well beyond the ſea as at home, vntill within 60. yeares paſt, there ſprang vp famous, and worthy perſons, of ſingular perſeuerance and learning: which from the courſe of auncient beginnings, ſet foorth the ſame that was formally written of, with additions aunſwerable to time preſent, vſing new Titles, wherevnto is added ſo much as hath bene brought to light by the trauaile of others, as Conradus Geſner of Tygure, Phiſition, writing of the nature of beaſts, birds, fiſhes, & Serpents, Fuchſius, Mathiolus, Theophraſtus, Paracelſus, and Dodoneus, theſe wrote of the natures, operations and effects of Hearbs, Plants, Trees, Fruit, Seeds, Metalls and Mineralls. Sebaſtian Munſter, Henry Cornelius Agrippa, and others of Aſtronomie and Coſmographie. Abraham Ortelius of Antwarpe for maps and diſcriptions: all which woorkes hath done great good in diuerſe and ſundrie Commonwealths. I haue therefore as an imitator of the learned, for the good will I bare to my countrie, collected forth of theſe aforeſaid Authors, the like deuiſes, which they in times paſt gathered of their elders, and ſo renuing the whole booke, as is apparant by additions, is brought home, the Master, the Pilot, and the profit thereto belonging, deſiring the wellacceptation of the ſame, and the friendly correction, if anye wordes ſhall happen to be miſtearmed in the imprinting (hoping that there will be found very fewe.) . . .

It is of considerable interest to find some of the writers, to whom reference has been now so often made, flourishing as Batman's authorities three hundred years ago.

To the list of the editions of this remarkable work I am able to add a notice of the translation into Dutch. It was printed at Haarlem, by Jacob Bellaert, in 1485, and forms a thick quarto volume, double columns, in black letter. It is adorned with a number of full-page illustrations, as of birds, beasts, fishes, the earth, the stars, human employments, and one or two others. These are interesting; spirited sometimes, sometimes humorous, and always characteristic. The copy I have contains 464 leaves; the first and last leaves, which are blank, are wanting. Hain (No. 2522) mentions the translation, but not having seen a copy gives no collation. There is one in the British Museum. The book is of course described by Campbell,* and is quoted by Panzer and others.

* *Annales de la Typographie Néerlandaise au XVe Siècle*, La Haye 1874, p. 67.

According to Campbell the large woodcut at the end represents the arms of the printer, but the Museum catalogue says the arms of the city of Haarlem.

12. Inasmuch as the treatise on practical medicine by Valescus de Tharanta is nowhere called "Secrets," it ought not, in accordance with the rules to which I have tried to adhere, to form part of these "Notes." But whereas it deals with the same subject, and very much in the same way as other works which, because they happen to bear the title of Secrets, have been included, it is attaching undue weight to what is after all the arbitrary limitation of a mere name, if it be passed over. Strictly speaking, summaries like the present belong to the history and bibliography of medicine, and not of the arts; but considering that they are sometimes distinguished as books of receipts, and embrace not only medical but technical and other secrets as well, it has not been possible to refuse them recognition universally. Anyhow the present one may be mentioned, if for no other reason than as a precursor of Varignana's treatise mentioned in the first supplement, *Trans.* (N.S.) II., p. 381.

The biography of the author is meagre, and is based mainly on his own Prologus. He there calls himself "Valescus, gallice Balescon de Tharanta,"* and says he was a doctor of arts and medicine, but in his humility he calls himself more than once "discipulorum medicinæ discipulus." At the close of the prologue he tells us that he began his book in the year 1418, the eve of St. Barnabas the Apostle, after thirty-six years spent in medical practice; and he then refers to certain contemporary events by way of confirming this date. He says this was in the first year of the pontificate of Martin, "remoto scismate."† King John was reigning in Portugal, constantly at war with the Saracens.‡ Of the

* Hain (*Repertorium*, No. *15244) puts him under the form Valastus de Tarenta, which occurs in some editions of his works.

† This is correct enough. Pope Martin V. was elected November, 1417; the Antipope Benedict XIII. was deposed the same year, and John XXIII., who had been deposed in 1415, and was kept in prison till 1419, recognized Martin V. at Florence as a condition of regaining his freedom. So that the schism was at an end for the time.

‡ John I. the Great; but, so far as I can ascertain, it was with Castile rather than with the Saracens that he was in conflict.

country of Foix and Béarn, Isabella, his mother, was countess and mistress.* "Carolus Albricus," "the most faithful king of France," was then reigning, "with wars and troubles which increased mightily throughout the whole kingdom."† The English were fighting in Normandy, and took possession of Rouen. These dates and events certainly coincide with the time at which Valescus was busy with his book. They were examined also by Astruc,‡ who viewed them as confirmatory of Valescus' accuracy as a contemporary chronicler.

Ranchin calls him a Portuguese, and says that during the "vacances" he visited his native country every year. Subsequent writers repeat this statement.

The date of his settling in Montpellier is not quite fixed; it may have been about 1380, but from the above remark in his prologue, he must have begun the practice of medicine in 1382. He pursued this profession "with marked success and distinction," and ultimately became one of the most distinguished ornaments of the medical school at Montpellier. Castellanus says he was chief physician to the king of France.

The earliest summary of his life I have met with is by the said Petrus Castellanus,§ which is worth quoting:

> Valescus de Taranta, Monspelii, circa ann. Domini CIↃCCCLXXXII. Medicinam laude magnâ, & multorum fructu insigni facere cœpit. In quo genere studij cum triginta sex annos continuasset, quæcunque & lectione, & quotidianis experimentis didicerat, in unum volumen congessit, quod *Philonij* nomine circumfertur. Tradunt etiam Archiatrum fuisse Regis Gallorum.

---

* See Ersch u. Gruber, *Encyclopädie*, XLVI., p. 119.

† Charles VI., *bien-aimé*, the well-beloved. I have not traced the name "Albricus." He had certainly numerous wars on his hands. Agincourt was fought in 1415. Rouen capitulated in January, 1419.

‡ *Mémoires pour servir à l'Histoire de la Faculté de Médecine de Montpellier*, Paris, 1767, 4°. See pp. 208-211, where an account of our author is given.

§ *Vitæ Illustrium Medicorum qui toto orbe, ad hæc usque tempora floruerunt.* Antverpiæ, 1618, p. 166. Castellanus, Pierre Duchâtel, was an antiquary and physician, and became professor of Greek at Louvain. This volume contains 180 brief biographies, alphabetically arranged, a list of which is given by Mercklin (*Lindenius Renovatus*, Norimb., 1686, p. 883). The reprint by Gronovius, *Thesaurus Græcarum Antiquitatum*, Lugd. Bat., 1701, X., col. 853, contains only the biographies of Greek physicians, and should not have been put forward as Castellanus' complete work. *See* Stolle, *Anleitung zur Historie der Medicinischen Gelahrheit*, Jena, 1731, p. 243.

Mercklin,* to the preceding, which he has simply copied, has prefixed the statement: "Natus est Monspelii circa A.C. 1382," and subjoined another: "obiit circa A.C. 1418, ætatis suæ 36," without apparently having observed the irreconcilability of the statements. Mangetus,† who quotes the passage from Mercklin in full, directs attention to the discrepancy, but makes no attempt to remove what is obviously a muddle on Mercklin's part. Freher ‡ takes his account from Castellanus and not from Mercklin, but he commits the same error of placing his death in 1418. He adds, however, from Caspar Schott,§ the wonderful story of the rejuvenescence of a prioress, which Valescus || uses as an illustration of that subject. Kestner ¶ gets round the difficulty by giving no dates at all, but he is misled by Mercklin into saying that he was a native of Montpellier. He corrects, however, one or two mistakes of other writers. Portal,** following Ranchin, calls him a Portuguese, and gives a favourable notice of his book.

Eloy †† calls him a native of Montpellier, and repeats that he became physician of Charles VI. "about 1400." This is a new date added on to the older accounts. He says he made some excellent observations, which were still founded on in Eloy's time. He adds that Valescus did not understand Greek, and wrote shocking bad Latin. In the later edition ‡‡ there is an entirely new and in every respect a more satisfactory account,

* *Lindenius Renovatus*, Norimb. 1686, p. 1036.

† *Bibliotheca Scriptorum Medicorum*, Geneva, 1731, II. ii., p. 355.

‡ *Theatrum*, Norimb. 1688, p. 1211.

§ *Physica Curiosa, sive mirabilia Naturæ et Artis*, Edit. altera, Herbipoli, 1667, p. 419. Schott calls him Valesius Tarentasius.

|| *Philonium*, Lib. VI., cap. 12. Rembert Dodoens, or Dodonæus, at the close of a small volume, *Medicinalium Observationum exempla rara*, Coloniæ, 1581, pp. 289-293, Hardervici, 1621, pp. 205-208, quotes a few striking cases from the *De Curandis Morbis* of "Valescus Tharantanus." Of course the prioress figures as one of them, p. 291, or p. 207. The story is told also by De Longeville Harcourt, *Histoire des Personnes qui ont vecu plusieurs siècles . . .*, Paris, 1715, p. 268.

¶ *Medicinisches Gelehrten-Lexicon*, Jena, 1740, p. 834.

** *Histoire de l'Anatomie*, Paris, 1770, I., p. 237.

†† *Dictionnaire Historique de la Médecine*, Liège, 1755, II., p. 431.

‡‡ *Dictionnaire Historique de la Médecine*, Mons, 1778, IV., pp. 458-9. Still there are slips; for instance: "Il se donne lui-même le nom François de Balescon de Tharare, dans la préface," etc. (p. 458).

which Dezeimeris has done little but copy and condense. Following Astruc Eloy praises the *Philonium* for its matter, but condemns it for its barbarous style. He does not know upon what authority Castellanus calls him the chief physician of the king, for his name is not found in the best lists of these physicians.*

Another writer who criticises his style is Albrecht v. Haller,† and he calls him "verbosus auctor," but credits him with discussing the opinions of others and expressing his own views. He gives a long but not very clear list of the editions of the *Philonium.*

Writers on the history of medicine during the present century have just repeated the older accounts. Dezeimeris' article,‡ for instance, is practically taken from Eloy, as I have just said. It is accurate, and as full as is possible within the limits of the book; it gives a list of editions and references to the authorities, though the latter are not sufficient, seeing that Eloy is omitted.

Upon this in turn is based the article by Gurlt,§ who has not attempted to add anything new.

A short notice of him was given by A. W. E. Th. Henschel,‖ in 1852, but this writer has not included a few quite legitimate facts, such as the

---

* While on this matter, it may be remarked that Dezeimeris says Tiraqueau, as well as Castellanus, Ranchin and Conring, gives him this title. He does not mention the passage, and it is possible, therefore, that I have overlooked it; but in one place Tiraqueau certainly does not give him the title: "Valescus Tarentinus, quem de Tarento vulgò appellant, cuius est Practica quam Philonium nuncupant, et alia quædam opera." *See* André Tiraqueau (Andreas Tiraquellus) *Commentarii de Nobilitate,* Lugduni, 1617, fol., p. 291, in the section in which he discusses whether the practice of medicine detracts from nobility. He gives a long alphabetical catalogue of physicians, in which the above appears.

† *Bibliotheca Medicinæ Practicæ,* Bernæ, 1776, Tom. I., pp. 454-5.

‡ *Dictionnaire Historique de la Médecine Ancienne et Moderne . . .* par MM. Dezeimeris, Ollivier (d'Angers) et Raige-Delorme, Paris, 1828-39, T. I., pp. 259-60. The author is called Balescon de Tarente, or de Tharare, in this work.

§ August Hirsch, *Biographisches Lexikon der hervorragenden Aerzte aller Zeiten und Völker,* Wien und Leipzig, 1884, I., p. 269.

‖ In his "Biographisch-litterarische Notizen, berühmte Wundärzte und Aerzte des XIII. und XIV. Jahrhunderts betreffend," article 34; published in *Janus, Central-Magazin für Geschichte und Literärgeschichte der Medicin . . .*" Gotha, 1853, II., p. 422.

undoubted popularity of the *Philonium*. He mentions only the Venice edition of 1502!

Haeser* furnishes a fuller review than Henschel, and has endeavoured to give a complete list of the editions. There is, however, a vagueness about it which renders me doubtful about several of them, especially as he does not seem to have any stronger support than the enumeration in Panckoucke's article.† He also is confused, for he says that Valescus lived at Montpellier from 1382 (which is not certain) and finished his treatise, the fruit of thirty-*eight* years' work, in 1418. These figures do not harmonize with those of Valescus himself.

A portrait of Valescus is contained in the *Ephémerides médicales de Montpellier*, for 1887.‡ I have not had the fortune to see it.

To a certain extent the confusion about the events of his life envelops the editions of the *Philonium* as well. It may, therefore, be convenient to have a digest of those which have been enumerated.

	Mercklin.[1]	Manget.[2]	Eloy.[3]	Panzer.[4]	Panckoucke.[5]	Haeser.[6]	Billings.[7]	Br. Mu.
Lugd.	—	—	—	1478	—	—	—	—
Lugd.	1490	1490	—	1490	—	1490	—	—
Lugd.	—	—	—	1490	—	1490	—	—
Ven.	—	—	1490	—	1490	1490	—	—
Lugd.	—	—	—	1496	—	—	—	—
Lugd.	1500	1500	1500	1500	1500	1500	1500	—
Lugd.	—	—	—	1501	—	—	1501	1501
Ven.	—	—	1502	1502	1502	1502	1502	1502
Lugd.	—	—	—	1516	—	1516	—	—

[1] *Lindenius Renovatus*, Norimbergæ, 1686, p. 1037.

[2] *Bibliotheca Scriptorum Medicorum*, Genevæ, 1731, II. ii., p. 355.

[3] *Dictionnaire Historique*, Mons, 1778, IV., pp. 458-9.

[4] *Annales Typographici*.

[5] *Biographie Médicale*, n.d., VII., p. 390.

[6] *Lehrbuch der Geschichte der Medezin*, Jena, 1875, I. 712.

[7] *Index Catalogue of the Library of the Surgeon-General's Office*, by John S. Billings, Washington, 1894, XV., pp. 560-1.

* *Lehrbuch der Geschichte der Medicin*, Jena, I., pp. 712, 713.

† *Biographie Médicale*, n.d., VII., p. 390.

‡ It is referred to in the *Index Catalogue of the Library of the Surgeon-General's Office*, by John S. Billings, Washington, 1894, XV., p. 561.

	Mercklin.	Manget.	Eloy.	Panzer.	Panckoucke.	Haeser.	Billings.	Br. Mu.
Ven.	1521	1521	1521	—	1521	—	1521	—
Lugd.	—	—	1521	—	1521	1521 (?)	—	—
Lugd.	1526	1526	1526	1526	1526	—	1526	1526
Ven.	1532	1532	1532	—	1532	1532 (?)	—	—
Lugd.	—	—	1535	1535	1535	1535	1535	—
Lugd.	—	—	1560	—	1560	1560	—	—
Ven.	—	—	—	—	—	1589	—	—
Francf.	1599	1599	1599	—	1599	1599	1599	1599
Lugd.	—	—	—	—	—	1599	—	—
Francf. Lpg. u. Jena,	1680	1680	1680	—	1680	1680	1680	1680
Leipz.	—	—	1714	—	1714	1714	—	—

I have examined, and may briefly describe the editions of 1501, 1502, 1526, 1599, 1680 in the British Museum; 1490 in the Hunterian Library; and 1501 and 1535 in my own collection.

The edition of 1478, Lugduni, per Johannem Cleyn, 4to, is quoted by Panzer (*Ann. Typ.* I. 532, 13), from *Catalogus Bibl. Jo. Frid Christii*, Lips. 1757, I., p. 234. He calls it *editio dubia*, and says that the date should probably be 1488, as the name of Cleyn as a printer does not occur earlier. Curiously enough, Hain (No. 15249), who did not know any copy, and simply repeats what Panzer says about the erroneous year, does not quote his remark about Cleyn's own date. As will be seen by the table, this edition was unknown to the older authorities.

For the so-called edition of 1401 see under 1501.

The first edition seems to be the following. There is a copy in the Hunterian Library, University of Glasgow.

f. [1] Title. Practica valefci de tharanta | que alias philonium dicitur
Verso blank.
f. [2] Tabula, ends f. 3 *verso*.
f. [4] *recto* blank. *Verso* Epistola.
a j Folium I. Incipit Prologus in practica3 | vfualem Ua- | lefci de tharāta etc.
Small folio. 4 ff. a to z, A to Y in eights; or, ff. [4] CCCLX numbered. Double columns. 51 lines. Black letter.
Yviij *recto* col. 2, colophon:
Preclariffimū op⁹ valefci de tharāta reuerēdiffi | mi mgrī necnō artis medicine doctoris famofif | fimi. Finit feliciter Impffum lugd. p Johānem | trechfel

alemanū. Anno nr̄e ſalut̄ Milleſimo | quadringētefimononageſimo Die vero decimo | nono menſis maij. Amen.
followed by Trechsel's device: I.T in a circle through which rises a line crossed near the top by two lines, one longer than the other; colour white on a red ground.

Trechsel's edition is described by Panzer, I. 541, 88, on the authority of Maittaire, p. 519, and *Catalogus Bibliothecæ Cœnobii Glunicensis in Austria*. Kestner, *Bibl. Medic.*, p. 601.

The second edition was printed at Lyons, "per magistrum mathiam huss alemanum," on the 20th of November, 1490. It is in gothic character, folio, 272 leaves. It is mentioned by Maittaire, p. 519, note 4; by Denis, *Suppl.*, p. 297; Braun, II., p. 203; Panzer, I., 541, 85; and Hain, No. +15251, who gives 267 leaves. I have not seen it.

Another, also dated 1490, is said to have appeared at Venice. I have not seen or heard of an actual copy, but it is noticed by Astruc, Panckoucke, Dezeimeris, and Haeser, who all, probably, simply copied Astruc's statement. Its existence seems doubtful.

The edition of 1496 appeared at Lyons, 14th Oct., in 4to, and was likewise printed by Johann Trechsel. It is mentioned by Denis, *Suppl.*, p. 418. Panzer (I. 549, 155) asks whether it may not be part of a collection entitled: "Amicus medicorum magistri Joannis Ganiueti." This has the same date, place, and printer as Valescus' work, and it is possible that they form part of one whole. Apparently Hain took this view, as he gives a reference to Ganivetus, No. *7467.

Hitherto I have not come across a copy of the edition printed in 1500 by N. Wolff at Lyons, in 4to. It is described in full by Hain (No. *15252), who says it contains 4 ff. not numbered and 364 numbered. Billings' account is as follows:

Practica, que alias Philonium dicitur, una cum Joannis de Tornamira introductorio. 3 p. l., ccclviiii ff. 4to. [*Lugduni.*] [*Ad finem:*] *Impressum per N. Wolff*, 1500.

It seems to be different from the following, which Panzer gives (IV. 73, 677):

Valesci de Taranta Praxis Medica. 1500. fol. Maitt. p. 738.

Mangetus quotes a Lyons edition of 1500, but without specifying the printer. Haller alludes incidentally to a Basel edition of 1500.

Next comes an edition of which I happen to have a copy:

Practica Ualeſci de Tharanta que als Philoniũ dicitur
vna cũ oĩb⁹ Antidotis tam ſimplicib⁹ q cõpoſitis vni-
cuiq3 egritudini a capite vſq3 ad pedes: appropria
tis Johãnis Meſue: per alphabetum: a magi-
ſtro Jacobo de partibus doctore Pariſiẽſi
celeberrimo: elegãter collectis. et cũ non-
nullis ſtomie canonibus cũ anatho
mia oſſium adiũctis planiſſime cũ
nũero ipſorũ ſc3. ccxlviij. abſq3
oſſibus ſiſaminis et oſſe lau-
de sub lingua: valde medi
co perutiliſſimis ex
pluribus doctori
bus emẽdatiſ
ſime excer-
ptis

Small folio. 4 leaves AA, a to z, A to Y, in eights, or ff. 12 not numbered, CCCLX numbered. Black letter, with numerous contractions; double columns, 51 lines; head lines and signatures. Title entirely in red.

Collation: 1st leaf, Title; *verso* blank.

2nd leaf, *recto*: ⁋ Incipit tabula hui⁹ libri . . . ends on the 4th leaf *recto*.

4th leaf *verso*. Dedication or address to the Chancellor of Montpellier University by Hieronymus Ferrera.

AA1 *recto*. Sũmula Jacobi de partibus, ends AA7 *verso*. 8 is blank.

a i *recto*. Text of the Philonium begins, and ends Y viij *recto*. ff. I-CCCLX. *Verso* is blank. At the end of the text is the colophon:

⁋ Preclariſſimũ op⁹ Ualeſci de Tharãta reue | rendiſſimi magr̃i necnõ artis medicine docto- | ris famoſiſſimi. Finit feliciter. Imp̃ſſum lugd̃. | p Johem Cleyn alemanũ. Anno noſtre ſalutis | Milleſimo quadringẽteſimop'mo, deci-moocta | no. kal. Decembris.

This is followed by Cleyn's device, white on a red ground, with the letters I C, which resembles Trechsel's.

Panzer (VII., 277, 10) quotes this edition, and considers the date a simple misprint, for after *quadringentesimo* he puts the word *sic*, and the book is placed under 1501. He quotes (VII., p. 163), *Catal. Bibl. Thottianæ* and *Bibl. Schw. iun.*

The copy in the British Museum, which I have examined since the preceding was written, is the same as my own. Cleyn's device, however, has been cut out, and there is added a folded leaf, equivalent to eight pages of the book, which contains on the half of one side the picture of a skeleton, with the bones marked with letters, and the corresponding names printed on the rest of the sheet round the picture. On the other side, occupying the whole sheet, is a picture of the veins, with a description of their functions in phlebotomy. This sheet, which was printed by Cleyn, as the type is the same as that used for the *Philonium*, is intended, I presume, to give the insight into the bones and veins referred to in the title. In this copy the tract of Mesue is placed at the end, instead of at the beginning as in mine. The date 1401 is considered a misprint for 1501, and the latter date is assigned to the book in the Museum catalogue.

In 1502 the *Philonium* was reprinted. It is a folio, double columns, gothic character, and contains A to Z, AA, BB, all in fours, and CC in 5 (as the register gives it) or ff. CCIX. There is no title page. At the end is this colophon:

> ¶ Preditiſſimū atq3 p̄clariſſimū opus Valeſci de tharan- | ta reuerēdiſſimi magiſtri necnō artis medicine doctoris fa | ma ſup ethera noti. vna cū introductorio celeberrimi viri | atq3 medicine doctoris Johānis de tornamira feliciter finit. | Impēſiſq3 nō mediocrit⁹ & cura ſollertiſſima petri liechtē | ſteyn colonienſis aggripine colonie: necnō emendatione | diligentiſſima eſt Impreſſu3 anno. ſalutis. M.ccccc.ij. duo- | decima Kal Martij Venetijs. Sūme & īdiuidue trinitati. | ſit laus: honor & gloria.

The above is from the copy in the British Museum (7306. k. 13 (1).).

This edition is given by Panzer (VIII. 356, 139), and he refers to Gruner's *Aphrodisiacus*, p. 28, not. l., and to Thott, l.c. VII., p. 54.

Under Lugduni, MDXVI. Panzer (VII. 314, 312) quotes the edition printed by Jacobus Myt, Anno ... MCCCCCXVI die vero XI mensis Augusti. 4. He refers to Gruner's *Aphrodisiacus*, p. 28, not. l. Haeser calls it a very elegantly printed edition—eine sehr zierlich gedruckte Handausgabe. I regret I have seen no copy.

Hallervordius* and Mercklin speak of an edition printed in folio at Venice in 1521: Apud hæredes Octaviani Scoti, but they give no particulars. Mangetus and Panckoucke copy from Mercklin. Haeser's edition in

---

* *Bibliotheca Curiosa*, Regiomonti, 1676, p. 347.

folio of the same date was printed apparently at Lyons, and Panckoucke also gives an edition, Lyons, 1521, in folio. Whether there be two such editions, or whether one of these authorities has made a mistake as to place or date I am unable to say, for I have seen no copy of 1521. I do not even know if Haeser simply quoted Panckoucke without having had access to a copy. This question, therefore, cannot be satisfactorily answered for want of the necessary data. Fabricius (*Bibl. lat.*, p. 873) calls this edition the first erroneously.* Billings describes it as follows:

> Philonium. Aureum ac perutile opus practice medicine operam dantibus, quod Philonium apellatur, consumatissimi medici domini Valesci de Tharanta. Novo ac diligenti examine correctum, novisque et pulchris marginalibus annotationibus exornatum. Introductorius etaim libellus ad practicam medicine partem domini Ioannis de Tornamira. 3 p. l., 223 ff. fol. [*Ad finem:*] *Venetiis impressum cura et sumptibus heredum ... Octaviani Scoti civis Modoetiensis ac sociorum.* 1521.

The edition of 1526 is one of those most frequently referred to. There is a copy in the British Museum (542. a. 2.) of which I can give an account:

> Philonium. | Aureũ ac perutile op⁹ | practice medicine ope- | ram dantibus: q̄d Phi | loniũ appellatur: Cõ- | ſumatiſſimi medici do | mini Valeſci de Tha- | ranta. Nouo ac diligẽ | ti examine correctũ. |
>
> Introductorius etiã | libellus ad practicã me | dicine partem domini | Joãnis de tornamira. | [Device.]
>
> 8vo. aa 4; a to z, A to Z, AA to GG in eights; or ff. [4] ccccxxiiij. Title in red, except the ornamental border and the device of a cross composed of six small crosses. Gothic letter, double columns, 52 lines. The word "Philonium' is outside the border.
>
> aai Title. On the reverse is a short address to the reader.
>
> aaij: Incipit tabula hui⁹ libri, ends aaiiij *recto*. *Verso* is blank.
>
> aj Text begins fo. i. but not numbered, and ends GGviiij *recto*, ſo. ccccxxiiij.
>
> The colophon is as follows:—
>
> ¶ Utiliſſimum atq̄ clariſſimum opus Vale | ſci de tharãta per q3 reuerẽdi magiſtr nec- | nõ artis medicine doctoris fama ſup ethe | ra noti. vna cũ introductorio celeberrimi vi | ri atq3 medicine doctoris Joãnis de torna | mira finit feliciter. Impreſſum lugduni per | ſedulũ virũ Jacobum myt. Anno virginis | partu. M.ccccccxxvj. die *no*. vj. mẽſis Martij. |
>
> The last *page* is blank.

---

* So says Kestner (*Medicinisches Gelehrten-Lexicon*, 1740, p. 834), but I have had no access to the edition referred to. The article in that which I have seen, Florentiæ, 1858, III., p. 570, is inaccurate, though this particular error has been expunged from it.

Myt's edition of 1535 is a reprint of this one.

Panzer, VII., p. 339, 524. He refers to Oertelii *Catalogus Bibl. Scholæ quæ Neustadii ad Aissum floret*, Norumb. 1787, 4to. P. 104.

Haller mentions an edition of 1531, and he is followed by Dezeimeris. No copy of that date has elsewhere come under my observation, and I am doubtful about it.

The edition of 1532 in fol., Venetiis, apud Hæredes Octaviani Scoti is mentioned by Mercklin, Mangetus, Eloy, and Panckoucke. Haeser quotes an edition of the same year, but judging from the arrangement of his list it would seem to have been printed at Lyons.

The first acquaintance with Valescus' book which I made was through a copy of the 1535 edition. The following is an account of it:

Philonium. | Excellētiſſimi ac | optimiſquibuſq3 Practicae Medicinae | ſtudioſis viris longe vtiliſſimum et neceſ- | ſariū opus : quod Philoniū inſcribitur : | preſtantiſſimi necnon conſummatiſſimi | Medici Domini Ualeſci de Tharanta : multoq hactenus fuerit nuper vigilan- | tius emendatum : cui acceſſit. D. Joan. de | Tornamira diligentiſſimi Galeni intro- | ductorius libellus varias morborū curas complectens. | 15 [Device] 35. |

¶ Uenūdātur Lugduni apud Scipio- | nem de Gabiano in vico Mercuriali ſub | inſigni Fontis. |

8vo. Signatures a to z, A to Z, AA to KK, all in eights; or ff. [i-iv] v to ccccxlvij, and one leaf not numbered. Black letter, double columns. The title page is printed in red and black, and is enclosed in four separate different pieces. The device is a fountain.

ai contains the title, on the reverse of which is a short address to the reader.

aij to aiiij *recto*: Tabula, or table of contents; *verso* is blank.

The text begins on a v numbered folio v, and goes down to KKvij *verso*, numbered ccccxlvij. At the end of the text is the register, followed by the colophon, which shows the estimation in which the book was held:

> ¶ Utiliſſimum atq3 clariſſimum opus Va-
> leſci de tharanta perq reuerendi ma-
> giſtri necnon artis medicine doctoris fa
> ma ſuper ethera noti. vna cū introducto
> rio celeberrimi viri atq3 medicine docto
> ris Joannis de tornamira finit feliciter.
> Impreſſum lugduni per honeſtū virum
> Jacobum myt Expenſis vero honeſto-
> rum virorum Scipionis de Gabiano et
> fratrū aſtantium Anno virginis partu.
> M.CCCCCXXXV. die *vo. XX. menſis Aprilis.

The last leaf, not numbered, has the device of a fountain with two cupids and the word LAVAMINI, and round the border the words:

Ego Sitienti Dabo De | Fonte Aquæ | Vivæ Gratis | Apoc. 21.

The *verso* is blank.

The previous edition of 1526 was also printed by Myt. The present one is given by Panzer, VII. 363, 754, who refers to Hirsch, *Millenarius*, II., p. 57; and by Panckoucke, who calls it a 4°. Billings' collation tallies with mine. There is no copy in the Museum.

In the note to the reader, apparently by the publisher, or whoever it was who saw the book through the press, it is stated that this edition had been printed in a compact form, so that physicians, who have to go from place to place, may have a survey of human diseases which they can carry about with them as a "veni mecum," as such a book is called. This therefore was a pocket cyclopaedia of the healing art, as practised at the beginning of the sixteenth century, and may be compared with the analogous works of De Dondis and of Varignana, the former in large folio, the latter in small octavo. All these works are of especial value as illustrative of medicine from 1300 to 1385. There is wanted a good critical account of the natural and medical science of the 14th century, of which less is known than of the 13th, in so far as the men of the later century were by no means so distinguished as their immediate predecessors.

Of the edition of 1560 Haeser, taking his information possibly from Panckoucke, says nothing but that it is in 8vo, and apparently came out at Lyons. I suspect that Haeser means the work of Guido Desiderius (Gui Didier) who was physician of the monastery of Saint Antoine de Vienne. It has the following title:

Epitome operis perquam utilis de morbis curandis Valesci de Taranta. Lyons, 1560, in 8vo.

Haller commends this epitome.

The Venice edition of 1589 was also in 8vo, and appears to be authentic. It, too, is mentioned by Haeser. I have not seen copies of either edition.

The edition of 1599 (B. M. 547, i. 1) is based on that of Guido Desiderius, amended and enlarged by Joannes Hartmannus Beyerus, and was printed at Franckfurt, "ex Officina Chalcographica Romani Beati, ſumptibus Nicolai Baſſæi Bibliopolæ, M.D.XCIX." It is a 4to in eights, and contains) : (in six, A to Z, Aa to Vv in eights, Xx in six; or pp.: contents [10, 2 blank], text 677, index [22, 1 blank]. It is a very commonplace book on thin soft paper, which has foxed.

Against the editions of 1560 and 1599 the reader is warned by Astruc.* Another critic says of the 1599 edition that it is curtailed and altered by Beyer, who has not hesitated to add to it a number of Paracelsian ideas.†

Long after, the book was reissued with a preface by Georg Wolfgang Wedel, at Franckfurt and Leipzig, "Sumtibus Joannis Adami Kästneri Bibliop. Jenæ, Typis Joannis NisI, Anno M DC LXXX." Wedel's preface is dated Jena, Cal. Mart. M.DC.LXXX. It also is a 4to, a, b, A to Z, Aa to Zz, Aaa to Zzz, Aaaa to Zzzz, Aaaaa to Vvvvv, or pp.: title, preface, and contents [16], text 871, index [27]. The title page is in red and black; but the book is an ugly seventeenth century quarto. [B. M. 546. g. 19.]

For the edition of 1714 in 4to, I have only Panckoucke's and Haeser's authority. Apparently it was a reissue of the previous edition of Wedel, 1680.

It is obvious that the book enjoyed a very great reputation. The authorities already referred to so often, commend Valescus for his record of original observations, and Conring,‡ a writer who is not lavish of his commendation, makes the following observation: "Valescus item Tarantanus, Regis Franciæ Protomedicus, auctor Philonii, docti ad praxin valde utilis scripti," and afterwards (p. 236) refers to it as "doctum librum."

A commendatory notice of him is given by Freind, who says that

---

* *Mémoires pour servir à l'Histoire de la Faculté de Médecine de Montpellier*, Paris, 1767, 4to, pp. 208-211, where in his notice of Valescus, he gives a brief summary of the *Philonium*, mentions the editions of 1490 and 1521, and, while condemning the above, commends the compilation of Rembert Dodoens, or Dodonæus. I do not see exactly for what reason, because Dodoens makes only three or four brief extracts, which can in no way be compared with a deliberate edition of the book, however badly executed. One would almost suppose that Dodoens' book had not been examined.

† *Dictionnaire Historique de la Médecine Ancienne et Moderne*, par MM. Dezeimeris, Ollivier (d'Angers) et Raige-Delorme. Paris, 1828-1839. 4 tomes, 8vo. I., p. 260.

‡ *In Universam Artem Medicam Introductio*, ed Schelhammer, Spirae, 1688, 4to, p. 103.

"Valescus de Taranta was almost the only one who, about 1400, wrote from experience, and not from books only. . . . He has left us a book call'd *Philonium*, wherein are several good observations relating to the practice of both *Physick* and *Surgery*. . . . And indeed this author interposes his own judgment very often, in some difficult points of Practice: a rare instance in these times, when none had yet scarce begun to think for themselves."* The author of the article in Hirsch's *Lexikon* states that he was one of the lights of the older school at Montpellier, and Portal † has a favourable opinion of Valescus as a practitioner and man of original powers.

The edition of 1680 has been called the best. As none of them are ever likely to serve the modern physician as a *vade mecum*, I should suppose that the early editions will still prove the most attractive from the collector's point of view, irrespective of Haller's insinuation: *Antiqua etiam est nitida, etsi Gothicis literis excusa, editio.* The paper, printing, and *ensemble* of a working quarto of the middle of the seventeenth century form one of the severest trials to which the taste of a bibliophile can be subjected.

13. The various editions of Gregory Reisch's *Margarita Philosophica* having been already enumerated by me (Part V., p. 433), it will be sufficient to tabulate those acquired since my first notice of it. They are—

Strassburg, 1504.	Strassburg, 1512.
Strassburg, 1508.	Strassburg, 1515.

Basel, 1535 and 1583.

Of the 1515 edition there are copies differing in details from one another, so that of that edition, at all events, there were at least two issues. Of the different editions, the first, 1503, is perhaps the most abundant, though the book in one issue or another is not uncommon. But to get a copy in fine condition, clean, unwormed, perfect, especially with the relative map or maps and the music plates, is very difficult indeed. The map has been almost always removed and other illustrations as well. As the bibliography of this book involves much detail,

* *The History of Physick*, London, 1726, Part II., pp. 295–7.

† *Histoire de l'Anatomie*, 1770, I., p. 237.

I intend to consider it more minutely in a separate paper, and at the same time I may possibly give some notion of its contents, as a summary of learning at the close of the fifteenth century.

14. A treatise contemporary with the *Margarita Philosophica* is entitled *Cueur de Philosophie.* Up to the present time this book has not appeared in my "Notes," and it has therefore the attraction of novelty. As to its theme, it deals pretty much with the secrets of Nature generally, not of the Arts, and it covers part of the ground cultivated by Reisch. Like him too, the author selects the catechetical form, and carries on the teaching by question and answer between a disciple and his master.

As the book, like the *Margarita,* deserves more than a mere passing glance, I shall confine myself, for the present, to the bare mention ot the editions which I have seen. Those which I have acquired are dated 1521 and 1524. They are both in small 4to, printed in black letter, for François Regnault, Paris. They have a few woodcuts. These are neither in the British Museum, nor are they noticed by Brunet. In the Museum are editions dated 1514 and 1520, and in addition to them Brunet mentions one by Antoine Verard about 1504, and others of 1521, 1529, 1534, and 1538. A copy of Verard's edition I have seen in the Municipal Library at Lille; and quite recently, in the Mazarine Library at Paris, I found an edition of 1530. As this one is not referred to by any bibliographer, I may give here the note which I made upon it:

Senfuit le Cueur de Philoſophie: contenant pluſieurs demandes et queſtions propoſees par le ſaige Placides au Philoſophe Tymeo: Et les reſponces cōtenuz en icelluy. Auec le traictie de Leſpere du mōde Et le Cōpoſt et Kalendrier touchant le cours du ſoleil et de la Lune des feſtes fixes et le biſſexte. Le tout figure: Tranſlate nouuellement de Latin en Francoys. xxx.

[Device]

+ On les vend a Paris en la grād Rue ſainct Jacques en Lenseigne de la Rose blanche courōnee par Philippes le Noir Libraire demourāt au dit lieu.

Small 4to. a in six, b to l in fours, m in eight; n to z, A, in fours; B in eight, C to F in fours, G in six; or ff. [6] CXXVI.

Black letter. Numerous woodcuts.

ai Title, in black and red. The device is Aristotle under a tree with an open book on his knee, holding a conversation with a draped female labelled Philosophie. This device has on the two sides a "billet" border alternately red and black, like those in the *Secrets de l'hystoire naturel.*

aii *recto*: La prologue, ends *verso.*

aii *verso*: La table, ends avi *recto.*

avi *verso*: Woodcut of a king seated, a queen (?), a kneeling female, and a man dressed as an ecclesiastic.

bi Text begins and goes to the end, Gvi *verso*. Then the colophon:

Cy finiſt Le Cueur de Philoſophie tranſlate de Latin en
Francoys a la requeſte de Philippes le Bel Roy de
France: Nouuellement Imprime a Paris pour
Philippe le Noir, Lung des deux Relieux iu-
rez: Et Maiſtre Imprimeur a Paris, De-
mourant en la grand Rue Sainct Jac-
ques: A lenſeigne de la Rose blanche
couronnee. Et a eſte acheue dym-
primer le XXXIII. iour de Mars (*Sic* for XXVIII?)
M.CCCCC.XXX.

15. Among the Italians who have compiled books of Secrets, I spoke in the first supplement of Guglielmo Varignana. He was one of a family of this name, several of whom were Professors at Bologna.* Bartolomeo, his father, died in 1318-19; he himself died in 1330, and he left two sons, Pietro and Matteo.

Guglielmo, who concerns us at present, wrote a handbook of medical practice, which was not printed till the sixteenth century, but thereafter met with such success as not only to eclipse the fame, but almost to obliterate the

* The following are mentioned among the doctors in various faculties at Bologna by Giovanni Nicolo Pasquali Alidosi (*I Dottori Bolognesi di Teologia, Filosofia, Medicina, e d'Arte Liberali dall' anno* 1000, *per tutto Marzo del* 1623, Bologna, 1623, 4to): p. 24, Bartolomeo; p. 47, Domenico; p. 77, Giovanni; p. 79, Guglielmo; p. 140, Matteo; p. 155, Pietro.

An account of Bartolomeo da Varignana and of the castle from which the family name arose is given by Mauro Sarti (*De claris Archigymnasii Bononiensis Professoribus a Sæculo XI. usque ad Sæculum XIV.* Bononiæ, 1769-1772, folio; Tom. I., Part I., p. 480). It is clearly shown that they were of Bolognese, and not of Jewish origin, as some writers have stated.

memory of his father, Bartolomeo, though by contemporaries the latter had been more highly esteemed as a physician. Bartolomeo wrote several treatises, but as they were not printed they had not the fortune to be known to subsequent times in the same way as the volume of medical receipts and secrets by his son.*

The earliest edition I have found mentioned is dated 1519. I have not seen a copy of it, but the title is given by Billings as follows:

> Secreta sublimia ad varios curandos morbos verissimis auctoritatibus illustrata additionibus nonnullis: flosculis item in margine decorata: diligentissime castigata: nusquam impressa feliciter incipiunt. 5 p. l., 129 ff. 1 l. 12° [*Impressum Papiæ, per Magistrum Bernadinum de Geraldis*, 1519].

The year after, an edition was printed at Venice, of which I have been fortunate enough to secure a copy.

> Guilielmi Varignane fecre | ta fublimia ad varios curandos | morbos veriffimis auctorita- | tibus illuftrata additionib⁹ | nõnullis: flofculis ite3 in margine decorata dili | gẽtiffime castigata: | nufq3 impreffa fe | liciter inci- | piunt ∴ | ✠ |
>
> Small 8vo. Signatures A–R in eights; or, ff. [8] 127, last leaf, blank (?) wanting. Black letter.
>
> Colophon: ¶ Impreffum Venetijs per Alexandrum de Bin | donis. Regnante Principe Leonardo Lau- | redano. Anno Domini. M.D.xx. Die | xx. Menfis Nouembris.

This is an attractive little volume of undoubted rarity. It is in Mercklin's† list, from which it is repeated by Haller.‡ There is no copy in the British Museum, but it is also described by Billings.§

The next edition, 1522, likewise occurs in Mercklin's and Haller's enumeration, and the copy which Panzer‖ knew long ago was in the library of the Royal Society.¶ There is none in the British Museum. The following account is given by Panzer, and for the collation I am indebted to Billings:

> Guilielmi Varignane fecreta fublimia ad varios curandos morbos veriffimis autoritatibus illustrata additionibus nonnullis flosculis. Item in margine decorata: diligentissime castigata: nusquam impressa feliciter incipiunt. Infigne Vincentii de Portonariis. In fine: Lugduni Impreffum per Jo. de Cambrey Anno a Virginis partu. 1522. Deo gratias. 4. [3 p. l., lvi. ff. sm. 4°.]

---

* Sarti, *Ibid.* p. 483. † *Lindenius Renovatus*, Norimb., 1686, p. 385.

‡ *Bibliotheca Medicinæ Practicæ*, Bernæ, 1776, I., p. 443.

§ *Index-Catalogue*, 1894, XV., p. 593.

‖ *Annales Typographici*, 1801, IX., p. 519, 465 b.

¶ It is so still. See *Catalogue*, 1883, p. 1083.

The edition which appeared at Lyons in 1526 seems to have been well known, as it is enumerated by Mercklin, Haller, Eloy, Panckoucke, Haeser, and is included in his list by Billings. His collation is: lxxxvii. ff., 3 l. Sm. 4° [*Lugduni, impressum per J. de Cambray*, 1526]. There is a copy of this edition in the library of the Royal Medical and Chirurgical Society, London.* Mangetus, who took his list from Mercklin, misprints the date 1626, as is pointed out below.

Several other editions are enumerated, but while some are authentic, others appear to me to be simply misdated. The following are those which I have observed:

1531 (?), 21 p. l., 477 pp., 12° Basileæ, per H. Petrum.

Billings gives this, but it seems to be identical with the undated edition in the British Museum, referred to by me in the previous supplement.

An edition of 1536 is mentioned in the *Biographie Universelle*, T. 42, p. 637. I have little doubt that this is a misprint for 1526.

In the first supplement I alluded to an edition of 1539, of which there is a copy in the Museum. It is mentioned by Mercklin, Haller, Eloy, and Haeser. Billings' collation is: lxxvii. ff., 3 l., 12°, Lugduni, J. Flaiollet; which agrees with the copy I have seen. There is a copy of the same date and place, but designated a 4°, in the library of the Faculty of Physicians and Surgeons of Glasgow.†

Haeser alone is the authority for an edition ot 1540, Venice 8°, but from its being placed between those of 1519 and 1526 in his chronology, I have little doubt it is a misprint for 1520.

The edition of 1545, Basel in 4°, mentioned by Mangetus and by Panckoucke, may possibly be the undated edition I described before. I have not noticed any other reference to such an edition.

The edition of 1597, Basel 8°, edited by Caspar Bauhinus, is that most requently spoken of. I have already mentioned the copy in the British Museum. Billings gives the following collation: 7 p. l., 375 pp., 16° Basileæ per S. Henricpetri. In the Museum copy there are eight pre-

* *Catalogue*, London, 1879, vol. ii., p. 486. The title there given is: *Opus de curandis egritudinibus*. goth. l. 4°. Lugd. 1526.

† *Catalogue*, 1885, p. 645.

liminary leaves. There are copies in the University library, Aberdeen,* and in the library of the Royal Medical and Chirurgical Society of London.†

Mangetus ‡ quotes an edition of 1626: "Lugd. apud Joh. de Cambray," in 4°. This again, from its coming between the editions of 1520 and 1539, and from its printer, is obviously the 1526 edition with the date misprinted. Besides, he admittedly quotes from Mercklin, who has no edition of 1626, but only of 1526.

Varignana has been commended by several authorities. Conring says of him, when discussing the notable men of the epoch: "Item Guilielmus Varignana, Judæus quantum apparet, vir magnæ eruditionis," § and again, "Ad quos tamen adhuc referendus est, qui quasi empiricam curandi morbos rationem tradidit *Guilielmus Varignana;* cujus liber titulo *Secretorum* verissimis auctoritatibus illustratus ad multorum morborum curationem valde juvare nos potest."‖ In a note to this passage (p. 252), Conring compares Fioravanti with Varignana, but gives the former a higher place as being a "rationalising" empiric, who is justified in styling his book *Compendium Secretorum rationalium.* Mangetus ¶ calls him "ſublimis et excellens medicus," and copies Conring as to his nationality. He gives a brief abstract of the book, and quotes the editions from Mercklin.

Portal ** devotes a paragraph to Varignana. He states that according to Conring he was a Jew, that he practised in Genoa, was highly distinguished as an author, and published two works, *Opera Medica* and *Secreta sublimia.*

Pretty nearly the same narrative is repeated by Eloy.†† He mentions the father, and says of Guglielmo that he taught for several years at Bologna, his native place, whence he was called to Genoa, where he wrote his two books, of which he enumerates the following editions respectively: *Secreta*

* *Catalogue*, Aberdeen, 1874, vol. III., p. 294.

† *Catalogue*, London, 1879, vol. II., p. 486.

‡ *Bibliotheca Scriptorum Medicorum*, Genevæ, 1731, II. ii., p. 458.

§ *In Universam Artem Medicam Introductio*, ed. Schellhammer, Spiræ, 1688, p. 102.

‖ *Ibid.*, p. 237.

¶ *Bibliotheca Scriptorum Medicorum*, Genevæ, 1731, II. ii., p. 458.

** *Histoire de l'Anatomie*, Paris, 1770, I., p. 204.

†† *Dictionnaire Historique*, Mons, 1778, IV., p. 479.

Medicinæ, Papiæ, 1519, 8°; Venet. 1520, 8°; Lugd. 1526, 4°; 1539, 8°; Basil. 1597, 8°. *Opera Medica de curandis morbis*, Basil. 1545, 4°; 1595, 8°; Lugd. 1560, 8°.

These older accounts were examined and criticised by Tiraboschi.* Relying upon the authority of Alidosi and Sarti, quoted above, he shows that his father, Bartolomeo, was a native of Bologna, and of a noble family, and is at a loss to know how he could be thought to be of Jewish extraction. He points out, moreover, that Conring does not say that he was of Jewish origin, as Portal makes him do, but only that he appears to have been Jewish. He does not know any authority for his practising in Genoa, and he believes that the two titles, *De Curandis morbis* and *Secreta sublimia*, refer really to one book, and not to two.

Henschel, a more recent writer than these, and a professed investigator of the history and biography of medicine, curiously enough refers to Conring's statement only as to Hebrew descent. He does not seem to have known about Sarti's inquiry. It looks as if he had better evidence for saying, on the strength of Varignana's own dedication, that his patron was named Meladinus, and was a nobleman of Croatia and Bosnia, and that the author lived latterly in Illyria.†

In Varignana's case, likewise, there seems to be a good deal of obscurity surrounding the events of his life. It may be remarked that the article by Weiss ‡ is based on Tiraboschi's account.

16. Whether Raoul du Mont Vert is the real or fictitious name of the author of the little volume, *Les fleurs et secretz de médecine*, I have not been able to ascertain. The book itself has the merit of originality; it is independent of similar collections. Like many other French books printed early in the sixteenth century, the editions are all rare and little known. I described formerly one of 1547 and a late reprint of 1602, besides enumerating the editions in the British Museum and those noted by Brunet.§ I have recently got other two not in the Museum and omitted

* *Storia della Letteratura Italiana*, Firenze, 1807, t. V., parte I., p. 257.

† *Janus*, 1853, II., p. 390. See full reference in note ||, p. 194.

‡ *Biographie Universelle*, Paris, 42, p. 637.

§ Part V., p. 441.

by Brunet. Both are undated, and both are in small 8vo, printed in black letter. One of them has no place or printer's name, the other issued from the press of Alain Lotrian and Denis Janot at Paris, and may therefore have appeared as early as 1520. This work requires fuller consideration than I can give it here at present, and I hope to subject it to a special examination.

17. A much-esteemed book must have been Vicary's *Englishman's Treasure*, which was edited and enlarged and supplemented by William Bremer. The earliest edition that I was able to submit formerly to the Society was printed by Thomas Creede at London, 1596, in black letter, Part V., p. 454. There was also a very dilapidated copy, certainly dated 1696, and purporting to be of the seventh edition, Part IV., p. 306. Another copy, however, of the seventh edition was dated 1626, and this agrees better with those now to be described.

The first edition seems to be that imprinted at London by John Windet for John Perin in 1586. It is a small black letter quarto: 5 preliminary leaves, B to L, A, Bb to Dd in fours, Ee in two, or pp. [10] 115 [1 blank]. It contains Vicary's Anatomy, Remedies for Wounds, various medical receipts, Urines, of the Bath of Baeth. In the later editions much matter was added to the original book.

The next copy I have to show is of the sixth edition; it was "Imprinted at London by Thomas Creede, 1613." It is a small 4to, in black letter, pp. [8] 224 [8]. It has the dedication to the Governors of St. Bartholomew's Hospital, a short note "To the Reader," Vicary's address to the Surgeons, and the picture of the human skeleton, Vicary's True Anatomy, Secrets of Chirurgerie, of Medicine and of Vrines, the English Baths, and a collection of medicinal preparations, waters, oyntments, plaisters, etc., by G. E., and the last eight pages contain an index. There is no plate of the blood-vessels in this edition as there is in that of 1641. This is rather a neat little book.

The eighth edition, a copy of which is in the Hunterian Library and in the British Museum (1039. g. 7), was printed at London by Alsop and Fawcet in 1633. It is exactly like those of 1626 and 1641, being in 4to and black letter, pp. [8] 264 [8]. It does not, however, contain the diagram of the skeleton of the 1626 or of the veins of the 1641 edition. In the 1641 edition there are receipts, pp. 246-272, which are not in that of 1633.

That of 1641, just mentioned, is the ninth edition, and it too was printed by B. Alsop and Tho. Favvcet. The two editions are in exactly the same style, in small quarto and in black letter. There are a few minor changes in the title page, the most important being that, in the seventh edition, the additions about waters and medicines, etc., etc., are ascribed to G. E., Practitioner, whereas, in the ninth edition, they are ascribed to W. B., who, according to the Museum catalogue, is William Boraston. There are a good many alterations in the arrangement of the two editions, and the ninth is undoubtedly considerably enlarged. It contains pp. [12, including the frontispiece], 292, [15, 1 blank]. The colophon runs thus, on a leaf marked Rr.:

London, | Printed by B. Alsop and Tho: Favvcet, | dwelling in *Grubftreet*, neere to the lower | Pumpe, 1641. |

This copy is in about as good condition as a book of its date and character could be—clean, crisp, with the frontispiece of the blood-vessels, perfect in every way, and in old sheep binding with clasps. Another copy of the same edition I have seen from Mr. Hailstone's library in equally fine state.

The first tract, Vicary's Anatomy, is the oldest treatise on the subject in English. It was originally printed in 1548, but apparently no copy of it has survived. It was reprinted in 1577, and the so-called unique copy of this edition has served for a reprint by the Early English Text Society in 1888. It is No. LIII. of the Extra Series. Vicary belonged to St. Bartholomew's Hospital, and he compiled his tract for the special instruction and guidance of surgeons. Thereafter it became a sort of St. Bartholomew text-book, and all the editions enumerated down to this one were dedicated to the Governors of the Hospital.

It is a collection of some importance, for it represents several sides of the practical and medical knowledge of the time, and it was considered authoritative, seeing that it continued in use for over a century. The tract on the plague, in the ninth edition, reveals the terrors under which people lay, before they had sufficient knowledge to discriminate between natural phenomena and their own superstitions. The seven prognostics of the plague are instructive from this point of view, and what is said may be compared with another little treatise on the same lugubrious theme: ΛΟΙΜΟΤΟΜΙΑ, *or the Pest anatomized*, by George Thomson, London, 1666, in 16°.*

* [January, 1897. These works might be considered of special interest just now, when precautions are taking to prevent an incursion of the plague from India to Europe.]

ERRATA IN THE FIRST SUPPLEMENT.

Transactions (N.S.) II.

Page 372, line 2, Reprint page 9, *delete* as follows.
Page 403, line 22, Reprint page 40, *for* seems *read* seem.
Page 403, line 24, Reprint page 40, *for* 1559 *read* 1599.

BIBLIOGRAPHICAL NOTES

ON

HISTORIES OF INVENTIONS

AND

BOOKS OF SECRETS.

THIRD SUPPLEMENT.

BY

JOHN FERGUSON, LL.D., F.S.A. LOND. AND SCOT.,

REGIUS PROFESSOR OF CHEMISTRY IN THE UNIVERSITY OF GLASGOW.

GLASGOW:

Printed at the University Press by

ROBERT MACLEHOSE & CO.

1898

Read to the Archæological Society of Glasgow, 15th March, 1894.
100 *Copies Reprinted.*

No. XI.

BIBLIOGRAPHICAL NOTES ON HISTORIES OF INVENTIONS AND BOOKS OF SECRETS.

(*THIRD SUPPLEMENT.*)

BY

JOHN FERGUSON, LL.D., F.S.A.Lond. and Scot.
REGIUS PROFESSOR OF CHEMISTRY IN THE UNIVERSITY OF GLASGOW.

(*Read at a Meeting of the Society, held on 15th March,* 1894.)

1. The present supplement, though numbered separately for reference, really consists of the third part of the paper submitted to the Society on the above-mentioned date. It embraces treatises in a variety of languages, on a variety of topics: receipts useful in different arts and sciences; secrets of the toilet; cures for ailments; wonders and prodigies of nature. It differs, therefore, from the previous supplements, the first of which was restricted mainly to Italian books, and the second to books on medicine.

During the time that has elapsed since the paper was written, it has been revised and entries made of a few newly discovered works and editions. The English books, however, which were originally included, have been removed, and reserved for separate treatment. The number of them has become so large, that I have thought it desirable to consider them by themselves, in order that the character and extent of this literature, published in this country, more particularly in the seventeenth and eighteenth centuries, might be judged of, apart from analogous foreign works. Moreover, so few of the English books mentioned happen to be translations, that the disjunction of these from the originals will not, I trust, cause inconvenience. Any reader, however, who may wish to have a view of the consecutive editions and translations of any single treatise, and who might be hampered by this disjunction, will have his wants supplied, as far as possible, by cross references from one set of books to the other.

I cannot refrain from saying once more that the amount of early technical literature, whether in respect of distinct works or of different editions, far surpasses what I believed to exist. Whether it was instructive or not, there was plenty of it current, and if our predecessors were ignorant it was not for want of books. The authors, however, seem to have been as ill-informed as their readers; how then could they impart instruction, no matter how many books they wrote and sold?

2. The work with which the present section opens might have appeared more appropriately in the second Supplement amongst medical writings. As its name denotes it contains prescriptions for every disease, the preparation of ointments and pills, the virtues of a number of plants, and, incidentally, certain cosmetic receipts. It may perhaps be regarded as too purely medical for the present list, but it may very well receive notice as a contemporary counterpart to the works of Valescus and Varignana on one hand and of the later *Treasure of Evonymus* on the other. How far it is the work of Galen himself I am not prepared to say, but if it be his in part there does not seem to be any corresponding separate treatise by him. It purports also to be translated from the Latin, but if that be so I have not seen any mention of a copy of the original. In the British Museum Catalogue, the present work is put in a place by itself, neither amongst the genuine nor the supposititious works of the author. While, of course, it may be authentic, it is possible that its name denotes no more than that it is a collection of the prescriptions which, at the beginning of the sixteenth century, passed under Galen's name, and were opposed to the chemical or spagirical remedies which were just beginning to show themselves. One thing is quite certain, that the medico-astrological introduction is modern, as well as various medicaments—not to speak of diseases—which were unknown to the ancients.

3. The two copies which I have got are interesting by the wide interval between their dates and by the consequent differences between them as far as externals go.

Recettario di Galeno Opti- | mo e probato a tutte le infirmita che accha | deno a Huomini et a doñe di dentro et di | fuori li corpi. Et cõ remedio de cõferuar la | fanita et prolõgar la vita: et con molte altre | cofe noue che in li altri prima ftampati non | erano: et con molte altre recette cõtra la pe | fte. Tradutto in volgare per maeftro Zua- | ne faracino Medico Eccellentiffimo ad | inftantia de lo imperatore.

Small 8vo. A to H in eights; or ff. 63 [1]. Black letter. Double columns.

Collation: A1 *recto*, Title page. The lower half is occupied by a woodcut representing a patient on a chair having something done to his right knee. Behind is the physician apparently occupied in dismembering birds, and there is an attendant with a bason. *Verso*, Tabula, ends A5 *verso*.

A6 *recto*, Text begins.

B2 *verso*, is a woodcut of the human body as under the signs of the Zodiac.

B4 *recto*, Woodcut of the Tabula Salamone.

H7 *recto*, the text ends. After the text is the Registro, and the Colophon:

¶ Stampato in Vinegia per Francefco di Alef | fandro Bindoni: & Mapheo Pafini: com | pagni: nel Anno. 1528. del me- | fe di Febraro.

H7 *verso*, blank.

H8 *recto*, blank. *Verso*, printer's device of Tobit and the Angel. This device came first under my observation in editions of the *Physonomia* of Michael Scotus, printed in 1533 and 1537 by the same printers, and it will recur once or twice in the present review. The difference between this and these other books is that this is in black letter in double columns, while they are in italics in long lines. The result is that the former looks antique, the latter modern, with an interval, in one instance at least, of only five years between the two.

4. The other later edition is as follows:

Recettario di Galeno. Approvato, Et Molto Vtile alle infirmità, à che sono sottoposti gli corpi humani. Con rimedij di conseruare la sanità, e prolongar la vita con altre noue ricette, che non erano negli altri prima ristampate, & lire da preseruarsi contra il mal contagioso. Tradotto in lingua volgare, per lo Eccell. Maestro Giouanni Saracino Medico, ad istanza della Cesarea Catholica, e Real Maestà dell' Imperatore.

In Venetia, M.DC.LXXXIII.

Apresso Sebastian Menegatti. Con Licenza de' Superiori.

18mo. A to G in twelves; or, pp. 154 [14]. On p. 14 is a woodcut of the veins.

This is a poorly got up little volume, intended apparently for everyday use. Neither of the preceding is in the British Museum. The copies which are there are dated respectively Venice [1508], 1518, 1545, 1640, 1645. The book, therefore, must have met some sort of want, else it could hardly have reached the ripe age of a hundred and eighty years.

5. In bibliography it is sometimes almost startling how books related to one another, and of unquestionable rarity, come together from quite different quarters. It is not so long ago since I got a copy of the 1520 edition of the *Secreta Secretorum*, and the description of it was hardly published when a copy of the no less rare German translation made its appearance.

Das aller edlest und bewer | test Regiment der gesundtheyt, Auch von allen ver- | borgen künsten vñ Künigklichen Regimenten Aristotelis. Das er dem Grossmech | tigen Künig Alexandro zůgeschriben hatt.* Auss Arabischer sprach durch Meister | Philipsen, dem Bischoff Vonn Valentia, der Stat Jerapolis, In das latein | verwandlet, Nachmals auss dem latein in das Teijtsch gebracht, Bey | Doctor Johañ Lorchner zů Spalt (So beyder Keyser Frydrichs | vñ Maximilians Löblicher gedechtnuss Rath vñ Mathe- | maticus gewesen) nach seinem tod geschribẽ gefunden | zů auffenthaltung vñ fristung yn gesundtheit | menschlichem lebenn zů gůtt; Durch Jo- | hann Besolt in Truck verordnet. | [Vignette]. M.D.XXXI.

Small 4to. A to N in fours; or ff. [4] XLVIII.

The Vignette represents Aristotle presenting his book to Alexander.

Aj contains the title; *verso*, Vorrede; Aij *recto* to Aiij *recto*, the Contents; Aiij *recto*, Aristotle's preface, ends Aiiij *recto*. Aiiij *verso* is taken up with a woodcut of Alexander in the dress of a royal warrior of the year 1531. B to Niiij *recto* contains the text. At the end is the colophon:

Gedruckt zů Augspurg durch Heynrich Stayner, Am. 1. | tag Martij, des M.D.XXXI. Jars |

The *verso* is blank.

So far as it goes this is a translation of the first part of the 1520 edition, embracing the *Secreta Secretorum* and omitting the tracts *Signa temporum*, *de Intellectu*, *de Universalibus* and others formerly enumerated.* But there are several omissions; for instance, a part of the chapter on Stones (German Translation, chap. lix.) and sections both at the beginning and the end of the book. It has not however been disembowelled like the English translation of 1702.†

* Supplement II., *Transactions*, III., 1897, p. 180. † Part IV., p. 326.

The copy of this translation in the British Museum (520, a. 12) was published by Stayner in 1530, in 4to. It seems to be different, for the collation given in the Catalogue is ff. 46, whereas my copy has ff. 48. There is a translation also in French, but I have not seen either of these editions.

6. Do as one will it seems impossible to escape from Albertus Magnus and his two books of Secrets. In the previous Supplement I thought I was done with him for the time being, but, since its issue, another has put itself in evidence. The two treatises are in one volume, printed uniformly; but as they have distinct title pages and separate signatures and pagination, they are virtually independent, and I do not doubt may sometimes be met with detached from each other.

Alberti | Magni | Specvlvm | Astronomiae : | Nvnc Primvm E | m.f. Codice in lucem edi- | tum. | Praemittvntvr Avtem | eiusdem Authoris libelli, | De Virtvtibvs Herba- | rum, Lapidum, & animalium quorun- | dam, Item de mirabilibus mundi, & de | quibusdam effectibus causatis à | quibusdam animali- | bus, &c. | Lvgdvni | Anno MDC.XV.

24mo. Signatures A to G in twelves; or pp. 168 (misprinted 166).

The *Liber Aggregationis*, which treats of plants, stones, and animals, with the *de mirabilibus mundi*, occupies to p. 115. The *Speculum Astronomiae* fills up the rest. This tract is a novel addition, and I have not met it before along with the *Liber Aggregationis*.

7. The other treatise has the following title:

Alberti | Magni | De Secretis | Mvliervm Libel- | lvs, Scholiis | auctus, & à mendis re | purgatus. | Cum Indice locupletissimo. | Lvgdvni | Anno M.DC.XV. |

24mo. Signatures A to G in twelves. G 11 and 12 are blank; or pp. 164 (misprinted 166), [4 blank].

There is nothing noticeable about this edition. There is a copy in the British Museum, 1175 a. 2 (2).

8. [May, 1898. In 1884 (Part III., p. 213, Reprint, p. 28) I referred to a book entitled *Ein Newer Albertus Magnus*, printed at Frankfurt, by Weygandt Han, without date. In a note the similarity between a section of it with its illustrations and the work of Thomas Raynalde, *The Birth of Mankinde*, was pointed out.

I have just got another edition of this work, printed by Han, but dated 1556. I have some difficulty in deciding which of the two is the earlier. Although similar in all respects they are not identical. The title page in black and red, the woodcut border containing in compartments the most prominent physicians of antiquity, the ornamental capitals, the woodcuts, the type, and the same number of leaves, are contained in both; but details of typography, the arrangement of the title page, the position of the signatures, the use of capitals, and so on, show that the books were printed at different times.

In the note referred to it is implied, erroneously, that the edition of 1540 of Raynalde's book is in the Hunterian Museum. The editions there are of 1545, 1565 and 1598. I have myself acquired copies of the two earlier editions. That of 1545 is in 8vo, printed in bold black letter. The signatures are: A in four, B to I, Hh, in eights, Hhh in six, K to Q in eights, R in four, S to X in eights, Y in ten. There are four leaves of engravings in this copy, but, two, if not four, are wanting. There is no copy of this edition in the British Museum.

The 1565 edition is a large 8vo, or small 4to, printed with two sizes of black letter. The signatures are: A in four, B to S in eights, T in six, and there are four leaves of woodcuts, which form a complete set. The copy I have belonged to Herbert, and is described by him in the *Typographical Antiquities*, London, 1785, I., p. 581.

This edition is in the British Museum.

The edition of London, 1598, is much inferior to both the preceding. It, too, is in black letter, but it is a poor fount, and the paper is soft and weak. It is in small 4to, A in four, B to O in eight, or pp. [8] 204. The woodcuts—poorly executed—are included in the signatures and pagination as well. There is a copy in the Hunterian Library, and in the British Museum. Other editions followed in the seventeenth century.]

9. A collection of receipts in French, professing to be translated from the Italian, has come into my possession. As it was obviously a chap-book of some time last century, I was sceptical about its being a translation, and thought that this statement might be a ruse to procure for the book a currency which it might not have enjoyed if it had been

announced as of French origin. I investigated the matter, and have been rewarded by the discovery in the British Museum of a cluster of receipt books, which have not hitherto been included in my lists. The collection was really first written in Italian, and the book enjoyed an intermittent circulation for upwards of two hundred years, as the following enumeration shows. It is more than probable, however, that there are editions besides those to which I shall refer.*

The list begins with the Italian of 1529, which is probably the very first of the set. The description is from the British Museum copy (7953, f. 32).

Opera Nvova | Intitolata Dificio De | Ricette nella quale ſi contengono tre | utiliſsimi Ricettari, | Nel primo ſi tratta di molte & diuerſe uirtu. | Nel ſecondo ſe inſegna a comporre uarie ſorti | de foaui & utili odori, | Nel terzo & ultimo ſi tratta di alcuni rimedi | ſecreti medicinali neceſsari in riſanar li | corpi humani, come nella tauola | qui ſequente ſi puo uedere. | M D X X I X

Small 4to, A, *b* to *f* in fours; or ff. [4] 20, but the numbers are irregular. Printed in italics. At the end of the text:

Stampato in Venetia per Giouanantonio et | fratelli da Sabbio. MDXXIX.

Ai, Title, surrounded by a broad border of white loops and twists upon a black ground.

Verso, Full page picture of a master in a chair, dictating to a clerk or scholar. There are books, a sphere, a compass, etc., and the picture is surrounded by a border of white knots on a black ground—different from the other.

Aii *recto*, Incomincia La Tavola. Ends Aiv *recto*.

Aiv *verso*, Picture of a monk in a laboratory, directing a servant who is pounding something in a mortar; boxes, an hour glass, etc., are on a shelf.

bi, Allo mio benigno lettore, and then the first receipt: Remedio contra ogni puzor et fetor di boca. This first part ends ei *verso*, or ff. 1-17, (*sic* for 13).

diii *verso*, At the top is a note to the reader within a white and black border.

eii, The second part begins with a title page enclosed in a white and black border like that round the *verso* of the first title. It ends e iii, fo. 19 (*sic* for 15).

eiv, The third part begins with a border like that round the first title. On the *verso*, top half, is a woodcut of a distilling apparatus. This ends fiv *recto*, or fo. 20. *verso* is blank. The borders and woodcuts give this book a very odd appearance.†

---

* May, 1896. Shortly after the above was written, and after the other editions had been described, I got a copy of the Italian of 1538, which proved that my surmise had been correct.

† The only copy I have seen for sale was in the catalogue of a Florence bookseller, but I failed to get it.

10. I may now enumerate the other editions in Italian and in French which are in the British Museum, and the two which are in my possession.

Opera Nvova Inti- | tolata Dificio Di | Recette Nellaqvale Si | contengono tre utilifsimi Recettarij. | Nel Primo fe tratta di molte & diuerfe uirtu. | Nel Secondo fe infegna à comporre uarie | forti de foaui et utili odori. | Nel Terzo et ultimo, fi tratta di alcuni rime- | di fecreti medicinali neceffari in rifanar | li corpi humani, come nella | tauola fe contiene. | [Device.] In Venetia. |

Small 8vo. A to D in eights, or ff. 30 [2]. Printed in italics, but the index in Gothic letter. The device is a draped female figure marked Fede, holding a cross with her left arm and a cup on her right palm. A hilly landscape, with a town in the background. Round two sides of the picture are the words: Senza di me l'hvom | fassi a Dio ribello. |

A1, Title. *Verso*: to the reader.

A2 *recto*, Text begins and ends D6 *recto*, ff. 2-30.

D6 *verso*, La Tavola, ends D8 *verso*. Then the colophon:

Stampato in Vineggia per Mathio Pagan | in frezaria, al fegno della Fede. |

No date. f. 24, woodcut of a Still. [B. M. 1038, d. 4.]

11. In another edition of the *Deficio* it is surprising that the title should have been radically altered by the omission of the very phrase by which the book was to be identified: *Intitolata Deficio di Recetti*. This copy came into my hands quite unexpectedly after the rest of the list was finished. It is not in the Museum.

Opera Nvova | Nella Qvale Si Con | tengono tre utiliffimi Ricettari. | Nel primo si tratta di molte | et diuerfe uirtu. | Nel fecondo fe infegna à comporre | uarie forti de fuaui, et | utili odori. | Nel terzo et ultimo fi tratta d'alcuni ri- | medi fecreti medicinali neceffari in | rifanar gli corpi humani. | Come nella Tauola | qui feguente fi | pouedere. | MDXXXVIII.

Small 8vo. A to H in fours, or ff. 32. Printed in italics. The title is enclosed in a woodcut border of scrolls, faces, and grotesque figures—a copy of that round Michael Scotus' *Physonomia*, 1533, 1537.

Woodcut of a distilling apparatus on f. 21 *verso*.

Collation. Ai Title. Aii *recto*, the text begins with a note to the reader, and ends Giiii *verso*. Then follows immediately the Table of Contents, ending Hiiii *verso*. The register comes next, and lastly the colophon:

In Vinegia per Guilielmo da Fontareto.

M D X X X V I I I.

12. B. M. 1168, e. 8 (2).

Opera Nvova | Intitolata Dificio Di Recette. Nellaqvale Si | contengono tre vtilif- fimi Recettarij. | Nel Primo fe tratta di molte & diuerfe virtu. | Nel Secondo fe infegna à comporre varie | forti de foaui & vtili odori. | Nel Terzo &

vltimo ſi tratta di alcuni | rimedi ſecreti medicinali neceſſari in | riſanar li corpi humani, come | nella tauola ſe contiene. [Device of Tobit and the Angel] M. D. L.

Small 8vo. A to D in eights; or ff. 30 (misprinted 29) [2]. Printed in Italics; the Index in Roman.

A1, Title. *Verso*: address to the reader.

A2 *recto*, Text begins, and ends D6 *recto*, ff. 2-30.

D6 *verso*, Incomincia la Tavola, ends D8 *verso*. Then the colophon:

Stampata in Vinegia per Franceſco Bindoni, & Ma- | pheo Paſini compagni. Nel anno del | Signore. M. D. L. | Del Meſe di Mazo. |

P. 24 is the woodcut of the Still.

Simply a reprint of the edition of 1529. It resembles Pasini's editions of Michael Scotus.

13. I now come to the copies of the French translation in the British Museum which I have examined. The earliest (1038, a. 1) is of date 1541, but it is obviously not the first:

Traicte | nouueau, Intitule | Baſtiment de receptes: Nou- | uellement traduict de Italien | en langue Francoyſe. | Contenant trois pe- | tites parties de re- | ceptaires. | Le contenu deſquelles vous voy | rez en la page ſuyuante. | Oultre la precedente impreſ- | ſion ont eſte adiouſtez aulcuns | ſecretz pour les maladies | des femmes, tant en tra- | uail denfant que aul- | trement.

Small narrow 12mo. A to L in sixes. No foliation. Small Gothic character. There is an ornamental device at the bottom of the title.

Ai, Title. *Verso*: contents of the three parts.

Aii *recto*, Dizain.

Aii *verso*, Au Lecteur, ends Aiii *recto*.

Aiii *verso*, Text begins, and ends Lv *recto*.

Lv *recto*, Colophon:

☞ Imprime en An- | uers par Matthieu | Crom. Lan. M. D. | xlj.

Lv *verso* and Lvj are blank.

The text of the Italian book comes down to Li *recto*. The extra medical receipts are contained Li *verso* to Lv *recto*.

This is a very curious little book. It is of the size and shape of the 1534 edition of Rabelais.

14. The next, in the Museum, is marked 1169, a. 3:

Bastiment | de Receptes, contenant trois | parties de receptuaires. | ¶ La premiere, traicte de diuerſes ver | tuz & proprietez des choſes. | ¶ La ſecõde, eſt de diuerſes ſortes d'o | deurs, & compoſitions d'icelles. | ¶ La tierce comprent aucuns ſecretz | medicinaulx, propres a con- | ſeruer la ſanté. | ¶ Auecq certains Remedes | contre la peſte. | Item | Le plaiſant Iardin | des Receptes. | ¶ En Anuers ches Iehan Richart | au Soleil dor. | 1555. |

Square 16mo. A to P (?) in eights; imperfect. No pagination. No Index.

Ai *recto*, Title.

Ai *verso*, Text begins, and goes to Pvij *verso*; all thereafter is wanting.

The Italian book ends Hi *recto*. On the same the extra receipts begin and end Hiiij *verso*.

Hv *recto* to Li *verso*, Aultres ſecretz adioutez oultre les precedentes impreſſions.

These are chiefly for beautifying the person, and at the end several pages are devoted to remedies for the plague.

Lij *recto*:

Le plaiſant iardin | des receptes, ou ſont plantez diuers | arbriſſeaulx & odorantes fleurs du | creu de Philoſophie naturelle, | cultiue par Medecins treſ- | expertz en phiſique ſpe | culation, contenant | deux parties: | ¶ La premiere traictera des reme- | des pour maladies furuenan- | tes au corps humain. | ¶ La ſeconde de pluſieurs ioyeu- | ſetez a faire en toute hon- | neſte compaignie. | Item pluſieurs receptes pour faire | pouldre a Canon. | Item la medecine de Maiſtre Grima- | che, contre pluſieurs maladies. | Item pluſieurs nouueautez compo- | ſez par Symon de Millan. | Traduict de langue Italique en | Francoys par Maiſtre Quil | lery de paſſebreue. |

Maiſtre Grimache's receipts are partly in verse, and they are for the cure of social and domestic troubles rather than for bodily ailments. They are not reprinted in the later editions.

15. A gap of a hundred and forty-four years ensues before the next edition with which I am acquainted appeared. One can hardly suppose that the book fell entirely into oblivion for that period, and was then revived; but if editions of it were published in the interval, no copy has come under my notice, whereas there are enough later ones to show that the book retained something of its old popularity. The first I have to mention is in the British Museum (1038, e. 37).

Le | Bastimens | Des | Receptes, | Traduit D'Italien | en François. | Augmenté d'une infinité de beaux ſe- | crets depuis peu mis en uſage. | Avec un autre petit traité des Receptes | intitulé le grand Jardin. |

A Troyes, | Chez Jacques Oudot, Im- | primeur ruë du Temple. 1699. |

Small 12mo. (A) to M in eights and fours alternately. No pagination. Very roughly printed on poor paper; a chap-book.

Signature A is omitted.

(A)i is the title.

(A)ij. The text begins and ends M4. There is no index or contents.

The Italian book goes down to Iij *recto*, the extra receipts from Iij *verso* to Iv *recto*.

"Le plaisant Jardin," with a sort of contents title, like 1555, from Iv *verso* to M4 *verso*.

This edition stops at "Pour faire quelque oyſeaux sembleront être mort, lesquels pourtant ne le seront pas."

The 1555 edition contains a good deal more.

The fly-title, prefixed to the *Jardin*, is as follows:

Autre | Petit Traité | Des Receptes intitulé le | plaiſant Jardin, cultivé | par Medicins trés experts | en Phiſique. | La premiere partie traitera des | remede (*sic*) pour les maladies qui | ſurviennent au corps humain. | La ſeconde pluſieurs joyeu- | ſetez à [*] en toute honnête | compagnie. | Item pluſieurs Receptes pour | faire poudre à Canon. | Traduit en (*sic*) Italien en Fran- | çois; par Maître Quillery | de Paſſebreuve. |

16. The second copy is in the Museum (1038, a. 2 (2)).

Le | Bastiment | Des Receptes, | Tradvit D'Italien | en François. | Augmenté d'vne infinité de beaux ſecrets, | depuis peu mis en vſage. | Auec vn autre petit traicté de Receptes, | intitulé le grand Jardin. | [Device.]

Imprimé a Troyes, *&* ſe vend | A Paris, | Chez Antoine De Raffle, | Imprimeur & Marchand Libraire, | Ruë de petit Pont, à l'Image | Sainct Antoine. |

Small 12mo. A to M in eights and fours; no pagination. Chap-book. No date, but Sæc. XVII.

Ai, Title. *Verso*: to the reader.

Aij *recto*, Text begins and ends M 4 *verso*. No Index.

The Italian book ends Hij *recto*. The extra receipts Hij *verso* to Hiv *verso*. "Le Plaisant Jardin" Ii *recto* to Miv *verso*.

17. The third copy is also in the Museum (1038, a. 2 (1)).

Le | Bastiment | Des Receptes. | Traduit d'Italien en | François. | Auec vn petit traicté de Receptes inti- | tulé le plaiſant Iardin. |

A Troyes. | Chez Denis Clement: | demeurant en la ruë du Temple. |

Small 12mo. A to M in eights and fours. No pagination. Chap-book poorly printed. No date, but Sæc. XVII. No Index.

Ai, Title.

Aij *recto*, Text, which ends Miij *recto*.

Miij *verso*, Av Lectevr.

Miv. A woodcut on both sides of men on a hunting expedition with bows and dogs—to represent the month of March.

On Cv *recto* is a picture of ladies riding, entitled: Mayvs.

On Gvj *recto* is another entitled November.

The Italian book ends Gvj *recto*. The extra receipts begin Gvj *verso* and end Hi *verso*. "Le Plaisant Jardin" begins Hij *recto* and ends Miij *recto*.

18. The latest is that in my own collection.

Le | Batiment | Des | Receptes, | Traduit D'Italien | en François. | Et augmenté d'une

---

* *Faire* is wanting in this copy. Compare 1555.

infinité de beaux | Secrets depuis peu mis en ufage. | Avec un autre petit Traité de Recepte | intitulé le grand Jardin, |

A Troyes, | Chez Jean Antoine Garnier, | Imprimeur-Libraire, rue du Temple. | Avec permiffion.

12mo. A to M in alternate eights and fours; or, pp. 140 [1, 3 blank]. The book is not dated, but the license is dated Paris, October 22, 1738.

The book is rudely printed on rough paper. It is not in the British Museum.

19. It was the acquisition of this copy which led me to the discovery of the Italian original of 1529, and of the other editions now enumerated. There could hardly be a greater contrast in appearance than what exists between the first edition, with its unusual ornamentation and handsome exterior, and these last chap-books, printed in very common style on miserable paper, without a vestige of a thought about their outer form.

It is difficult to say what kept it alive for over two hundred years, from 1529 to 1740. As it stands, it is a miscellaneous gathering, not arranged on any principle, and there is not even a list of contents! It was not, therefore, meant to be consulted in an emergency. It contains medical receipts, practical receipts, receipts for the toilet, perfumery, dyeing, cooking, preserving, legerdemain tricks, and so on, but it is impossible to find what one wants without hunting through the whole book. Withal the book is interesting in respect of its age and persistence.

20. In connection with personal decoration, and as anticipating the very rare work of Jean Liebaut * I happened to allude to the book of André Le Fournier, so long ago as 1883.† When I quoted it I knew that it was most uncommon, and I never anticipated having a copy for exhibition. This, however, is one of the shy books that come occasionally out of their retreats, and the obtaining of it has caused me to look rather carefully into its history, especially as Brunet on the subject is not more satisfactory than usual. The edition of 1530 was what I mentioned in 1883.

It seems to be the first, and is in the British Museum (1174, d. 2). It is as follows:

> ⁋ La decoration | Dhumaine nature, et aornement des | Dames, Compile et extraict des tres excellēs docteurs, et plus ex | pers medecins, tant anciens | que modernes par Maiſtre | Andre le fournier docteur | regent en la faculte de | medicine en luniuer | ſite de Paris Nou | uellement impri | me et non veu | par cy de | uant. | Auec Priuilege |
>
> ⁋ On les vend a Paris par Jehan | ſainct Denys & Jehan Longis, au | palays a la gallerie, par ou on va a | la Chancellerie. | Md.xxx.
>
> Small 8vo. Signatures: a in eight; aa, four. A to G, all in eight. a and aa are not numbered, but A to G are numbered: Fo. i to Fu (*sic*) lvi, which is correct. The title is printed in red and black. Black letter with a few ornamented capital letters.
>
> Collation:
>
> ai *recto*, Title. *Verso*: License to print, granted to Pierre le ver, dated Septr. siziesme, 1530, which ends aii *recto* (misprinted aiii).
>
> aii *verso*, Woodcut. A man in robes, with ermine tippet and round cap, holding a hand mirror (?) in his left hand and pointing with the forefinger of his right. Behind his head is a star, over his head five stars, in front two stars; in the background a river, and spires and other buildings; in front a tree and a flowering plant, close beside him what looks like a small lion.
>
> aiii *recto*, Andreas le Fournier's Dedication to Johannes Morin, in Latin, ends aiiii *verso*. At the end is a facsimile of Le Fournier's signature.
>
> av *recto*, Le prologue, ends avii *recto*.
>
> avii *verso*, Woodcut of a man in gown and cap on a chair, with a book lying open on a desk before him. This woodcut is enclosed in a border of four quite different pieces and over it in large letters:
>
> Cy finiſt ce Pre | ſent Prologue |
>
> aviii *recto*, The Table, which ends aaiiii *verso*.
>
> Ai *recto*, The text begins, and ends Gviii *recto*, ff. i to lvi *recto*. The *verso* is blank.
>
> Gviii *recto*, After the text is the colophon, in black:
>
> ⁋ Cy finiſt ce preſent liure de la decora- | tion dhumaine nature. Nouuellemēt im | prime a Paris, par Pirrre (*sic*) lever Et fut | acheue le. xviii. du moys doctobre. Mille | cinq cens trente. | *

In the later editions Le Fournier's dedication and the license to print are omitted. The first edition has wider margins, and is in a handsomer form than the others.

---

* Brunet spells the word "presant." He condenses the colophon, omits the misprint, and writes *Leber* for *lever*. *Manuel*, 1862; III., col. 932.

In it the subject is divided into the following heads:

La decoration et ornement des dames.
Des pouldres odoriferantes et eaue muscat.
Des Huilles odoriferes.
Leaue de Tartre.
Oingnemens bons pour la personne.
Pour faire ypocras.
Pour la grauelle.
Pour faire bon basme.

To this in the 1531 edition a few receipts are added.

21. The 1531 edition is also in the British Museum (1174, d. 1).

La decoratiō | Dhumaine nature, et aornement | des Dames, Cōpile et extraict des | tres excellēs docteurs, et pl⁹ expers | medicins, tāt anciens q̄ modernes | par Maiſtre Andre le Fournier do- | cteur regēt en la faculte de medeci- | ne en luniuerſite de Paris. |

¶ Item par hōneſte maniere, mon- | ſtre a toutes gēs pluſieurs belles et | ſouueraines Receptes tāt en lart de | medicine que pour faire ſauōs poul | dres et pommes redolentes en ſen- | teurs Auſſi pluſieurs eaues deli- | cieuſes & amoureuſes a lauer & net- | toyer tāt les corps que les habille- | mēs, lesquelles preſeruent de toute | corruptiō. Imprime Nouuellemēt | par Claude Veycellier. |

¶ On les vend a Lyō en rue mercie | re a lenſeigne ſainct Jehā baptiſte. |

Small square 8vo. Signatures: A to G in eights, ff. [i] ii—l numbered, vi not numbered. The title is in red and black, enclosed in a border of four pieces. Black letter, with a few ornamented capitals.

Collation:

Ai *recto*, Title.

Ai *verso*, The Prologue, which ends Aiii *recto*, or f. [i] *verso*, to f. iii *recto*.

Aiii *verso*, The text begins, and ends Gii *recto*, or f. iii *verso* to f. l *recto*.

Gii *verso*, The table, which ends Gvi *verso*.

Gvii *recto*, Senſuyuent aulcunes receptes | de nouueau adiouxtes |, ends Gviii *recto*. *Verso* blank.

On Gii *recto*, or f. l *recto* after the text is the colophon:

¶ Cy finiſt ce preſent liure intitu- | le la decoratiō dhumaine nature. | Nouuellement imprime a Lyon, | par Claude veycellier libraire et | imprimeur demourant en rue mer | ciere a lenſeigne ſainct Jehan ba- | ptiſte. Et fut acheve le p̄mier iour | de Mars mil cinq cēs trēte & vng. |

The difference between this and the 1530 edition is a couple of pages of additional receipts.

22. The next edition is that of which I can show a copy.

La decoratiō | Dhumaine nature, & aornemēt | des dames, Cōpile & extraict des | treſexcellēs docteurs et plus ex- | pers medecins, tant anciens que | modernes par Maiſtre Andre le | Fournier docteur regent en la fa| culte de Medecine en Luniuer· | ſite de Paris. |

¶ Itẽ par honeſte maniere, mon | ſtre a toutes gẽs pluſieurs belles | et ſouueraines Receptes tant en | lart de medecine q̄ pour faire fa- | uons pouldres et põmes redolen- | tes en ſenteurs. Auſſi pluſieurs | eaues delicieuſes et amoureuſes | a lauer & nettoyer tãt les corps q̄ | les abillemẽs, leſquelles prefer- | uent de toute corruption. |

¶ On les vend a Lyõ en rue mer | ciere a lenſeigne. S. Jehã bapti. |

Small square 8vo. Signatures: A to G in eights, or ff. [i] ii—l numbered, vi not numbered. Title red and black, enclosed in a border of four pieces, black letter, with a few ornamental capitals.

Collation:

Ai *recto*, Title.

Ai *verso*, The Prologue, which ends Aiii *recto*, ff. [i] *verso* to iii *recto*.

Aiii *verso*, The text begins, and ends Gii *recto*, or ff. iii *verso* to l *recto*.

Gii *verso*, The Table, which ends Gvi *verso*.

Gvii *recto*, Senſuyuent aulcunes receptes | de nouueau adiouxtees. | ends Gviii *recto*. *Verso* is blank.

The colophon follows the text on Gii *recto*, f. l:

¶ Cy finiſt ce preſent liure intitu | le la decoration dhumaine natu- | re. Nouuellemẽt imprime a Lyon | en la maiſon de Claude veycellier | en rue merciere a lenſeigne ſainct | Jehan baptiſte. Et fut acheue le | xxiiii. iour de Auril. Mil cinq cẽs | trente et deux.

The type used in this edition is well shaped and vigorous, but as the margins are narrow the page has a plethoric look. The title page, with its alternating red and black ink and ornamental border, has a rather rich appearance. I have met with no other copy, but it is mentioned by Brunet.

23. The last of the copies which I have seen is in the British Museum (7441, a. (2)):

La decora- | tion Dhumaine nature, & aorne· | ment des dames, Compile & ex- | traict des treſexcellens docteurs | et expers medecins, tant an· | ciẽs que modernes, par Maiſtre | Andre le Fournier docteur regẽt | en la faculte de Medecine en Lu | niuerſite de Paris. |

¶ Itẽ pluſieurs ſouueraines Re | ceptes tãt en lart de medecine que | pour faire ſauons, pouldres, & põ | mes redolentes. Auſſi pluſieurs | eaues prouffitables a lauer & net | toyer tant les corps que les abil- | lemens, leſquelles preſeruent de | toute corruption. |

¶ On les vend a Lyon par Thi | bault Payen. |

Small 8vo. Signatures: A to F in eights; or ff. xliij numbered from the beginning, *i.e.* A—Fiij; the remaining five leaves are not numbered. Title all in black, with woodcut border. Black letter of somewhat angular appearance.

Collation:

Ai *recto*, Title.

Ai *verso*, The Prologue, ends Aiij *recto*, or f. iij *recto*.

Aiij *recto*, The text begins, and ends Fv *recto*, which is not numbered, though it is in reality f. xliiij.

Fv *verso*, The table, ends Fviij *verso*.

Immediately after the text Fv *recto* is the colophon:

¶ Cy finiſt ce preſent liure intitu-|le decoration dhumaine natu-|re. Nouuellemēt imprime a Lyon | Par Thibault Payen. Et fut | acheue le ſeptieſme iour de moys|de Aouſt. Mil. v. cens trente & ſept.|

In this edition the extra receipts, which in the 1531 and 1532 editions follow the table, are printed before the table, but their titles are not included in it.

The most noteworthy circumstance connected with the editions is that though the book first came out at Paris, these subsequent editions were all printed at Lyons.

24. Besides the editions of 1530, 1531, and 1532, Brunet* enumerates the following: 1533, Paris, Pierre Leber; 1536, Lyons, François Juste; 1537, Lyons, Morice Roy and L. Pesnot, but with 1536 and the name of Juste at the end, so that this would seem to be the previous edition with a new title page; 1541, Lyons, Huguetan; 1545, Lyons, Chaussard; and 1582, Lyons.

In his account of the 1530 edition he makes some mistakes: "Petit volume composé de 10 ff. prélim. . . . , et de lvj ff. chiffr. pour le texte, sign. aa et A—G . . ." There are in reality 12 preliminary leaves, and there is another signature: a.

In the *Supplément*† the account of the first edition is fuller and more accurate, mention being made of the woodcuts and of the contents of signature a. The edition of 1536 is styled "charmante." The edition of 1537, Lyons, Thib. Payen, which is in the British Museum, is added to

---

* *Manuel*, Paris, 1862, III. 932.

† *Supplément*, Paris, 1878, I. 818.

the list, and of the 1541 edition, Lyons, Gilles et Jacques Huguetan, the only copy alluded to is that in the Méon catalogue.*

Brunet mentions an undated Lyons edition by Veycellier, as well as that of 1531, but it is possible that these two are the same, and that the date, which happens to be at the end, has been overlooked. An undated edition, however, quite possibly exists.

Brunet's first article, with a few verbal alterations, was reprinted, blunders and all, by Graesse. The supplemental article of course could not be included.

25. [May, 1898. Among those of Italian origin in my first Supplement (*Transactions*, N. S., II., p. 382) I drew attention to a very early treatise on dyeing by Gioanventura Rosetti, called *Plictho*, and said that only one edition had appeared. That is not correct, for there is a 4to reprint of 1611, and I have just got another.

Libro di Tentoria Intitolata Plicto, Che infegna à tenger Panni, Lana, Tele, Bambafi, Aze, Sede, . . . Et l'Arte di conzar, e colorir Corami, & camozzar Pelle, . . . Col modo di far Aque per fcriuere e miniare, colorir Ofsi, Legni &c. Nouamente arichito di belliffimi Secreti vtiliffimi á ogni qualità di perfone. . . .

In Venetia. Preffo li Zattoni. A Rialto . . . 1672.

Small 12mo; sigs. A to K in twelves; or, pp. [1-5] 6-228 [12]. It is got up in the same size and style as the *Recettario di Galeno* of 1683 (see § 4 above), of the *Secreti* by Venturini, 1672 (see Supplement I. § 41) and others. It was the fashion of the time.

Compared with the original it exhibits a great falling off, but, on the other hand, it is enlarged in one or two places.

It is surprising how regularly books degenerate after their first appearance.]

26. No nation of Europe surpassed the Italians in the production of receipt books, of one sort or other.

Here is one more example, which I found in the British Museum (1168, e. 8 (1)).

* *Catalogue des Livres Précieux, Singuliers et Rares, tant imprimés que Manuscrits, qui composaient la Bibliothèque de M* * *, . . . A Paris, chez Bleuet jeune, Librarie, place de l'Ecole, No. 45. 1803, 8vo. P. 108, No. 901. The Méon collection contained a number of very rare books of Secrets.

Opera Nova | Piacevole Laqvale | Insegna Di Far Varie | compositioni odorifere per far | bella ciafchuna donna. | Et agiontoui molti fecreti neceffarii | alla falute humana como in la ta- | bula fe contiene. Intitulata | Venvsta. |

[Device of Tobit and the Angel, as in the *Dificio*, 1550, by the same printers. That in the *Recettario*, 1528, and in Michael Scotus' *Physonomia* is different.]

In Vinegia appreffo di Francefco Bin- | doni, & Mapheo Pafini. | M. D. L. |

Small 8vo. A to D in fours; or, ff. 15 numbered, and 1 not numbered. Roman character.

A1, Title. *Verso*: Evstachio Celebrino alli Lettori.

A2 *recto*, The Text begins, and ends Dii *verso*.

Diii *recto*, Tabula begins, and ends Div *recto*. The *verso* is blank.

At the foot of Div *recto* is the colophon:

¶ A laude de lo eterno Dio: & vtilita del feme humano, | Finiffe la eccellentisfima opera intitulata Venufta | Stampata ne la inclita Citta di Vinegia per | Francefco Bindoni, & Mapheo Pafi- | ni compagni. Nelli anni del Si- | gnore. M. D. L.

It is a small collection of cosmetic secrets: to make unguents, washes, soaps, perfumes, wines, to beautify the skin, hair, teeth; and there are others on subjects similar to some in Albertus. A work on the theme of the present volume appeared in a French translation in 1561, of which there is a notice below (§ 43). I have not been able to compare the two, and do not know if the whole or any part of the Italian was included in that translation. In any case, the French book is larger than the Italian one, and the author must therefore have had other sources to draw upon.

27. [May, 1898. In the first Supplement (§ 21)* there was a notice of a little book on the art of perfumery printed at Venice in 1555. I have just got another edition of the same tract.

Notandissimi | Secreti De d'Arte | Profvmatoria, | Per far Ogli, Acque, Pafte, Balle, Mofcardini, Vccel- | letti, Paternoftri, e tutta l'arte intiera, | come fi ricerca, | Cofi nella città di Napoli del Reame, come in | Roma, e quini (*sic*) in la città di Vinegia | nuouamente ristampati. | Con Privilegio. | [Device: Two cupids holding laurel wreaths, flying above a cloud, with the motto ET ANIMO ET CORPORI; enclosed in a scroll border.] In Venetia. |

Small 8vo. Signatures: A to K in eights; or, ff. 74, and table [6]. After the table the colophon:

In Venetia appreffo Francefco Rampazetto, | M. D. LX. | Printed in italics.

At the foot of f. 73 *verso* is the note: E tutte quefte cofe fon compofte per Gio- | uanni Ventura Rofeto Veneto. |

---

* *Transactions*, N. S., II., p. 386.

I suggested that this might be the author of the work on dyeing, entitled *Plictho*, printed by Rampazetto in 1540. There is everything in support of such a supposition.

After having compared them, I have found, as I anticipated, that the 1555 and 1560 editions of the present book are not identical.]

28. Books of medical secrets, from the very nature of man's interest in health, have been prominent in previous lists. They are not, however, always constructed on the same lines, but present considerable variety in their discussion of the theme. While, for example, the works of Varignana and Valescus de Taranta * deal with diseases and their treatment and cure, the *Treasure of Evonymus* contains the methods of preparing the remedies for use in treating diseases.

This work, *Thesaurus Evonymi de Remediis Secretis*, had a considerable circulation, not only in Latin, but in the vernacular of several nations. It was certainly written by Conrad Gesner; for, although his name does not appear anywhere in the first edition, 1552, it does so in the second, 1554. It is, moreover, distinctly assigned to him by Placcius † and other writers on anonymous and pseudonymous books, and, if it needed any confirmation the title of the second division of the work would supply it amply.

29. A copy of an undescribed edition, absent too from the Museum Catalogue, has to be added to the roll of the issues of this book. It is a small square 16mo, pp. 567 [41], and bears date, Venice, 1556. I have some doubt about the date of this volume, and a very great deal about the place. From the whole fashion of the book it seems to have been produced at Lyons, like the formerly-described editions of 1554 and 1555.‡

---

* Supplement II., *Transactions of the Archæological Society of Glasgow*, 1897, III., p. 191 and p. 206.

† *Theatrum Anonymorum et Pseudonymorum*, Hamburg, i. 1708, folio; Part 2, p. 266, Number 981. I can hardly think Gesner intended any concealment. If he did so it soon came to an end, for he appears as the author in Simler's Epitome. See *Bibliotheca Instituta et Collecta Primum a Conrado Gesnero Deinde in epitomen redacta ... per Iosiam Simlerum*. Tiguri, 1574, small folio. The first part of Evonymus occurs on p. 138, col. 2, line 42; the second part on p. 142, col. 1, line 57. The same references apply to the edition of the Bibliotheca: *Amplificata per Iohannem Iacobum Frisium*, Tiguri, 1583.

‡ Part V., p. 444. Part I., p. 188. Part III., p. 218.

It has no colophon and no printer's name, which in itself is a suspicious circumstance. The woodcuts of plants and apparatus are of inferior execution, so that it has all the look of a surreptitiously-printed edition, wherever it was done.

30. In my first reference of all to Gesner (Part I., p. 188) the second part of the *Treasure* was mentioned, but at that time only the English translation and Liebaut's adaptation or paraphrase in French could be shown. At last, however, a copy of the Latin original has made its appearance. Its title leaves no doubt as to its authorship:

> Conradi Gesneri . . . Euonymus, ſiue de Remedijs ſecretis, pars ſecunda: nunc primum opera & ſtudio Caspari Wolphii . . . in lucem editus.

The colophon states that it was printed at Zürich, by Christopher Frosch (*sic*), in 1569. It is in small octavo, ff. [8] 247 [16, and 1 blank?]. The copy I have is in poor condition, but a book, one example of which occurs once in twenty years, cannot be prudently rejected by the bibliographer because it is not exactly in the state a fastidious collector demands.* There is a copy of this first edition in the Museum, but that is the only other I know of at present.

31. Of the translation into French by Jean Liebaut I have got two editions. One was printed at Lyons in 1593; it was mentioned in my first list,† but I had no example. The other has had the date unfortunately cut off, but it may have been printed at Paris, and, so far as one can judge prior to 1593. Two Paris editions are mentioned: 1573 and 1578 *chez Jacques Dupuys*. In the absence of a date, this edition can be identified only by a full description. The title, which is printed in black and red and has a vignette of a furnace and distilling apparatus, is as follows:

> Quatre Livres | Des Secrets | De Medecine, Et | De La Philosophie | Chimiqve. Faicts Francois Par | M. Iean Liebaut Dijonnois, Docteur | Medecin à Paris.

---

* However, I am not quite sure that there is not a special interest attaching to a volume of this kind, which has obviously been maimed in the discharge of the duty which it was created to perform; but I admit that I should have preferred a clean copy, uncut if possible.

† Part I., p. 188.

Efquels font defcrits plufieurs remedes finguliers pour | toutes maladies, tant interieures qu'exte- | rieures du corps humain. | Traictees bien amplement les manieres de diftiller eaux, huiles, | & quintes effences de toute forte de matieres, preparer | l'Antimoine & la poudre de Mercure: faire les ex- | tractions les fels artificiels, & l'or potable. | [Vignette of a still.]

Small 8vo. ā, A to Z, Aa to Qq, all in eights (but āviij and Qqviij are wanting in this copy) or ff. [8, 1 wanting], 297, [15, 1 wanting]. The woodcuts of apparatus all very closely copied from the original, but are not identical with them. Further, as showing that some care has been employed, they are not reversed.

Liebaut's preface is dated 1573. In it he says that, having perused the second part of *Evonymus*, edited by Wolphius, he had paraphrased and supplemented it for the use of apothecaries, so that they might make their distillations better. Comparison, however, with the original Latin shows that Liebaut has not departed materially from Gesner's text.

Both parts may be regarded as forming a treatise on the preparation of substances by distillation, with their uses in chemistry and medicine, and, in fact, the work to a large extent deals with pharmaceutical chemistry.

32. I could now give notices of a considerable number of editions of the Secrets of Alexis which have come before me from time to time. Among these is one printed at Pesaro, "per Bartolomeo Cesano," 1559, containing the first and second parts. The first part is a reprint of what appeared at the same place in 1557, already referred to. The 1559 edition is not mentioned in any catalogue I have seen. There is a copy of the first French translation of the first part, printed at Antwerp by Plantin in 1557; the first Dutch translation also by Plantin in 1561; an Italian edition of the whole four parts, Venice, 1639; a French edition of the whole four parts, Lyons, 1639; another in 16mo, Rouen, 1627, and in 8vo, Rouen, 1691. The later French editions contain the translation of the *Kunst-buch* to which reference has been made so often.* It is not contained in the Italian. As I intend an account of these editions to be included in a separate paper † on Alexis the preceding will suffice on this occasion.

---

* Part I., p. 192; *Proceedings of the Philosophical Society of Glasgow*, 1888, XIX., p. 126.

† The first part, dealing with the biography of Alexis, was read to the Society of Antiquaries, London, February 11th, 1897. The second part, devoted to the Bibliography, was read, February 10th, 1898, to the same Society.

33. An author, whose name occurs now and again in catalogues attached to a collection of prodigies of nature, is Pierre Boaistuau. Two of his productions have fallen into my hands, the *Histoires Prodigieuses* and the English translation of the *Theatre du Monde.* Though quite familiar to me by name I had not had the curiosity to look at his writings till I had got these copies. No sooner, however, had I examined the former than I recognized both in the text and illustrations the closest resemblance to Fenton's *Secret Wonders of Nature*, and a brief comparison soon proved that Fenton's book was a translation of Boaistuau's, and not an original work as I formerly supposed.* This fact has obliged me to include it now as a book on the secrets of nature, analogous to those of Polydore Vergil, Obsequens, Jonstonus, and others.

With regard to the *Theatre du Monde* it fortunately does not belong to the literature now under review, for it is one of the most sombre books it has been my lot to come across for many a day. The pessimism of the passing moment is sunshine compared with the gloom of Boaistuau's outlook.

34. Of the author, what little is known is recorded by La Croix du Maine and Du Verdier.† The orthography of his name has been a puzzle to Frenchmen and foreigners alike; indeed I am not convinced that the bearer of it was himself quite sure about it. In the editions of the *Histoires Prodigieuses* of 1566, 1576, 1594, 1598 he is called Pierre Boaistuau; of 1571, Boaisteau. La Croix du Maine calls him Pierre Boistuau; Du Verdier, Boaistuau. The forms Boystuau and Boastuau are said to occur, and Lenglet Dufresnoy uses the forms Boiaistuaux,‡ Boiestuaux,§ and Boistuaux.‖ In Dutch it is Bosteau; in Spanish, Bouistau. Alday calls

---

* Part IV., p. 306.

† *Les Bibliothèques Françoises de la Croix du Maine et de Du Verdier, sieur de Vauprivas; Nouvelle Edition, . . . revue corrigée et augmentée . . . par M. Rigoley de Juvigny, . . .* Paris, 1772-73, in six vols. 4to. Du Maine's article is in Tome II., pp. 254-256; Du Verdier's, Tome V., p. 237. The extracts from Boaistuau, Tome V., pp. 238-242.

‡ *Recueil de Dissertations, Anciennes et Nouvelles, sur les Apparitions, les Visions & les Songes*, Avignon & Paris, 1752, T. IV., p. 228.

§ *Ibid.*, T. I., p. 131; IV., p. 244. ‖ *Ibid.*, T. IV., p. 296.

him both Boaistuau and Bouaystuau. The natural outcome of this bewildering variety is that the *Biographie Universelle* contains two quite distinct notices of him! One is under the spelling Boaistuau, and is in Tome IV., 1843, p. 477, by A. Beuchot and De Noual la Houssaye. The other is under Boistuau de Launai, in the same volume, p. 607, and is by La Fontanelle. Both are founded on what is said by La Croix du Maine and Du Verdier, and contain practically the same statements, though they differ in certain details. But the funny thing is that the two have been gravely inserted in perfect good faith and in apparent ignorance of one another!

35. Boaistuau, which seems to be the best authenticated spelling, called Launai, or Launay, or Lawney—even here there is want of unanimity—was a native of Nantes in Brittany. He may have been born towards the close of the fifteenth century or in the early years of the sixteenth. He died at Paris in 1566, and was buried in the "Cimetiere des Écoliers," near the Church of S. Estienne du Mont.

La Croix du Maine calls him "homme très-docte & des plus éloquens Orateurs de son siècle, & lequel avoit une façon de parler autant douce, coulante & agréable qu'autre duquel j'aye leu les écrits." * He is praised for his French style, which is in marked contrast with that of Belleforest, who wrote a continuation of the *Histoires tragiques* as well as of the *Histoires prodigieuses*. This merit, if no other, seems to have carried some of his books through numerous editions. A list of them is given by both the writers mentioned. They include: *Théâtre du Monde*, of which there were more than twenty editions, besides translations into Latin, English, German, and Spanish. In the British Museum are editions of 1561, 1562, 1571, 1577, 1595; in Latin, 1589; in German, 1606; in Spanish, 1593; in English by John Alday, 1566(?), 1574, 1581, 1663. The copy I have is of the 1581 edition. It is a small 8vo volume, printed at London by Thomas East for John Wyght; and it is in black letter. It has Wyght's device of a man in a doctor's gown and cap with the punning motto, instead of East's crest of the black horse, as in

* Ubi sup., II., p. 254.

Batman's Bartholomew Glanville's *De Proprietatibus Rerum*, 1582.* The title runs thus:

> Theatrum Mundi. The Theatre or rule of the World, wherein may be seene the running race & course of euery mans life, as touching miserie and felicitie, . . . wherevnto is added a learned and pithie worke of the excellency of man, written in the French and Latine tongues by Peter Boaistuau, Englished by John Alday, & by him . . . corrected and amended, . . .

36. *Histoires tragiques*, from the Italian of Bandello, went through an "infinity of editions." *Les Amants Fortunés*, is a translation apparently from the Heptameron. *Traité de la paix et de la guerre; Traité de l'Eglise Militante; L'Histoire de Chelidonius*, 1559; *Visions prodigieuses*, in Lenglet Dufresnoy's work above quoted, 1752, T. I., are also by him.

Some works were left in manuscript, including a translation of St. Augustine's *Cité de Dieu* and a *Traité des pierres précieuses*, in which are discussed the nature of genuine precious stones, and the tricks and impostures practised with false ones. I should have liked much to see what he had to say on this last subject and the sources from which he drew his information.

37. His *Traite de l'excellence & dignité l'homme* was printed along with the *Théâtre du Monde*, and it forms the second part of Alday's translation above mentioned. Two or three extracts from these are given at the end of Du Verdier's article,† by which some notion may be formed of his style. It seems certainly simple, easy, and dignified.

But De la Monnaye has a different estimate of him. "Il a passé (he says) dans son temps pour un beau parleur, avoit quelque lecture, du reste fort superficiel, ne sachant absolument point de Grec, & n'entendant qu'assez médiocrement le Latin." One example is given from his "Excellency of Man," where he takes Clazomenae, the native place of Hermotimus, for Hermotimus himself!

Du Verdier's article is nothing more than a list of Boaistuau's writings.

38. The *Histoires prodigieuses*, the work which, in consequence of

---

* Supplement II., *Trans. Archæol. Soc.*, Glasgow, III., 1897, p. 189.

† Ubi sup. T. V., p. 237.

Fenton's having adapted and adopted it, now claims our attention (as it seems to have done that of the author's contemporaries), ran through a number of editions. I find mention of the following:

1556, apparently the first edition, 1557, 1560, 1561, 1564, 1566, 1567, 1568, 1571, 1574, 1575, 1576, 1594, 1598.

Of these there are in the British Museum:

1566. Paris, Vincent Norment & Jeanne Bruneau, 8vo. It contains only Boaistuau's work, and has clear woodcut illustrations.

1571. Paris, Jean de Bordeaux, 16mo. It contains Boaistuau, Tesserant, and Belleforest, with quite small cuts.

1576 (misprinted 1756 in the B. M. Catalogue). Paris, Charles Macé, 8vo. A page for page reprint of the 1566 edition, with Belleforest's histories added. The cuts are very poor.

1594. Anvers, Guislain Janssens, 12mo. It contains Boaistuau, Tesserant, Belleforest, and Hoyer. It is very closely printed, and the woodcuts are much reduced in size.

1598. Paris, la Vefue Guillaume Cavellat, in six small volumes, square 16mo. To those in the 1594 edition are added Arnaud Sorbin (translated by Belleforest) and "I. D. M.," from various authors. It is a neat collection, with passable woodcuts.

39. The copy which I have is not in the Museum.

Histoires Prodigievses, Extraictes de Plvsieurs fameux Autheurs, Grecs & Latins, sacrez & Prophanes, diuisees en deux Tomes. Le premier mis en lumiere par P. Boistuau, surnommé Launay, natif de Bretagne. Le second par Claude de Tesserant, & augmenté de dix histoires par François de Belle-Forest Comingeois. Auec les pourtraicts & figures. [Device.]

A Paris, chez Iean de Bordeaux au cloz Bruneau, à l'enseigne de l'Occasion. 1574. Auec priuilege du Roy.

Two volumes 16mo. Vol. I. ff. [8], 1-216; Vol. II. ff. [4], 217-278, [3], 279-396 [4]. The woodcuts are small, but tolerably sharp. This seems to be simply a reprint of the edition of 1571, by the same publisher.

40. The Spanish translation was made by Andrea Pescioni from "Bouistau," Tesserant, and Belleforest. It appeared at Medina del Campo, 1586, small square 8vo, and Madrid, 1603, in the same form. These contain no cuts, and are poorly printed. What little there is to say about the English translation will be included in the next supplement.

41. The Dutch translation [B. M. 12430, aa. 17] is entitled:

Het wonderlijcke Schadt-Boeck der Historien... Dordrecht, 1592, in small 8vo. It is in black letter, and the woodcuts are small and blurred. It is in four parts, with separate pagination, but the signatures are continuous.

There is another work [B. M. 7002, a] entitled: *Toonneel der Gedenkwaardigste Wonderen, Geschiedenissen en Vreemdigheden, uyt de Heer Bosteau en anderen dus in beknopt Duyts gestelt door Jan van Duisberg, Boek-verkooper*, Amsterdam, 1657, 18mo. It is a collection in 23 chapters of remarkable things, mainly in natural history. The subjects are the following: Stones, the load-stone, the satyrs of Bosgoden, burning mountains, wonderful and terrible deaths, examples of avarice, memorable changelings (mis-births), strange banquets, the lion and the slave, strange lives, strange snakes, visions and ghostly appearances, plants and roots, wonderful cruelties, strange properties of waters, peculiarities and acts of dogs, the cause of monsters, plants and stones which resemble certain animals, bees and ants, the likeness and unlikeness of men, wonderful fish and water animals, earthquakes and their causes.

This is an interesting book, which might well go along with the *Secret de l'hystoire naturelle*, but it is not a translation of Boaistuau, though some of the prodigies may have been taken from him.

42. The work which follows is a general collection of receipts, and though the book is not bulky, it is so closely printed that it contains a great amount of matter. About the author there is nothing to say, except that some writers hint at his identity with Hieronymo Ruscelli, who claims to be the true Alessio of Piedmont. But, so far as I am aware, there is no evidence in support of this notion.

It may be observed that Bariletto was also the publisher of the *Secreti* of Isabella Cortese. That book appeared in 1561,* and has the same device on the title page. The demand for such books was apparently so great that a publisher was not afraid to send out possible competitors for public favour on the same lines.

Della Svmma | De' Secreti Vniversali | In Ogni Materia | Parte Prima. Di Don Timotheo Rossello, Si | per huomini & donne, di alto ingegno, come an-

* Supplement I., p. 390.

cora per medici, & ogni forte di artefici in- | duftriofi, & a ogni perfona vir- | tuofa accommodate. | Con gratia, & priuilegio. | [Device.]
In Vinegia Per Giouanni Bariletto. M D LXI. |
Small 8vo. Parte Prima. Signatures: *, A to T in eights; or, ff. [8], 152. The device is a draped female holding a hand mirror.
At the end of the text, 152 *recto*:
In Venetia appreffo Giouanni Bariletto. | M. D. LXI. |
152 *verso*: the device repeated, as in the *Secreti* of Isabella Cortese, 1561.

The second part has a separate title.

La Seconda Parte | De Secreti Vniversali | In Ogni Materia Di Don | Timotheo Rossello. | Ne' Qvali Si Contengono I | rimedi vtili à l'infermità, & altre cofe bellifsi- | me, ingeniofe & artificiofe, come nella | tauola veder potrai. | Con gratia & priuilegio. | [Device.]
In Venetia Per Giouanni Bariletto. | M D LXI. |
Signatures: †, A to T in eights; or, ff [8], 152. At the end of the text, 152 *recto*:
In Venetia appreffo Giouanni Bariletto. | M.D.LXI.
152 *verso*: the device is again repeated.

43. Another rare work is concerned, like Fournier's and so many others, with secrets of the toilet. One of the copies is in the British Museum (547, e, 8(2)), but I have not been successful in finding the Italian original.* There is no reason for doubting that this, like the *Batiment des receptes*, is really from the Italian. The copy which I have is of a later edition, and it will be observed that the title pages are different, and that both the fact of its being a translation and the initials of the translator—if that is the meaning of S. E. S. X.—have been deleted. It would seem, too, that the house and shop occupied by Victor Sertenas in 1561 were inhabited in 1566 by Victor Norment and Jeanne Bruneau, who have already appeared as issuing an edition of Boaistuau that same year. Was this a new firm, or had Sertenas changed his name to Norment? It would perhaps be considering too curiously to discuss the matter now, especially as I have no facts bearing on it; but it is a curious question. This is the copy I have seen in the British Museum.

Recveil De | Plvsievrs Secretz | trefvtiles, tant pour l'ornement que la | fanté du corps humain, tirez des | plus excellens auteurs, tant | Grecz (*sic*) que Latins. | Avqvel

* See what is said on this subject under *Opera Nova Piacevole*, 1550 (§ 26).

Avons Adiovste | vn traicté des deſtillations, contenant pluſieurs | receptes d'eaues Imperiales, d'Ange, | Naffe & autres ſemblables :| Novvellement Tradvict | d'Italien en François, | Par S. E. S. X.| [Device.]

A Paris, | pour Vincent Sertenas Libraire, demeurant en la | rue neuue noſtre Dame, à l'image ſainct Iean l'E- | uangeliſte. Et en ſa boutique au Palais, en la gal-| lerie par ou on va à la Chancellerie.| M.D.LXI.| Avec Privilege.|

8vo. Signatures A to H in eights; or, ff. [1] 2-59 [5].

Ai title. Aij to Hiij Text. Hiiij to Hviij Table of Contents.

The first part contains receipts for beautifying the face, hands, and body.

The second for making soaps, violet powder, oil, odoriferous water, and various medical receipts.

Part three treats of distilled waters of various kinds, medical, fragrant, &c.

44. The next copy is in my own possession.

Recveil | De Plvsievrs | Secrets Tresexcel- | lens Et Admirables, | pour l'ornement & embelliſſement | de la face, & autres parties du corps.| Plvs, | Vn Traicte De La Maniere | de diſtiller eaux de ſenteurs, faire parfums, | pommes d'odeurs, & autre choſes | excellentes, par cy deuant | non cogneuës.| [Device.]

A Paris.| Pour Vincent Norment & Ieanne Bruneau, en la rue neu- | ue noſtre Dame, à l'image S. Iehan l'Euangeliſte, & au | Palais, en la galerie par ou on va à la Chancellerie.| M.D.LXVI.

Small 8vo. A to H in eights; or ff. 59, and 5 of table of contents. At the end: Acheve D'Imprimer | Le Cinqviesme Iovr | De Mars. M.D.LXVI.|

The device consists of two right hands out of clouds, clasping each other and holding a sheaf of arrows with the points downwards. In the ornamental border are the words: Vincenti Non Victo Gloria.

45. Still another Italian book on the same subject of personal adornment has to be noticed:

Gli | Ornamenti | Delle Donne | Tratti dalle Scritture d'una Reina Gre- | ca per M. Giovanni Ma- | rinello, & diuiſi in | quattro libri, | Con due Tauole, una de' Capitoli, e l'altra d'al- | cune coſe particolari. | Opera utile, & neceſſaria ad ogni gentile perſona. | Con Privilegio. | [Device.]

In Venetia Appreſſo Franceſco de' Fran- | ceſchi Seneſe. M.D.LXII. |

Small 8vo. *, A to Z, AA to RR in eights; or, ff. [8] 319, [1 blank]. The device is Peace, a draped female holding up an olive branch in her right hand and an inverted torch in her left, enclosed in a scroll-border. On the *verso* of RRvij are the "Registro" and colophon: In Venetia | Appreſſo Franceſco de Franceſchi Seneſe, Al ſe- | gno de la Pace. M.D.LXII. |

The four books—the contents of which are not very strictly discriminated—contain respectively methods of removing defects which interfere with the symmetry of the body, the hair and everything connected with its preservation, improvement, and colour, the face and various secrets for making it "beautiful for ever," and, in the fourth book, the rest of the body. It is comprehensive and elaborate, and it is to be hoped that the Italian ladies duly appreciated the efforts of the author, which were put forth for the express purpose of pleasing them and assisting them in the great work of making themselves even more beautiful than they were already.

46. The acquisition of several copies of Levinus Lemnius' curious collection of wonders and secrets carries me back to my first paper on the present subject.* At that time I did not know the date of the first edition.

The earliest which I have as yet seen, came into my hands a little time ago.

Levini Lem- | nii Medici Zirizaei | Occvlta Natvræ Mira- | cula, ac varia rerum documenta, probabili | ratione atque artifici coniectura duobus | libris explicata, quæ ſtudioſo aui- | doq́; Lectori non tam vſui | ſunt futura, quàm ob- | lectamento. | Elenchus operis, horum omnium | gustum exhibebit. [Device.]

Antverpiae, | Apud Guilielmum Simonem. | Cum Priuilegio. | 1561. |

Small square 8vo. Sigs. ✠, A to X in eights, Y in four; or ff. [16] 164. The device is a lad who having plundered a bee-hive is set upon by bees; with the motto: Dvlcia Mixta Malis.

Notwithstanding its date I am nearly certain that there must be an earlier edition. The author's epistle dedicatory is dated December, 1558, and the privilege April, 1559; so I should suppose the first edition came out about the middle of 1559. It contains only two books; the later editions have four.

47. It was reprinted by Plantin at Antwerp in 1564, in 12mo, and it must have been in demand, for another edition came out in 12mo, in 1567. These two editions I have not seen, but I have here a copy of the edition of 1574, which was also published by Plantin. It is an 8vo, pp. [16] 566 [for 582] [33, 1 blank], printed in italics, with the device of the compass on the title page. The Dutch words which occasionally occur, are printed in a peculiar black-letter, which is much more striking than that employed in

* Part I., p. 187. [Reprint, 1896, p. 10.]

the 1581 edition formerly described. The present copy is also in contemporary binding of fine smooth light-brown calf, with the initials D. M. and the date 1575.

48. Even more interesting are the three earliest editions of Gohory's French translation:

> Les | Occvltes | Merveilles Et Se- | cretz De Natvre, Avec Plv- | fieurs enfeignemens des chofes diuerfes tant | par raifon probable que par coniecture arti- | ficielle: expofées en deux liures de nõ moin- | dre plaifir que proufit au lecteur ftudieux.| Par | Leuin Lemne Medecin Zirizeen, & nouuellement tra- | duit de Latin en François, par I. G. P. | Avec deux tables, lune des argumens | des chapitres, l'autre des fingu- | lieres matieres d'iceux.|
>
> A Paris, | Par Pierre du Pré, Libraire iuré en l'vniuer- | fité, au premier pillier de la grand | falle du palais.| M.D.LXVII.| Auec priuilege du Roy.|
>
> It is a small 8vo. Sigs. A to Z, and Aa to Dd, ã, ẽ, all in eights; or, pp. [28] 402 [2] [32]. Gohory's preface dated Paris, 2 February, 1567, would show that this is the first edition. The present one is the Yemeniz copy, and it has the book plate.

49. This was reprinted at Orleans by Pierre Trepperel in the following year, 1568, in a square 16mo. The copy in the Hunterian Museum was described in my very first paper on the subject, but a little time ago a copy of this very rare volume came into my possession. It agrees with the description already given.*

50. Another edition, probably the second, was published at Paris in 1574. The title is the same as the other, but the page is differently arranged; there is a vignette of a galley with oars and sails and shields and a covered poop, and the book was printed at Paris: Pour Galiot du Pré, Rue S. Iaques, à l'enfeigne de la Gallere d'or. This also is in 8vo, signatures A to Z, Aa to Cc in eights, Dd in four, ✠, *, in eights, * in four; or ff. [1], 212, [20].

Galiot du Pré was doubtless some relative of Pierre, who printed the first edition. In this later edition the numbering of the leaves has been introduced; an archaism all the more singular that in the first edition it is the pages which are numbered. All three editions are attractive.

---

* Part I., p. 187, but *Paris* is incorrectly given as the place instead of Orleans.

51. In the first paper * of all on this subject I gave an account of a tiny volume entitled *Künstbüchlin*, and afterwards showed that, with certain modifications, it appeared in whole or in part in numerous subsequent editions, and was translated into French, Dutch, Danish, and English. The name, however, was used by other compilers, and was given to books of receipts differing from the series referred to, although resembling them in subject and in treatment.

One of these has come before me, and it may be described, to distinguish it from the others.

Kunstbüchlin. | Wie man auff | Marmelstein, Kupffer, Mes- | sing, Zihn, Stal, Eisen, Har- | nisch vnd Waffen, etc. Etzen, vnd | künstlich vergülden sol. | Mit vorgehendem Bericht: | Wie man Dinten, Pre- | silgen, vnd alle Metalfarben zum | schreiben, Mancherey (*sic*) Farben, Parge- | ment vnd Federn zu ferben, Alle Metallen aus | der Federn zu schreiben, Golt vnd Silber Fun- | damentlein, vnd Goltwasser auff allerley | Ballerey, vnd dergleichen mehr, ma- | chen vnd Temperiren sol. Zu dienst vnd Ehren allen | Schreibern, Auch den vnerfarnen der | Etzkunst, zusamen bracht | Durch Andream Helmreich, Rechen- | meister zu Hall. | Von newem vbersehen, gemeh- | ret vnd gebessert. 1574. |

Small 8vo. A to E in eights, F four, or, ff. [44]. Title red and black.

Fiiij *recto*, the colophon:

Gedruckt zu Witteberg, | Durch Lorentz | Schwenck. | 1574.

The detailed title page saves one the trouble of describing the contents of this little tract.

52. It is no easy matter apparently to exhaust the editions of Wecker's Secrets. Here are some new ones.

De Secretis Libri XVII. Ex variis authoribus collecti, methodicéque digefti, & aucti per Ioan. Iacobvm Vveckervm, Bafilienfem, Medicum Colmarienfem. Accefsit Index locupletifsimus. Cum Gratia & Priuilegio. [Device.]

Basileae, Ex Officina Pernea CIƆIƆXXCIIX.

Small 8vo. Signatures:)(in eight, a to z, A to Z, Aa to Il in eights, KK2; or, pp. [16], 865, and Index [30].

Wecker's epistle dedicatory to Baron Lazarus Svendius is dated: Colmar, August 1st, 1582.

The edition of 1701, already mentioned, is a simple reprint of this.

---

* Part I., p. 192. *Proceedings of the Philosophical Society of Glasgow*, 1888, vol. xix., p. 126, and Supplement, 1894, vol. xxv., p. 224.

Wecker's collection has been taken from all the old Secret writers, and is an excellent handbook to them.

Whether the present is the first edition or not, I am not certain, but considering the date of the dedicatory epistle, as well as of the French translation, I should suppose there was one earlier edition, if not more.

53. The translation of the book into French seems to have been taken in hand at once by Pierre Meyssonier, for the dedication of it by him to Hierosme de Chatillon is dated: Lyons, January 1st, 1584. This translation might, of course, have been made from a MS., but there is no evidence of that, and, anyhow, it is simpler to suppose that he had a printed edition before him. This would involve a Latin edition of 1582 or 1583, corresponding, therefore, to the date of Wecker's dedication, but I have not seen one so early. Nor do I know for certain when this translation was first published, but I should suppose that it must have been in 1584, the year of the dedication. I should also infer, from what is said on the title page of the following French copy, which is the earliest that I have seen, that the book passed through several editions in the ten years subsequent to the date of the dedication.

The title referred to is as follows:

> Les Secrets Et Merveilles De Nature, Recueillis de diuers Autheurs, & diuifez en XVII. liures, Par Iean Iacqves Vuecher, de Bafle, Medecin de Colmar. Traduicts en François, & corrigez outre les Imprefsions precedentes, felon la traduction Latine: Auec vne Table tres-ample. [Device.]
>
> A Lyon, pour Thomas Soubron, & Moyfe des Prez. 1596.

This is an octavo of pp. [16] 1086, 2 blank, [91] of index, and 5 blank. As the copy I have happens to be as clean and fresh as when it was printed, it looks well enough, but the paper is of the soft, porous kind so common in the cheap Lyons books of the time, and the page is not attractive.

54. Another followed in 1606, but here again I am unable to say if there were any in the interval. I should suppose there were.

> Les Secrets et Merveilles de Natvre. Receuillis de diuers Autheurs, & diuifez en XVII. liures, Par Jean Jacques Vvecker, de Bafle, Medecin de Colmar. Traduits en François, felon la traduction Latine: Reueu, & corrigé de nouueau. Auec vne table tres-ample. [Device.]

A Tovrnon, Par Clavde Michel, & Thomas Sovbron. 1606.
8vo. Signatures: *, A to Z, Aa to Zz, Aaa to Zzz, Aaaa to Ffff, all in eights; Ffff 7 and 8 blank. Pagination: [16] 1246, [49, 5 blank].

The translation is Meysonnier's. The dedication to Hierosme de Chastillon is reprinted with the date: Lyon, January 1st, 1584. There is no difference between this and the earlier edition.

55. Another which I have is got up in the same style, but it professes to be improved:

Les | Secrets | Et Merveilles | De Natvre. | Recueillies de diuers Autheurs, & diuifez | en XVII. Liures. | Par Iean Iacqves Wecker | de Bafle, Medecin de Colmar. | Liure non feulement neceffaire aux curieux, ains à tous ceux qui | font profefsion des Arts Liberaux, & fubtiles inuentions: | tant pour l'exercice de l'Art Militaire, qu'autres | de quelques vacations qu'ils foient. | Reueu, corrigé, augmenté, & enrichy de plufieurs figures, & mis en | meilleur ordre que les precedentes Impreffions. | [Device.]
A Roven, | Chez Clavde Le Villain, Libraire & Relieur | du Roy, ruë du Bec, à la bonne Renommee. | M. D C. XXVII. |
8vo. Signatures: †, A to Z, Aa to Zz, Aaa to Vvv; or, pp. [16] 1012 [41, 1 blank]. Title red and black.

This is on very soft paper, and has a dingy look. Notwithstanding its assertion I do not see any difference in the contents between this edition and that of 1596. The Dedication is left undated. Perhaps the publisher thought it was time to forget the fact that it went back to 1584.

56. When the English translation of Wecker was under consideration before (Part III., p. 225), editions of 1660 and 1661 were both mentioned. Having the 1661 edition only, I was unable to settle whether it was a "genuine second edition, or the original edition with a new-dated title page." I have now got the 1660 edition, and find that the second alternative is the correct one, namely that the 1660 title page has been cut out and the 1661 title page substituted, the book otherwise being the same; and further that both title pages bear the name of Dr. R. Read; so that Watt and Lowndes are wrong in calling him *Alexander*.

57. Almost contemporary with Cardan flourished Jean Bodin, known by his works on the study of history and on the Republic. He is also responsible for a book on Demonology in which he gave fair scope to

his superstitious credulity, his ignorance, his fear, and as a consequence his cruelty.

Not to be behind he too blossomed out as the expounder of creation. Having displayed such intimacy with the mysteries of the realms of darkness, he was Hegelian enough to know that he must balance it by a corresponding explanation of the world of light.

The *Theatrum Mundi*, as he called his book, first appeared in 1596, so far as I am aware.

Vniversae Natvræ Theatrum, In Qvo Rervm Omnivm effectrices caufæ, & fines quinque libris difcutiuntur, Avtore Ioan. Bodino,
Lvgdvni, Apvd Iacobvm Rovssin,
Typographum Regium. M. D. XCVI.

8vo. Sigs. ã4, A to Z, Aa to Qq in eights, Rr in four. Pp. [8] 633 [1 blank].

This book is decently printed, with liberal margins and spacing, which give it a handsome look.

58. Another edition came out later:

Hanoviae, Typis Wechelianis apud Claudium Marnium, & hæredes Ioann. Aubrii. M. DC. V.

8vo. Sigs.):(, A to Z, Aa to Qq in eights, Rr in four. Pp. [16] 633 [1 blank].

This is very nearly a line for line, page for page reprint of the 1596 edition. But the paper is not so good, the printing is inferior, and the margins are curtailed, so that the volume looks commonplace beside the other.

The conclusion is striking:

Finis Theatri Naturæ, quod Io. Bodinus Gallia tota bello ciuili flagrante confcripfit.

Both copies are in Sir William Hamilton's collection in the University Library.

59. The copy I myself have is of the French translation by François de Fougerolles. Like several of these encyclopædic surveys it is in the dialogue form. It is divided into five books: Elements; Earth, Air, Water; Plants and Animals; Souls and Capacities of Men; The Heavens, the Planets, etc. It comes in here, then, less as a work on technical secrets as on the secrets of nature and of natural history, and it goes along with such books as Cardan's just mentioned, and a good many others.

A book by a man like Bodin is of some note in the history of knowledge, and especially of physical knowledge, for it reveals what was thought and believed about nature in the latter half of the sixteenth century. I am doubtful however as to his knowledge, and not the least doubtful as to his dogmatism.

The following is the title of the French translation:

Le Theatre De La Natvre Vniverselle De Iean Bodin Ivrisc. Auquel on peut contempler les caufes efficientes & finales de toutes chofes, defquelles l'ordre eft continué par queftions & refponces en cinq liures. Oeuure non moins plaifant que profitable à ceux qui voudront rendre raifon de toutes queftions propofées en Philofophie. Traduict du Latin par M. François De Fovgerolles Bourbonnois Docteur aux Arts & en Medecine.

A Lyon, Par Iean Pillehotte, à l'enfeigne du nom de Iesvs, M. D. XCVII. Auec privilege pour dix ans.

8vo. Signatures †, †† in eights, ††† in four, A to Z, AA to ZZ, AAA to NNN in eights, but NNN 7 and 8, doubtless both blank, wanting, or pp. [40] pp. 917 [25] [4 blank? wanting]. Woodcuts in the Text.

In the later pages are contained a set of tables in which the contents are systematically and logically classified. This is perhaps the most valuable part of the whole. They are not in the Latin editions which I have examined.

60. The popularity which certain of these books enjoyed has been attested by the editions of them which have been enumerated. One of the most popular must have been the *Libro de Phisonomia Natvral*, seeing that it was in vogue from 1609 till 1858. These were the earliest and latest editions I knew of in 1885, but I have since found one still earlier, which seems to be actually the first. On the chance of its being so, it may be fully described:

Libro | De Phisono | mia Natvral, Y | varios fecretos de naturaleza: el qual | contiene cinco tratados de materias | diferentes, no menos curiofas | que prouechofas. | Compuefto por Geronimo Cortes, natural | de Valencia. | Dirigido al Maeftro Chriftoual Colon, | Vifitador general en el Arçobif | pado de Valencia. | [Device.] |

En Madrid, | En cafa de Pedro Madrigal. | Año. 1598. | A costa de Miguel Martinez. | Vendefe en las librerias de Palacio. |

Small 8vo. Signatures A to P in eights, or ff. 120. The device is a hooded falcon perched on a left hand out of a cloud, with the motto: Post Tenebras spero Lucem. It may be pointed out that a similar device of a hooded falcon, perched on the branch of a tree, and with the same motto, occurs in the *Kunst Boeck* of Symon Andrée, 1549.*

The book is divided into five parts: physiognomy, of romero, of ardent water, secrets of natural things, secrets of the four elements, the heavenly bodies and their effects.

An interesting section of the book is the list of authorities upon whom the author has relied for the five tracts above mentioned.

Thus, for the first, he quotes Tisnerio, Escoto (Michael Scotus), and Pedro de Ribas. For the second and third: Arnaldus, Pliny, Herrera, "el maestro Zapata Cirurgico" (as to whom something has been already said in Supplement I., § 27), Galen, and many others, as Dioscorides, Theophrastus, Serapio, and Crecentinus. For the fourth: Mizaldus, Pliny, Cardan, Florentino, Albertus, Aranda, Porta, "y don Alexios Piamontes." For the last: Pliny, Alfraganus, Aristoteles—Principe de los Philosophos, with "an infinity of other authors."

This list also shows how books like the present were mere compilations and condensations from other writings, and contained no original matter and no criticism. They are, therefore, valuable now chiefly as commonplace books and as indices to their predecessors.

61. After existing for a century and a half, Cortes' book, harmless enough one would have thought, fell under the ban of the Inquisition. Here is a copy of such an edition:

Fisonomia, | y | Varios Secretos | De | Naturaleza, | Compuesto | Por Geromino Cortés, | natural de la Ciudad de Valencia. | Están Expurgados, Segun | lo manda el Decreto de la Santa | Inquisicion de 13 de Junio | del año de 1741. | Con Licencia. | Barcelona: Por Joseph Giralt Impresor, Pla- | za de Santa Ana. |

Small square 8vo, Signatures A to Q in eights, R four, or pp. [4] 260. The book has no date, but, of course, must be after 1741. After a rough comparison with the 1598 edition I do not see any evidence of expurgation. It may, however, be in the modification or removal of words or phrases that were thought objectionable.

* Part II., p. 263. *Proceedings of the Philosophical Society of Glasgow*, 1888, Vol. XIX., p. 137.

62. One of the most attractive of all the collections is the following, which is all the more so because such books in Dutch are not common:

Secreet-Boeck | Waer in vele diversche Secre- | ten, ende herrlicke Consten in veelder- | leye verscheyden materien, wt seker Latijnsche, | Fransoysche, Hoochduytsche, ende Nederlandt- | sche Authoren, te samen ende by een ghebracht | zijn: Waer van den meestendeel der voor- | schreven Authoren namen daer by | gheciteert worden. | [Device.]

Tot Dordrecht. | By Abraham Canin, Int | Jaer ons Heeren, 1601. |

Small 8vo, A to Z in eights, A to D in eights, and E, 1 leaf, or 4 pp. prelim. and text, pp. 419, so numbered, but pp. 180-189 (inclusive) are numbered twice. Black letter. The device is a lion holding a book.

The preliminary pages contain title, privilege, and author's preface.

The work contains secrets referring to almost every imaginable topic, and the names of the authorities are frequently given at the end of each. Among these are Mizaldus, Albertus, Fallopius, Cardanus, Fioravanti, Porta, Alexius, Isabella Cortese, Roscellus, Theophrastus (which?), Silvius, and others.

There are secrets of the fire, air, water, earth; of the mind, as to strengthen the memory; of the body; against divers sicknesses, in general, and in detail; cosmetics, wine; of sleep (opium was used); the secrets of beasts, horses, sheep, pigs, dogs; of wild animals, deer, hares, wolves, moles, weasels; ants, scorpions, fleas; serpents, and so on; of fishes; of birds; of flies; of plants, trees [and grafting] (a good deal of this from Mizaldus); of the vineyard; of root plants, radishes, etc.; flowers, fruits, seeds; metals, iron and steel, etc.; soldering; silvering and gilding; of sulphur; salt, glass; gardening; secrets of clothes, how to clean them; secrets of writing, of ink, of illuminating; of varnish; of gums and resins; Aurum Musicum; colours of various kinds—green, blue, white, etc.; Lutum Sapientie; to dye thread or linen brown; to dye skin, bone, and horn; secrets of the air and weather; signs of snow; signs of wind; of thunder and lightning; the cause of earthquakes; signs of great tempests; signs of a fruitful and unfruitful year; secrets of conserves and comfits, marmalade, etc.; secrets of cookery; secrets of compound distilled waters; secrets of metalline waters (nitric and other acids, etc.); secrets of balsams; of oils from seeds and from spices; of turpentine; of aqua vitæ.

It is a comprehensive and entertaining collection of old physical, chemical, and technical knowledge. The only copy of this edition which I have seen is in the British Museum, 1036, c. 3 (2).

63. Another, which I have, appeared in 1609, but whether it is the second or not, I am unable to say. The following is its title:

Secreet-Boeck | Waer in vele diversche Secretē | ende heerlicke Consten, in veelderleye | verscheyden materien, wt seker Latijnsche, Fran- | soysche, Hoochduytsche, ende Nederlandtsche Authoren, te | samen ende by een gebracht zijn: Waer van den mee- | stendeel der voorschreven Authoren namen | daer by gheciteert worden. | By een vergadert door D. Doctor Carel Batin. | [Device.] Tot Dordrecht. | By *Joris Waters. Anno. 1609.*

Small 8vo. Signatures: A to Z in eights, pp. [4] 356 [7, 1 blank]. Black letter. The device is a Triton blowing on a conch; round the woodcut are the words: Stvdys Immortalitem Acqvirimvs.

So far as one can judge by a comparison of the table of the contents of the previous book as given with the present, this is a simple reprint. The most important point in this edition is the adding of the author's name. I cannot find it, however, in the biographical dictionaries so that it is possibly a pseudonym.

64. The following tiny chap-book I found in the British Museum, 1048 c. 4 (3):

Petit Traicté. Contenant plufieurs Secrets naturels, bien experimentez par l'Epagerie Tramontane, lefquels il en faict participant à tous ceux qui font curieux de leur fanté & qui defirent voir des belles curiofitez.

A Aix, Par Iean-Courraud, & Philippe Coignat,
Imprimeurs ordinaires de ladite ville.
M. DC. VI.

Sm. 8vo, pp. 14.

Besides a few medical receipts there are some miscellaneous ones: the tree of Diana; to clean pearls; to dye white hair black with silver nitrate; to freeze wine with ice and saltpetre, and some more like these.

65. There is another chap-book similar to the preceding and to those already described in the first supplement:

Thesoro | Di Secreti | Natvrali | Difpenfato da me Benedet- | to, detto il Perfiano. | [Vignette]
In Roma, | Appreffo Giacomo Mafcardi. | M.DC.IX. | Con licenza de' Superiori.

8vo. Signature A, 4 leaves. The title is surrounded by a border of separate scrolls. The vignette is an old bearded man holding his cheek with his hand. It may be a portrait of the author in meditation, but from the expression which the limner has put into it, I judge that he means it as a vivid delineation of the pain of toothache, and the desirability of buying the pamphlet for the sake of the cure.

This pamphlet contains only brief directions for the cure of some common diseases: boils, toothache, bleeding at the nose, the bite of a mad dog, scrofula, and so on. The remedies are for the most part derived from plants. It is a very singular assemblage of secrets.

66. As I have often had occasion to mention Neri's *Art of Glass*, first published at Florence in 1612, in 4to, I may refer to another edition with which I became acquainted a little time since. It shows how much esteemed the book was, as giving sound and accurate information on its subject, that it was printed so often as a sort of standard work. The edition to which I refer was included in the Italian translation of Baumé's *Chimie Expérimentale*. I have not seen this translation, but I have a copy of Neri's part, of which copies apparently were put separately in circulation. It is a small 8vo, pp. cxliii. [1 blank] with only a short title: *L'Arte Vetraria di Prete Antonio Neri*, without place, printer's name, and date. The appended licence to print the whole work of Baumé is, however, dated 1787, and is granted to "Francesco di Niccolò Pezzana, Stampator di Venezia." There is nothing of interest in the book itself, apart from its being another and comparatively late edition.

67. Godefroy Roussel, if such be his real name, is not noticed in any work that I have consulted, and a book of receipts is all that I know him by:

Les | Secrets | Descovverts | Des Arts, Tant De | Pharmatie que de celuy de Distiller, | vulgairement nommé Alchemie, | ou Spagirie: par le moyen desquels | l'on paruient à la perfection; tant par | Theoricque que practique, à rẽdre | l'Or Potable, succinctement dé- | duicts en forme de Dialogue. | Oeuure non moins necessaire que profitable | à tous Amateurs de science. | De l'inuention de Godefroy Rovssel, | Maistre Appoticaire à Paris. | Dedié Av Roy. |

A Paris, | Et se vendent par l'Autheur, ruë de la Sauon- | nerie, entre la pierre au laict & la place aux | veaux: Chez Guillaume Baudeau, | Maistre Chandellier. 1613. | Auec Priuilege de sa Majesté. |

Small 8vo. Signatures: †, †, ††, in fours, A to I in eights, K in four; or pp. [1]-7 [8-22] [2 blank], 138 (misprint for 147) [5].

The epistle to the reader is a defence of alchemy, which according to this author seems to be identical with chemical pharmacy. The first part contains an exposition of the general principles of pharmacy, classifications of the properties of drugs and their uses. The second book deals with the secrets of distillation, treated from a very lofty theoretical standpoint. An account of some preparations is given, and a good deal is said about potable gold and its "mystical" preparation. At the end there is a section on the sympathetic relation of man to the great world, and the doctrine of signatures is introduced.

The notable thing about this work is that the doctrines are expounded by an aspirant to the mastership in pharmacy, in the form of a disputation or maintenance of theses, before a master who commends his expositions and suggests new themes for him to discourse upon.

68. The inclusion of the following little volume in the present list requires possibly some justification, for it is in reality on chemistry, rather than on practical arts. However, as a section of it is concerned with the preparation of substances and mixtures to be used in medicine and in the arts, I have decided to insert it:

Den Nederlandtsche Slevtel van t'Secreet der Philosophie, in welck grondelijc bewesen wert, d'aert, so in't generael, als in't bysonder aller Metallen, als Goudt, Silver, Coper, etc. En dis gheheele Alchijmie, met haer verborghendtheden. Midtsgaders d'eerste materie der Philosophen, Dat is: Quinta essentia des Wijns, met haren volcomen ghebruyck, en verclaringhe aller duystere woorden des voorsz. Consten, op dat Paracelsus claerlijck in alles mach werden verstaen. Door Hermannvs Follinvs, der vrije Consten, ende der Medecijne Doctor, [Vignette of two furnaces.]

Ghedruckt t'Haerlem, by Adriaen Rooman, voor Daniel de Keyser, Boeckvercooper, woonende op't Sant, in't vergulden A. B. C. Anno 1613.

Small 8vo. A to I in eights, K four, or ff. 76. Black letter.

Woodcuts in the text.

Follinus was the author of a few other tracts.*

* For notices of Follinus and lists of his works, consult: Mercklin, *Lindenius Renovatus*, Norimbergæ, 1686, p. 408; Joh. Franc. Foppens, *Bibliotheca Belgica*, Bruxellis, 1739, I., p. 474; Christian Wilh. Kestner, *Medicinisches Gelehrten-Lexicon*, Jena, 1740, p. 303; *Biographie Médicale*, Paris, Panckoucke, n.d., IV., p. 177 (a depreciatory notice); *Biographie Nationale publiée par l'Académie Royale . . . de Belgique*,

69. Those who are interested in the deterioration of books, due to repeated publication, as illustrated by the *Dificio de Ricette* already noticed (§ 13), will find another example to their hand in the work of Etienne Binet.

Under the name of René François he produced a book called *Essay des Merveilles de Nature, et des plus nobles Artifices*, which is said to have appeared at Rouen in 1621, in 4to. The second edition is in the British Museum (835, h. 6), with the imprint: Rouen, Jean Osmont, 1622. It forms a respectable 4to, widely and legibly printed on fairly good paper, and it is decorated with an elaborate engraved title by J. Briot, and with passable woodcut illustrations. Altogether for a Rouen book of that date one might call it handsome. It contains pp. [2] [14] 592 [1, 1 blank].

In the Museum there is nothing between the second edition and the ninth (12202, aaa. 6). This is another Rouen edition, Charles Osmont, 1644. It is a closely packed octavo, with no engraved title, on poor soft spongy paper. It was no longer a book to be made look well, but it had to be turned out in as cheap and shabby a condition as possible, and Osmont has been quite successful in his effort. It contains pp. [16] 630 [2].

The copy that I have myself procured introduces a puzzle which, without having more editions, I cannot undertake to solve. It is of the tenth edition and was printed at Lyons by Nicholas Gay, in 1643, in small 8vo, pp. [16] 607 [1 blank]. Various explanations of this anomaly in the dates of the ninth and tenth editions may be suggested, but without more evidence they would be futile. This is a wretched production, on dingy porous paper, with a few miserably executed illustrations.

The thirteenth edition, Paris, chez Henry Le Gras, 1657, in 8vo, is in the Museum (58, b. 6). The printing, though very close, is legible, and the margins are ampler, so that on the whole this edition looks somewhat better than the preceding. It contains pp. [16] 607 [1 blank].

The collection is rather unusual; the themes of the different sections have very little to do with each other. Still the book is not without its

---

Bruxelles, 1880-1883, VII., col. 181; August Hirsch, *Biographisches Lexikon der hervorragenden Aerzte*, Wien und Leipzig, 1885, II., p. 398. The *Sleutel* is not mentioned by these authorities, which, perhaps, is another reason for including it here.

merits, and gives curious information on some out-of-the-way topics. The "most noble artifices" practically include the arts, as gardening, printing, painting, sculpture, dyeing, architecture, heraldry, music, etc. The aim of all this instruction, however, is not to enable the reader to understand or to practise any of the arts, but to talk about them! The author says so. It was meant for the fine gentleman of the time, who wished to be able to converse on any topic with an appearance of consummate skill and learning, though he may have had no practical knowledge whatever.

70. The author was born at Dijon, 1569, and died at Paris, 4th July, 1639. He joined the Jesuits in 1590, and became rector of the chief houses of the order in France. He wrote and published a great deal, but one critic says that he had more zeal and piety than talent, and that of his numerous publications only two or three are worth remembering, of which the *Essay des Merveilles*, now under consideration, is one. It went through twenty editions in a hundred years, and another writer says, this "work is not read now-a-days, and it does not deserve such neglect."* He called himself René Francois, in allusion to his name Binet, bis natus!

A long list of his works is given by Backer,† amongst which occurs the present *Essay*. The editions quoted are these:

Rouen, Osmont, 1621, 4to. Apparently the first edition.
Rouen, Osmont, 1622, 4to. The printed title calls this the fifth edition.
Rouen, Osmont, 1625.
Rouen, Osmont, 1626, 4°. The sixth edition.
Rouen, Osmont, 1629, 4°. The seventh edition.
Paris, 1646, 8°.
Rouen, Osmont, 1648, 8°. The ninth edition.
Rouen, Viret, 1657, 8°. The tenth edition.

But the editions I have seen and described above furnish forth a second list:

Rouen, Jean Osmont, 1622, 4to. The second edition.
Rouen, Charles Osmont, 1644, 8°. The ninth edition.
Lyons, N. Gay, 1643, 8°. The tenth edition.
Paris, Le Gras, 1657, 8°. The thirteenth edition.

---

* *Biographie Universelle*, Paris, 1843, IV., 347, article by Weiss.

† *Bibliothèque des Écrivains de la compagnie de Jésus*, par Augustin et Alois de Backer, Liége, 1853. Première Série, p. 95, No. 7. Here only the edition of 1621 is mentioned. The subsequent list is given in the Septième Série, Liége, 1861, p. 108, No. 8.

It is hard enough to reconcile the dates 1644 and 1643 of the ninth and tenth editions, but who will arrange the chaos of dates and editions of the first list and the second? Is all this disorder an outcome of the author's Order?

71. On pyrotechny, in the modern sense of the term, there is a treatise that ought certainly to be included in an enumeration of old books on practical arts and applied science.

> La Pyrotechnie de Hanzelet Lorrain ou ſont repreſentez les plus rare & plus appreuuez ſecrets des machines & des feux artificiels, Propres pour aſſieger battre ſurprendre & deffendre toutes places.
>
> Av Pont a Movsson par I. & Gaſpard Bernard. 1630.
>
> 4to. Signatures: *A*, A to Z, Aa to Kk in fours. Pagination: [8] 264.
>
> This book has an engraved title-page, and is enriched by a large number of engravings, of which the firework ship is the most elaborate.

The book is occupied mainly with fireworks of all kinds for war, including artillery, especially for siege purposes, and a portion of the work really deals with siege operations. Only a small part at the end is devoted to fireworks for amusement. It may be compared with the works of Bate, White, and Jones.

The author, Jean Appier Hanzelet, as he signs his name at the end of the dedication, was the son of an engineer who was in charge of the fortifying of Nancy. He was born in Lorraine in the sixteenth century, and was a printer and engraver. In 1620, at Pont-à-Mousson, he published, in 4to, a treatise on military machines and fireworks for war and pleasure, with one hundred plates executed by the author. The *Pyrotechnie* is an entirely new work. The plates are bold and instructive, and were presumably by the author.

The copy of the book which I have bears the book plate of "William Penn Esqr. Proprietor of Pensylvania. 1703."

72. In my "Notes" (Part VI., p. 11) an account was given of the English translation by Rowlands, London, 1657, of the notable work on natural history by Joannes Jonstonus. The first edition, in Latin, appeared at Amsterdam in 1632. I have not seen it, but I have got a copy of the next edition, the title of which runs thus:

> Iohannis Ionstoni | Thavmatographia | Natvralis, | In Claſſes decem diviſa: | in quibus | Admiranda | Cœli, | Elementorvm, | Meteororvm, | Fossilivm, | Plantarvm, |

Avivm, | Qvadrvpedvm, | Exangvivm, | Piscivm, | Hominis. | Editio Secunda priore auctior. | [Device of acorns and fleurs-de-lys.]
Amstelodami, | Apud Ioannem Ianssonivm. | Anno M. DC. XXXIII. |
24mo. Signatures: (?) in two, A to Z, Aa in twelves, Bb 2; or pp. [6] 578 [2].

This volume ranges with the Elzevirs. This copy has no frontispiece or engraved title, if the book was issued originally with such a decoration.

73. More than once have the works of Domenico Auda been cited.* Here is another edition:

Breve Compendio | Di Maravigliosi | Secreti | Approuati con felice ſucceſſo nelle | indiſpoſitioni corporali. | Diviso in Qvattro Libri. |
Nel I. Si tratta di Secreti medicinali.
Nel II. Di Secreti appartenenti a diuerſe coſe.
Nel III. Di Secreti Chimici di varie Sorti.
Nel IV. D'Aſtrologia Medicinale.
Con vn trattato per conſeruarſi in ſanità | Dato in Lvce Dal Signor | Fr. Domenico Avda | Capo Speciale nell' Archihoſpedale | di S. Spirito in Roma. | Con nuoua Aggiunta dell' iſteſſo Autore. | Dedicato | Al Molt' Illuſtre Sig. Pron Col. il Sig. | Fabritio | Lavro. |
In Milano, | Per Federico Franceſco Maietta. | Stampatore al Bottonuto in P. R. | Ad inſtanza di Carlo Gioſeppe Quinti.
24mo. ✠, A to N in twelves; or pp. [24] 304.

The introduction is signed by Gioseppe Quinti, and is dated Milano, li 23 Decembre, 1682.

74. Auda was a native of Lantosca, in the county of Nice, cultivated botany with success, and travelled much. He settled in Rome and practised pharmacy in the monastery of the "Minimi Osservanti" of St. Francis, to which religious order he belonged. The *Breve Compendio* was published at Rome in 1655, 1660; Venice, 1663, 1676, 1692, 1716; Turin, 1665; Cuneo, 1666; Milan, 1666.† Of this book editions of Rome, 1663;‡ Torino, 1665;‡ Venice, 1673;‡ Bologna, 1673;‡ Venice, 1676‡ and 1692,§ have been already mentioned. The *Practica de' Speziali*, his other

---

* Part IV., p. 322. Supplement I., *Transactions of the Archæological Society of Glasgow*, 1894, II., p. 399.

† G. G. Bonino, *Biografia Medica Piemontese*, Torino, 1824, 8vo, Vol. I., p. 396.

‡ Supplement I., *Transactions of the Archæological Society of Glasgow*, 1894, II., pp. 398, 399.

§ Part IV., p. 322.

book, came out at Cuneo, 1666; Venice, 1683; with a tract *delle confettioni nostrane*, and a *nuova aggiunta de segreti*. The edition of this work which has been included in these Notes is dated Venice, 1674.*

75. The introduction throws light upon what I have pointed out before is obscure, namely, the relation of Auda to Quinti. In 1711 there was printed at Venice, and sold at Liege, a French work by Quinti which is simply a translation of Auda's *Secreti.*† Subsequently‡ I described an Italian edition of the same places and date, also with Quinti's name attached to it. The explanation of this now presents no great difficulty. In the Milan edition, say of 1682, Quinti appears as the patron and editor. His address to an unnamed "Molt' Illustre Sig." begins: "Li secreti Medicinali dell' Auda, ancorche repplicati in moltissime copie uscissero dalle Stampe di Venetia, tutta via tanta stima, e prezzo si guadagnorno, che al desiderio comune sono riuscite pochissime . . . ." If this be compared with the corresponding introduction to the edition of 1711, it will be seen how the processes of absorption have resulted in the practical assimilation of the *Compendio* by Quinti: "Ces secrets de medecine que je fair paroître au jour, ont été si bien reçûs du Public," etc., etc.§ Auda presumably was dead by this time, and there may have been nobody to challenge Quinti's taking possession of the book and getting the credit of being the author. To display the evolution of the process would require a fuller set of the intermediate editions than I have at command, but in light of the result it cannot differ from what is technically called by the wise—"conveying."

76. Under the following title there appeared a volume of secrets:

Novveaux | Secrets | Rares & Cvrievx. | Donnés Charitablement au publiq | par vne perſonne de | Condition. | Contenant | Divers Remedes Eprovvez, | vtils & profitables pour toutes ſortes | de Maladies. | Et diuers Secrets pour la conſeruation de | la Beauté des Dames: Auec vne nou- | uelle maniere pour faire toutes ſortes | de confitures, tant ſeiches que liquides. |

A Paris, | Chez Iean Baptiste Loyson, | ruë Sainct Iacques prés la Poſte, | à la Croix Royalle. | M. DC. LX. | Avec Privilege Dv Roy. |

Small 8vo. Signatures: ã 8, ẽ 2, A to R in eights, S in four. ã to ẽ are not paged, but A–S are paged [1-2] 3-280.

---

* Supplement I., p. 400.

† Part IV., p. 323.

‡ Supplement I., p. 400.

§ Supplement I., p. 400.

ā and ē contain the title; the dedication to Mons. de Siffredy, which has no date, but is signed P. Erresalde; the bookseller to the reader; the table of contents; an extract from the privilege, dated 24 November 1659 and signed Dv Bisson, and a sonnet on the merits of the book by Du Pelletier. Then comes the text, preceded by a bastard title.

This, the first edition, which is in the British Museum (1038, c. 19), is well printed in large type on an uncrowded page with nice wide margins, and is a much more attractive book than the second edition.

77. Of this also there is a copy in the Museum (1038, c. 20), but I have one here which will show its chap-book character. The following is an account of it:

Novveavx | Secrets | Rares et Cvrievx. | Donnez charitablement au public | par vne perſonne de Condition. | Contenant | Divers Remedes Eprovvez, | vtils & profitables pour toutes ſortes | de Maladies. | Seconde Edition. | Augmentée de Remedes tres-ſouuerains | pour ſe penſer de la Maladie contagieuſe, | & ſe preſeruer d'icelle. | Auec diuers Secrets pour la conſeruation de | la Beauté des Dames, & vne nouuelle | maniere pour faire toutes ſortes de confitures, | tant ſeiches que liquides. |

A Paris, | Chez Iean Baptiste Loyson, au cinquiéme Pillier | de la grand' Salle du Palais, à la Croix d'Or. | M. DC. LXIX. | Avec Privilege Dv Roy. |

12°. ā in twelves. A to K in twelves, or pp. [24] [1-2] 3-240.

ā; Title. āij epistle to M. de Siffredy, signed P. Erresalde. āiiij *verso*, the publisher to the reader. āv *recto*, Table of contents, ends āxj *recto*. āxj *verso*, Extract from the privilege, dated 24 November 1659 and signed Dv Bisson. āxij *recto*, Sonnet on the book by Du Pelletier, but not signed in this second edition. Aj bastard title. Text, Aij-Kxij *verso*, pp. 3-240.

As is apparent from the bulk of the two editions respectively, the second is more closely printed, and is thereby made to contain more. The contents of the first edition, ending with "Pour la migraine," occupy only 225 pages; the remaining 15 contain receipts about the plague, which are not in the other.

This collection is not arranged on any definite plan. The first section contains remedies for all sorts of ailments, the second contains "Secrets to satisfy the curious," consisting mainly of receipts for beautifying the face, followed by a number of similar receipts, and for making "waters" from certain plants. The last section contains directions for making preserves of various kinds of fruits.

A peculiarity of the arrangement of this book is that the table of contents gives one order and the book itself contains the receipts in a different order. This makes it somewhat confusing.

78. So far as the title page goes, the book is anonymous, and the only indication of the author is the name, P. Erresalde, at the end of the dedication. Not satisfied that this was not a pseudonym I consulted a few authorities, but could not find mention of any such name. In the Museum Catalogue the book is entered under that heading without any question or cross reference. It seemed, however, to me as soon as I saw it, that it was not a real name, but an anagram or transposition, and that it might stand for De la Serre. In confirmation of this view I have found no one called Erresalde, which somehow does not impress one as genuinely French, but there is an author, Jean Puget De la Serre, who lived at the end of the seventeenth century, and was a voluminous miscellaneous writer. This treatise (like others) is not mentioned along with his principal works, but there is nothing in the theme to prevent him having been the author. At present I am disposed to ascribe the work to him.

79. During the whole course of my investigations into this literature I have not come across any book more curious, less known, and scarcer, than one which was written by George Horn. His name has been familiar to me for years as the editor of the works of Geber, but until the present little volume came into my hands I had not thought of him as an author.*

GeorgI HornI | Arca Mosis | Sive | Historia Mundi. | Quæ complectitur | Primordia Rerum Natura- | lium omniumque artium | ac ſcientiarum. | [Device.]
Lugd. Bat. & Roterod. | Ex Officina Hackiana. | A°. cIↃIↃcLXVIII. |

18mo. Sigs. *12, **6 (of which the sixth is blank), A to K in twelves, pagination [36] 220 [20]. The device is a flying eagle having a ribbon in its beak with the word *movendo.* *1 is an engraved title representing the finding of Moses. There are four female figures, two on each side of the Nile, which at this point is barely eighteen inches wide. To impart verisimilitude to the scene a crocodile about a couple of feet long is having a quiet stroll along the bank, regardless of the "invention" of Moses, which is just taking place. The background is occupied by the dense leafage of a luxuriant forest.

---

* He was the author also of a work on the history of philosophy: *Historiæ Philosophicæ Libri Septem*, Lugd. Bat. 1655, small 4to. As was to be expected from the author's proclivities it contains notices of alchemy.

Then come the printed title, Horn's dedication to Maurice of Beverward, dated: Lugd. Bat. 15 Aug. 1668; the preface, and a discourse on the first chapter of Genesis upon which this work is erected. The text goes from A to K2, and K3 to K12 contains an alphabetical index.

So far as I have observed, there is no copy in the British Museum either of this book or of another called *Arca Noæ*, 1666, 1674, which Graesse mentions, but which I regret I have not seen.

80. This is a book for which it is very difficult to find a place. It is on the secrets of creation, and includes a survey of natural history, and of the arts and sciences. The material which it embraces if properly expanded and elucidated would fill a volume a hundred times the size and more, in fact does fill the *Speculum Naturale*, *Morale* and *Doctrinale* of Vincent de Beauvais. The way in which Horn has managed to pack so much into such trifling space is by confining himself to names and the very briefest of definitions. Indeed the book is more a collection of notes to be elaborated than a treatise upon the subjects embraced. Beginning with the creation and general notions of space, form, and substance, he discourses on the elements and ultimate parts of matter, and on the great variety of their products and functions both in animate and inanimate nature. He, too, recognizes the sympathy and antipathy which apparently pervade the different parts of nature, and under the influence of that idea deals with disease.

Of the arts and sciences the only one which gets any attention is (as might be expected from the editor of Geber) chemistry, and to it is devoted a large proportion of the whole book, pp. 176-220, or exactly one-fifth.

For those who are interested in the history of that science, this is a valuable section, for in it Horn describes the apparatus, the operations, and the products, and thus gives an excellent *résumé* of the theoretical chemistry of the middle of the seventeenth century.

Altogether the *Arca Mosis* is one of the most original books which I have met with. If the *Arca Noæ* be as curious, it must be a desirable little volume.

81. The Secrets which passed under the name of Sir Kenelm Digby enjoyed some popularity at home, but were in still greater esteem on the

Continent. They appeared in Dutch, French, and German, and several editions have been quoted already, but the number is not exhausted.

I mentioned formerly* an edition of the *Remedes Souverains*, printed at Paris in 1684. As the privilege is dated 1668, there may have been several issues of the book prior to it; but there is nothing on the title page of that edition suggestive of an earlier one. It might have been the first. The licence also bears that the book was translated from the Latin and English, but I have not seen the original collections.

Another edition, similar to this, appeared in 1689:

Remedes Souverains et Secrets Experimentez, De M. Le Chevalier Digby, Chancelier de la Reine d'Angleterre. Avec plufieurs autres Secrets & parfums curieux pour la confervation de la beauté des Dames. Nouvelle Edition.

A Paris, Chez Guillaume Cavelier, au quatriéme Pillier de la grande Salle du Palais, à la Palme. M.DC.LXXXIX. Avec Privilege Dv Roy.

Small 12mo. ā in two, A to Z, Aa to Dd in alternate eights and fours; or pp. [4] 300 [28].

82. Another book, in its sixth edition, containing the preceding along with other tracts, dated 1700, was referred to at the same time.† I have got another edition of it also, called the seventh:

Nouveaux Secrets Experimentez, Pour conferver la Beauté des Dames, Et pour guérir plufieurs fortes de Maladies. Tirez des Mémoires de M. le Chevalier Digby, . . . Avec fon Difcours touchant la guérifon des Plaies, par la Poudre de Sympathie. . . . Septiéme Edition, revûe, corrigée & augmentée d'un Volume.

A La Haye, Et fe vend A Bruxelles, Chez Jean van Vlaenderen, vis à vis l'Eglife de S. Jean. M.DCC.XV.

Small 8vo. Tome I.: 4 pp.; *, **, in fours; A to M in eights; or pp. [20 including the frontispiece] 192. Tome II.: 4 pp; * in four; A to I in eights; K in six; or pp. [12] 156.

The *Remedes Souverains* occupy pp. 1–172, and the *Secrets pour conserver la Beauté des Dames*, pp. 185–192 of the first volume of this edition. The second volume contains the discourse on the powder of sympathy, and a number of receipts from various authors.

83. Of the Dutch translation I have two editions to describe. The book first came out in 1681,‡ and then it was revised and issued again.

* Part VI., p. 19. † *Ibid.*, p. 19.

‡ Prefatory Note to the Second Edition, by the Publishers.

Theatrum Sympateticum, ofte Wonder-Toneel der Natuurs Verborgentheden Behelſende Een uitſtekende Oratie, over het gebruik des Poeders de Sympathie, . . . door Kenelmus Digby, . . . Benevens Twee waardige Vervolgen, van veele zeldzame Antipathien en Sympathien . . . met meer andere Geheimeniſſen der Natuurkunde, Door N. Papinius, en A. Kircherus. Den Tweeden Druk, vermeerdert en verbeetert.

Te Leeuwarden, by Hendrik Rintjes, en t'Amſterdam, by Jan ten Hoorn, Boekverkoopers. M DC XCVII.

Small 8vo. * in four; A to U, Aa to Oo, in eights; or pp. [8, including the engraved title] 320; 196 (misprinted 169) [20].

The fourth edition is a reprint of this, and was published at Amsterdam by Nicolaas Ten Hoorn, in 1727. It is in 8vo; * in six; A to Z, Aa to Kk, Nn, in eights; Oo in four; or pp. [12, including the engraved title], 528 [24]. The signatures and pagination are continuous. This translation contains the following tracts:

Digby's Discourse on the powder of Sympathy.
Laurens Strauss's remarks thereon.
An account of the divining rod.
Papin and Rattray on Sympathy and Antipathy.
Kircher on Magnetic influences in Nature.

A translation of the *Remedes Souverains* mentioned above forms the second division of the Dutch book. I do not know if other editions of any of the preceding collections were subsequently published.

84. The following curious little book has, at least, the merit of novelty, and treats of a topic which seems to have very little literature devoted to it:

Ecole De La Mignatvre, Dans Laquelle On peut aiſément apprendre à Peindre ſans Maître. Avec Le Secret De faire les plus belles Couleurs, l'Or Bruny, & l'Or en Coquille.

Se vend à Paris, Chez Christophe Ballard, ſeul Imprimeur du Roy pour la Muſique, ruë S. Jean de Beauvais, au Mont Parnaſſe. M.DC.LXXIII. Avec Privilege du Roy.

Small 8vo. ã, ẽ, A to S, in fours; or pp. [16] 134 [Table, 10].

The dedication to Mlle. Fovcqvet is signed C. B., and is couched in the phraseology of the time, compounded about equally of inordinate flattery, vanity and humility.

The book is different from any which have hitherto found a place in these notes. It is an exposition of the art of miniature-painting, in which the way of drawing the picture, the preparation of the vellum, the colours and the mixtures of them, and the manner of laying them on, are described. The latter part is devoted to the painting of the skin, of the hair, of ornaments of gold, silver, and precious stones, of fire and smoke, of landscapes, and, lastly, of flowers. At the end is the method of preparing ultramarine, carmine, green, and gold.

85. One must suppose that the art of painting in miniature was a popular one, for the book passed through a number of editions, as the list following will demonstrate:

1672. Traité de la mignature.
Barbier.* This is seemingly the first edition.

1673. Ecole de la Mignature. Paris, Christophe Ballard. 8°.
Copy described above.

1674. Traité de la mignature.
Barbier. In a note it is said: "Les premières éditions, qui portent le titre de: 'Traité de la mignature,' furent imprimées en 1672 et 1674." The author did not know that the 1673 edition was already called *École*; always presuming that the respective treatises are identical, which I have not yet ascertained.

1679. École (l') de la mignature, dans laquelle on peut apprendre aisément à peindre sans maître, avec le secret de faire les plus belles couleurs, l'or bruni et l'or en coquilles. Nouvelle édition augmentée.
Lyon, François Duchesne, 1679, in 12, XII. 164 pp. et 1 f. de priv.
Barbier.

1684. Traité de Mignature, pour apprendre aisément à peindre sans maître, *etc.* [With a dedication subscribed C. B., *i.e.* Christophe Ballard.] Paris, 1684. 12°.
Under "B. C." in the British Museum Catalogue, 1043, d. 8.

1688. — Dernière édition, *etc.* [With a dedication subscribed C. B., *i.e.* Christophe Ballard.] La Haye, 1688. 12°.
Under "B.C." in the B. M. Catalogue, 1043, a. 13.

1696. — Troisiesme édition . . . augmentée. Paris, C. Ballard. 1696. 8°.
Under "B.C." B. M. Catalogue, 787, c. 24. Barbier, under the heading "École."

1697. — 4e éd. Paris, C. Ballard, 1697, in 12.
Barbier, under "École."

1708. Traité de la peinture en mignature pour apprendre aisément à peindre sans maître. Ouvrage corrigé et augmenté . . . auquel on a ajouté un . . . traité

---

* *Dictionnaire des Ouvrages Anonymes* par Ant. Alex. Barbier. Troisième édition revue . . . par Olivier Barbier, René et Paul Billard. Paris, 1874. Tome II., col. 15.

de la peinture au pastel, . . . la maniere de laver . . . toutes sortes de plans, le secret de faire le plus belles couleurs . . . avec une explication . . . de tous les termes propres au dessein et à la peinture. La Haye, 1708. 12°.
Under "Traité," B. M. Catalogue, 1043, d. 17.

1711. Traité de mignature. Nouvelle édition. Paris, 1711. 8°.
Under "B.C." B. M. Catalogue, 1253, a. 37 (2).

1724. École (l') de la mignature. Rouen, P. Dumesnil, 1724, in 12.
Barbier.

1759. — Bruxelles, J. Moris, 1759, in 12.
Barbier.

1769. — Paris, J. B. G. Musier, 1769, in 12.
Barbier. B. M. Catalogue, 7857, aa. 36, under "École."

1782. — Paris, Musier, 1782, in 12.
Barbier.

1802. — Paris, Moutardier, an XI., 1802, in 12.
Barbier. B. M. Catalogue, 7855, aa. 17, under "École."

1817. — Paris, Bachelier, 1817, in 12.
Barbier.

There are altogether sixteen editions in this list, but it is probable there are others. Barbier enumerates eleven; there are seven in the British Museum, of which three only are mentioned by Barbier, whereas eight in his list are not in the Museum. My own copy, 1673, picked up on the Quai at Paris, is neither in the Museum nor in Barbier.

86. From the preceding *résumé* it appears that the book was published with two different titles: *Traité* and *École.* As I have already said, I have not yet been able to settle by actual comparison whether the two treatises are otherwise identical or not; but from the way they are arranged in both Barbier's and the Museum lists I presume they are so.*

With regard to the author whose initials C. B. are at the end of the dedication, he was apparently at one time supposed to be Claude Boutet, and in my copy there is a manuscript note affirming this.

Barbier,† however, subsequently attributed the work to Christopher Ballard, the publisher, founding on a note by L. T. Hérissant. He says:

> L'auteur, à la fin d'un épître à Mademoiselle Fouquet, s'est désigné par les lettres C. B. qu'on retrouve encore dans l'extrait du privilége, qui termine ce volume.

---

* See note at the end of this paper.

† *Dictionnaire des Ouvrages Anonymes et Pseudonymes.* Seconde édition. Paris, 1822. I., p. 355.

Le privilége entier, qui est dans l'édition de 1672, où se trouve déjà l'ép re dédicatoire à Mademoiselle Fouquet, signée également C. B., porte en toutes lettres *Christophe Ballard*, et semble indiquer le vrai rédacteur de cet écrit, qui sans doute n'avait pas avoué ce travail, parce qu'il n'était qu'amateur dans la peinture. Il avait été réçu libraire-imprimeur en 1666.

The inference is plausible, but it is not conclusive. In my copy only an extract of the privilege is given, and only the initials C. B. appear, so that the full name was dropped immediately.

87. In the Museum Catalogue, as a consequence presumably of the cataloguing rules, the work appears under four headings. Thus under "École" are placed the editions of 1769 and 1802, and the author's name is given inaccurately as C. *Bullard.* This, of course, is useless as a cross reference, for this name does not appear in the Catalogue, and but for Barbier one could not have known that the correct form was *Ballard*, which *is* contained in the Museum Catalogue. Under this heading, again, there is another cross reference to "B. C." under which are placed the editions of 1684, 1688, 1696, and 1711. Finally, and without any cross reference at all, on looking under "Traité" in the Catalogue, one finds the edition published at the Hague in 1708. The result is that I am not certain that I have yet ascertained all the editions of this little treatise which are in the Museum.

88. Notwithstanding its title, the following is a general collection:

Recueil des plus beaux Secrets de Medecine, Pour la Guerison de toutes les maladies, bleſſures, & autres accidens qui ſurviennent au Corps humain; Et la maniere de preparer facilement dans les Familles les remédes & médicamens qui y ſont néceſſaires.

Comme auſſi pluſieurs Secrets curieux ſur d'admirables effets de la Nature & de l'Art.

Avec un Traité des plus excellens Preſervatifs contre la Peſte, Fiévres peſtilentielles, Pourpre, petite Verole, & toutes ſortes de maladies contagieuſes.

Le tout experimenté, recueilli, & donné au Public par une perſonne trés-habile & charitable. Divisé en Deux Parties.

A Paris, Chez Thomas Guillain, M. DC. XCV.

12mo. Signatures: A to S in twelves; or pp. [1-6] 7-406 [23, 1 blank].

A12 is an etching of an oriental physician (?) examining drugs brought to him by four attendants in oriental costume. Although included in the signature A, it is dropped in the pagination. The title is in red and black. The first six pages are occupied with title and note to the reader. Then the text, and the last 23 pages contain the Tables.

The first part is devoted to secrets of medicine (7-258).

The second part begins with secrets for beautifying the body—scents, etc. Then follow secrets about minerals, colours, ink; secrets of gardening and a number of miscellaneous secrets of medecine, magic, and so on.

It is rather an interesting collection. The author is said to be Michel Mourgues. [B. M. 1038, c. 28.]

89. A collection of secrets in B. M. 777, g. 11, is entirely new to me. It is not a translation of the preceding.

Thesaurus Secretorum Curiosorum In Quo Curiosa Non Solum Ad Omnes Corporis Humani cùm internos, tùm externos Morbos curandos, fed etiam ad Cutis, Faciei, aliarumque Partium ornatum, formam, nitorem, & elegantiam conciliandos continentur Secreta. Quibus Insuper Quamplurima Varii Generis non minùs Curiofa, qùam utilia addita funt Secreta. Cum Indice Partium, Capitum, & Materiarum quæ in toto Opere continentur.

Coloniæ Allobrogum Sumptibus Societatis. M. DCC IX.

4to. Signatures: §, §§, A-Z, Aa-Zz, Aaa-Zzz, Aaaa-Bbbb, in fours; or pp. [15, 1 blank] 668. Title red and black.

This is a great collection of medical prescriptions for all sorts of diseases, divided into twenty parts. Then part 21 contains secrets for beautifying the skin; and part 22 contains 13 chapters: 1 of secrets of the fire, 2 of water, 3 of the mind, 4 of food and drink, 5 of love, 6 of animals, 7 of fishes, 8 of birds and insects, 9 of metals, 10 of glass and gems, 11 of clothes, 12 of writing, 13 of painting and colours.

Some of these are taken from the *Secreet-Boeck*, and from Albertus Magnus, etc. They furnish an additional illustration of the persistence of these old notions, and also how books are compiled. The author conceals his name, and he gives no authorities.

A copy which I have procured coincides with the above, but the index, which in the Museum copy is at the end, is here placed at the beginning in the order of the signatures.

90. Disappointed once or twice in my attempts to get the following, I must take my description from the Museum copy (7953, a. 59).

Le | Parfumeur | François, | Qui Enseigne | Toutes les maniéres de tirer | Les Odeurs Des Fleurs; | & de faire toutes fortes de | compofitions de Parfums. | Avec le ſecret de purger | Le Tabac En Poudre; | & de le parfumer de toutes ſortes d'Odeurs. | Pour le divertiſſement | De La Noblesse, | & l'utilité des Baigneurs and Perruquiers. | Troisieme Edition | augmentée. |

A Amsterdam, | Chez Paul Marret, dans | le Beurs-ſtraat, à la Renommé. |

Small 12mo. A11, B to K in twelves; or pp. [44] 174 [12] [6]. Title, red and black. Frontispiece of a perfumer's shop, with a lady and a gentleman making purchases. This frontispiece, I presume, is the twelfth leaf of Signature A. Not dated, but marked 1700? in the Museum Catalogue.

The preliminary matter consists of an address to the reader, a list of the chief ingredients used in perfumery, with notes upon them, and a description of the principal compositions. The text (pp. 1–174) contains the mode of preparing the various perfumes, etc., arranged according to their uses. This is followed by the table of contents, and at the end (K9 to 11) is a catalogue of new publications by Marret. K12 is wanting; it was probably blank.

This edition contains a few receipts additional to those in the English translation, which will be described in the next supplement.

91. Of all the books which have been mentioned, not one could belong more strictly to the series under discussion than that which appeared under the following title:

Secrets | concernant | Les Arts | et Metiers. | A Paris, | Chez Claude Jombert, Quay des | Auguſtins, à la deſcente du Pont-Neuf, | à l'image Nôtre-Dame. | M. DCCXVI. | Avec Privilege Du Roy. |

12mo. ã8, ẽ4, ĩ8, õ4, ũ2, A to Z, Aa to Zz, and Aaa to Ddd in 8 and 4 alternately, Eee in 6; or pp. [52] 610 [2].

No author's name is given, but there is an address from the publisher to the reader, from which it would appear that he himself had been the compiler.

It is divided into 18 chapters, which deal respectively with the following Secrets: Etching and graving on wood or metal; metals, including transmutation, casting, mirrors; varnishes; mastic, cements, wax; glass and imitation gems and pastes; colours and painting; gilding, staining of wood, ivory, bone; making moulds for casts; inks; making, preserving, and retoning wine; vinegar; liqueurs and essences; sweetmeats, jelly, syrups, conserves; tobacco; removing spots and stains; enriching casts, sculptures, etc.; miscellaneous secrets, such as bleaching wax, making soap, fire that is not put out by water, a stone which catches fire with water, how to make real phosphorus from urine, which method the author says has never been so fully described before.

The chapters are followed by additional receipts supplementary to some of the sections.

This is a thorough-going technical receipt book or book of Secrets, which it would be worth while comparing with some of the older collections to ascertain how much had survived and how much was quite modern.

92. Based on the preceding is an enlarged and altered edition, which is in the Museum (1042, c. 4, 5).

Secrets | Concernans | Les Arts | Et | Metiers. | Nouvelle Edition, revûë, corrigée & | confidérablement augmentée. | Tome Premier. | (View of Nancy dated 1721.)

A Nancy, | Chez Jean de la Riviere, Marchand | Libraire à la Cour de S. A. R. fur l'Efpla- | nade, au Tems perdu. |

12mo. Tome I. 2 leaves; ã8, b4, c8, d4, A to Z, Aa to Xx in eights and fours alternately, Yy in four, Zz in 2; or pp. [4, 48] 540. In front of the text are two engraved plates in compartments representing the different arts and crafts dated 1721. At the end of the preface and also at the end of the book the little picture of Nancy is repeated and then the imprint:

A Nancy, De l'Imprimerie de Pierre | Antoine, proche l'Hôtel de Ville.

Tome II. 2 leaves. ã in twelve; A to Z, Aa to Tt in eights and fours alternately, Vu in six; or pp. [4, 24] 516. A third engraving prefixed to the text. The title pages are in black and red.

Jean de la Riviere himself seems to have been the compiler of this collection.

On the title page of Tome second the work is said to be: Divisez en IV. Parties, and instead of Nancy there is the publisher's device, a right hand out of a cloud holding a sceptre surmounted by an eye; two branches of olive (?); and a serpent biting its tail, with: Jean. | De La | Riviere | 1721 | inside.

The first volume of this work appeared in English with the title "Valuable Secrets in Arts and Trades," to which I shall refer when I come to English books.

The second volume is a treatise on dyeing, and deals with the subject under four heads: 1. Generalities, with a vocabulary of terms; 2 and 3.

The making of colours and other requisites, and dyeing; 4. Dyeing of leather. This is an important book for the history of this subject.

93. With a title similar to the preceding, there followed long after what may by courtesy be called a new edition. Whether any others appeared in the seventy years' interval I do not know, but it is creditable to the time that it was not a hard and fast reprint, like some which were produced about that period. The copy I have seen is in the British Museum (63, i. 3-6).

Secrets | Concernant | Les Arts Et Métiers; | Ouvrage utile, non ſeulement aux Artiſtes, | mais encore à ceux qui les emploient ; | nouvelle Edition, conſidérablement | augmentée. | Par une Société d'Artiſtes. | Tome Premier. [Troisieme.]| [Vignette of a small landscape with a castle.]

A Paris, | Chez Bossange & Compagnie, Libraires | & Commiſſionnaires, rue des Noyers, | N°. 33. | 1791. |

12mo. I. 1 leaf. A to Z, Aa to K in twelves, L11, L12 probably blank, wanting; or pp. [2] 814.

II. 1 leaf. A to Z, Aa to Nn in twelves, Oo in six, or pp. [2] 876.

III. 1 leaf. A to Z, Aa to Ii in twelves, Kk in eight, or pp. [2] 784.

IV. 1 leaf. A to Z, Aa, in twelves, or pp. [2] 576.

Vols. I. and III. have the title as above. [III. has a flower for the vignette.]

In Vols. II. and IV. the title is as follows:

Secrets | Concernant | Les Arts Et Métiers. | Onvrage (*sic* in II. corrected in IV.) utile, non-ſeulement aux Artiſtes, | mais encore à ceux qui les emploient. | Tome Second. [Quatrieme.]|

A Paris, | Chez Moutard, Imprimeur-Libraire | de la Reine, hôtel de Cluny, rue des | Mathurins. | 1790. |

This is a much more elaborate work than that of Nancy, 1721, and although it is based upon the first volume to a small extent, it is so much enlarged and altered that it must be considered a different book altogether. The second volume of the older edition about dyeing is omitted, and entirely new chapters are introduced about metals, enamels, and other technical subjects.

94. According to the date there was published at Frankfurt in 1719 a volume entitled:

Jo. Nicol. Martii, Med. Doctor und Practici zu Braunſchweig, Unterricht von der wunderbaren Magie und derselben medicinischen Gebrauch; auch von zauberischen und miraculosen Dingen: Sympathie, Spagyrik, Astrologie &c. Welchem beygefüget Ein Neu-eröffnetes Kunst-Cabinet in 178 Artikeln und Antonii Mizaldi Hundert Curieuse Kunst-Stücke. Frankfurth und Leipzig: Verlegts Christoph Gottlieb Nicolai. 1719.

The copy, however, of which the above is the title was certainly not printed in 1719, but, I should suppose, was reprinted by Scheible, 1820-30. It is a small 8vo volume containing 368 pages, of which the last 32 are occupied with very full advertisements of books on occult subjects, similar to the present. Amongst them is a reprint, parallel to this volume, of Fischer's *Albertus Magnus der Andere*, to which reference is made in this paper under the year 1790, and Tenzel's *Medicina Diastatica*, a Paracelsian book of which I have given a notice elsewhere.*

In the present volume there is some interesting matter, both practical and impracticable. The *Kunst-Cabinet* gives a number of baking and cooking receipts, drinks, wines, cosmetics and washes, sympathetic powder, remedies for a number of minor disorders. Next come all sorts of miscellaneous receipts, some of the arts, some of metallurgy, and some which are of the occult and useless kind. The author of the collection calls himself a "German Philotechnist," and he does not give us his authorities. The receipts taken from Mizaldus are much less practical, and the reasons alleged are sometimes of the quaintest and most superstitious description.

The most valuable part of the book is the first division, which is a dissertation, partly historical, partly critical, or at least descriptive, on the nature of Magic, and especially on the magical origin and treatment of disease. It deals pretty freely with Sympathy, and contains a number of useful references to the older literature of the subject. It will be examined more minutely in my paper on the Powder of Sympathy.

95. I have so very few absolute novelties to refer to that those which belong to that category deserve extra attention. Among these is the following:

Magia | Natvralis, | Sive | Exacta Declara- | tio fecretorum vulgi captum | fuperantium, ex varijs au- | thoribus & experientia | petita. | Lvgdvni | M.D.XXVI. |

---

* *Bibliographia Paracelsica*, Part III., 1890, p. 26 and p. 42.

This is a tiny 24mo, size 3½ by 2 inches, containing pp. 266 [30, 4 blank], poorly printed on poor paper. I have not observed any mention of it, or seen a copy of it before. It may have quite well emanated from Lyons, as it professes, but the date is certainly wrong. For (1) if it had been printed at Lyons in that year, it would, if I may judge from other specimens of about that time,* have issued from the press in a different style; and (2), if it had been printed in that year the author or compiler could not have quoted, as he has done, from Mizaldus, Cardan, George Agricola, Lemnius, Baptista Porta, Isabelle Cortese, Alexis, Bodin; not one of whom published anything for thirty years later. The truth is, the date is either late sixteenth or early seventeenth century, and it may be simply a misprint as it stands for MDLXXVI. or for MDCXXVI.

Comparison of the little volume with the two different 16mo editions of Albertus Magnus *de Secretis*, which bear date: Lvgdvni, 1566 (see Part III., pp. 210–211), makes MDLXXVI. a probable enough date for it. But that, too, is impossible; for, among the authorities quoted, are Castor Durans whose work appeared in 1595; Quercetanus, most of whose books were published subsequent to 1600, although one or two came out in 1575 and 1576; and Julius Cæsar Baricellus, whose writings appear to be all between 1600 and 1623. Quotations from any one of these makes either MDXXVI. or even MDLXXVI. impossible as the date of the book, and I have no doubt that the true date is MDCXXVI., no matter whether MDXXVI., the date on the title-page, be an intentional or unintentional error.

There is no author's name, no printer's name, no prefatory matter of any kind, but the text begins right off with an extract from Albertus Magnus. The collection is divided into two books, for no particular reason that I can see, for there is very little attempt to arrange the paragraphs, either according to subject or any other plan whatever. The contents consist of quotations of marvellous secrets from the authors above specified, and a few from some others of earlier date, such as Dioscorides, Galen, Pliny, Vitruvius, Cato, Palladius, Albertus Magnus, Paracelsus; but the great majority are from the secret-mongers of the latter half of the

* Say the works of Symphorien Champier.

sixteenth century, with a few unsigned paragraphs, which may be original. It is difficult, however, to see what was the motive for culling these secrets in particular and printing them by themselves in this minute volume. Anyhow, it is curious for its size, its rarity, and for displaying what sort of books could find currency some 250 to 300 years ago.

96. Albertus Parvus, the Small Albert, counterpart or burlesque of Albertus Magnus, the Great Albert (if indeed he can be burlesqued), has had the book of Secrets which goes under his name alluded to more than once.* The book, as to its contents, is worse than worthless, but the fact of its numerous editions and circulation as a chap-book shows that it has been believed in to a great extent. The editions already quoted are all got up in the shabbiest style, with fictitious place and person of publication. A copy I have recently got is no exception to the squalor of the others, but the title is slightly different:

> Le Solide Tresor des Merveilleux Secrets de la Magie Naturelle et Cabalistique du Petit Albert, Traduit exactement fur l'Original Latin, . . . A Geneve, Aux Dépens de la Compagnie.

It is a small 12mo, of pp. [8] 175 [1 blank] with 5 separate plates and several woodcuts in the text. In the later editions certain of the pictures, such as of Hippomanes and Enula Campana, which were supposed to engender love, the hand of glory, and some others, have separate plates devoted to them, but as far as poverty of execution goes there is nothing to choose between them.

This so-called Geneva edition has no date, but it belongs to the end of the seventeenth or early years of last century.

The only difference between this and the editions of 1729 and 1803 is that it wants a short supplement of eight or ten leaves containing "Secrets Curieux Eprouvez trouvez dans le Cabinet d'un Curieux de la Nature," which occurs in those editions. The most interesting of these, perhaps, is the description of a magnetic table or board for communicating with a friend at a distance. Round the board are the letters of the alphabet. On a pivot in the middle a magnetic needle can be mounted

---

* Part IV., p. 333.

when required. The persons who wish to communicate have two exactly similar boards. The message is sent by moving one needle to successive letters, when the other needle moves correspondingly. After the communication is finished, the needles must be removed, wrapt in cotton, and carefully protected from rust. The only thing left unexplained is the cause of the simultaneous motions of the needles.

This, however, was not a new idea, for it was started before these books were printed. A similar notion will be found spoken of by Strada,* and in a tiny volume by William Cooper there is an idea of the same kind.† Something very like it too was made use of by Lord Lytton in his story called "The Haunted and the Haunters," but it is there modified to suit the exigencies of horror.

The remainder of the supplement presents nothing else of interest. It is unnecessary to refer to this queer book unless to know what kind of secrets passed under the designation of magical, and what nonsense they were, or, at least, what perverted notions they expressed.

97. In 1885‡ I referred to a collection of secrets which, under the name of *L'Albert Moderne*, was intended to be more rational and more decent than the "great and little Albert." The book was first published at Paris in 1768; again in 1769, and a reprint at Basle in 1770. Each of these was in a single volume. I have not seen the third edition, which was issued about 1782, but a little time ago I obtained a copy of the fourth edition, in three volumes.

> L'Albert Moderne, Ou Nouveaux Secrets Et Procédés, Utiles Ou Curieux, Pour l'entretien de la beauté & de la ſanté; La guériſon des maux & maladies; La conſervation & les diverſes préparations des alimens & des boiſſons; Les diverſes parties de l'économie, tant civile que rurale; Les Arts & Métiers des Villes & des Campagnes. Quatrieme Édition. Augmentée de Conſeils & Avis pour apprécier, ces Secrets, & d'un 3me Volume. Tome Premier.
>
> A Paris, Chez Duchesne Libraire, rue Saint-Jacques, No. 47. 1793.

---

* *Prolusiones Academicæ*, Oxonii, 1746, p. 232. The *Prolusiones* were first published at Rome, in 1617.

† *Aurifontina Chemica*, London, 1680, p. 65.

‡ Part IV., p. 329, *Transactions*, N.S., I., 1888.

12mo. Tome I.: pp. [4] 479 [1 blank]; Tome II.: pp. [4] 424; Tome III.: pp. 8, 468.

The title really gives a very good summary of the contents.

The first volume, arranged similarly to the first edition, contains most of it, with certain omissions and additions. The second and third volumes are also arranged under the same main divisions: the health of man; the health of domestic animals; domestic economy, including the preservation and preparation of foods; useful practical receipts. Under each division the matter is arranged alphabetically. The drawback to the work is that the matter is distributed through the three volumes, so that they have all to be consulted for information on any single point. It is difficult to see why, when the book was issued as a whole, the contents were not brought together and properly systematized. However, as the book is extinct and never likely to be consulted for practical purposes, it is futile to criticise the shortcomings of the compiler now.

98. The jumble by Eberhardt Heinrich Fischer, which he calls *Albertus Magnus der Andere und Wahre*, was reprinted some fifty or sixty years ago. The edition mentioned before was that of 1797.* The present one has the following title:

Albertus Magnus der Andere und Wahre; das ist: Geheimnisse der Natur und Kunst, auch der raresten Wunderwirkungen in Hunderten hoch nützlicher, curioser, magischer, sympathetischer &c. Vorschriften.

Altona und Leipzig | bey J. H. Kave, 1790.

16mo, pp. 352. [The last sixteen pages contain the contents of a book: *Der geheimnissvolle Tausendkünstler*, by way of an advertisement.]

The date here given may have been that of the edition from which the present one was copied, but it is certainly not that of the present book, which looks like one of Scheible's issues about the year 1820-30. Anyhow it shows that there are people who have not forgotten the book, and think it worthy of being perpetuated. The volume contains a few useful receipts, but there is much nonsense in it as well. The authorities he quotes are the standard ones upon the subject: Cardan, Mizaldus, Wecker, Porta, Alexis and others. It can easily be seen, therefore, that the result

---

* Part IV., p. 331.

will be a mixed one, and Fischer either had not or refused to exercise the critical faculty in selecting materials for the work on which he has conferred the august name of Albertus.

If 1790 be the date of the first edition, as seems to be the case, it gives us an insight into the state of popular practical knowledge that a book could then be compiled, not from the works of the contemporary chemists, but from the secret-mongers of two centuries earlier; and if it is still purchased, not as a curiosity but for practical purposes, it shows how much has still to be done for the propagation of sound scientific knowledge among certain sections of the population in Germany.

99. There are few arts which are more dependent upon receipts than that of dyeing, and none therefore upon which one would expect to find more books. But apparently the old dyers were as reticent about their secrets as the foreman of a modern colour house, for the earlier literature is very meagre, and I have had only one or two distinct treatises to mention,* although I have pointed out repeatedly that many of the general collections contain secrets of colours and receipts for dyeing. This scarcity makes the two following books the more important:

Neu-ausgefertigtes und mit vielen Raren Kunst-Stücken geziertes Farben-Buchlein, Oder Gründlicher Unterricht, wie man Baumwollen-wüllen und leinen Garn, dessgleichen halb wüllen und halb leinen Gewebes, oder auch ganz wüllen oder ganz leinene Waaren in allerhand schönen Farben abfärben und zurichten solle; [vignette] Allen Färbern, Parchend-Webern, Tuch- Rasch- und Zeugmachern, Meselan, Warff- und Leinwebern, wie auch vielerley Handwerkern, Hausvättern und Hausmüttern zu dienlichen Nutzen am Tag gegeben, und mit vielen neuen Kunst-Stücken Leder zu färben, auch ein und andere schändliche Flecke durch sonderbare Erfindung aus wüllen Tuch und seidnen Zeug zu bringen.

Im Jahr Christi 1748.

Small 8vo. A to H in eights, but H8 blank, is wanting. No pagination. Title red and black. The Vignette represents the interior of a dye house with the workman prosecuting his business. As the title page almost implies or suggests, this is a chap book.

A practical treatise on dyeing which is a hundred and fifty years old—in fact any genuine technical treatise of such an age—cannot fail to be of

* See § 92, *Secrets Concernans les Arts et Metiers*, 1721.

interest to any one engaged in the same art in our own day. In the present case we have the materials employed—colours, mordants, leys, and so on; the quantities to be used for a given amount of stuff to be dyed; the process, the length of time the action is to go on; the precautions necessary to obtain a really good result. Everything is carefully detailed. These were the methods followed, when, apparently, cloth and articles of clothing were dyed at home, before dye works on the large scale were started. The attraction of such a book to the scientific chemist lies in the chemical compounds which were used, and in the empiricism of the processes involving actions which are beginning only now to be understood.

100. Curiously enough another book on the same subject had come under my notice a couple of months earlier than the preceding, showing, as I have before remarked, how related books come together quite unexpectedly.

L'Art de faire les Indiennes, Et de compofer les plus belles Couleurs, bon teint à cet ufage; de peindre les Étoffes de foie & en Mignature; de laver les Deffeins, Plans, Cartes Géographiques, &c.; de teindre le Bois, la Paille, le Crin, les Fleurs artificielles, &c., avec plufieurs fecrets pour faire toute forte de Couleurs qui n'altérent point les Étoffes, & qui font à l'épreuve du grand air & du foliel. Par M. Delormois, Deffinateur du Roi, & Colorifte. [Vignette.] A Paris, Chez les Libraires Associés. M. DCC. LXXX.

Small 8vo. *, A to E in eights, F in four, G in two; or pp. xvj, 84, viij.

This is a treatise on calico printing and dyeing. It describes the designing of the patterns, the method of printing, and the receipts for the colours. In the second part are given receipts for liquid colours, or, as they are called, "the secrets of the composition of the colours." This book therefore brings itself under the general category of secrets. Like the preceding it interests the calico-printer and chemist alike, but it deals with printing rather than with dyeing.

There is apparently no copy of this book in the British Museum.

101. One of the delights of a chap-book lies sometimes in the contrast between the resounding announcement of the title-page, with the puny echo of the text. Some examples have occurred in previous parts, and I shall have one or two more to produce hereafter, but the following is a typical instance:

Le Secret Des Secrets De La Nature, Extraits tant du petit Albert, que d'autres Philosophes Hébreux, Grecs, Arabes, Chaldéens, Egyptiens, et de beaucoup d'autres Modernes. Enrichi de plusieurs rares secrets de Cornelius Agrippa, Marac Trimegiste, d'Arnosa, de Villeneuve de Cardon, d'Alexis Piémontois, et plusieurs autres. Vu et corrigé par M. Mallemans. [Vignette of a cock pecking at a bunch of grapes.]

Limoges, Chapoulaud, Imprimeur-Libraire.

12mo. A, B, in six, C in four; or pp. 32. No date, but about 1800.

The performance corresponding to this vast promise is striking. There are secrets for procuring love, for silencing a dog, for dyeing the hair, secrets about sheep, secrets for amusement (preferably at somebody else's expense) quite dull and futile, for catching birds, and taking foxes, moles, hares, rats, for killing vermin, for curing a few ailments, for etching on iron, for changing lead into tin, some sleight of hand tricks,—and that is all. The author has, unfortunately, omitted the names of the philosophers of antiquity and of modern times who could authenticate the receipts.

102. The age of "Secrets" is not yet past. Witness a work printed within the last thirty years, the title of which might have introduced a book of the sixteenth century on the same topic. [B. M. 7944, aaa. 25.]

Secrets | Des Anciens | Et | Recettes Nouvelles | Pour la Toilette, la Propreté et les Vêtements | l'Hygiène et les Aliments | les Meubles, les Appartements, les Maisons et les Jardins | la Destruction des Animaux nuisibles | *suivis de* | Remèdes et Conseils pour la Santé |

Avignon | Amédée Chaillot, Editeur | Place du Change, 5 | 1865 |

12mo. pp. 197 [1 blank].

The Toilette includes removal of wrinkles and spots, the teeth, hands, baths, the hair, almond paste, pastils, scent bags.

Preservation of clothes—washing.
Preserves of fruits, olives, eggs, wine, etc.
Cleaning articles of vertu, bronzes, armour.
Cleaning of rooms—damp, paints.
Gardens—Destruction of insects, wasps, moths, ants, of moles, or worms.
Receipts for the preservation of health.

The three supplements now ended contain the books in foreign languages which have come under my observation during the last few years, and the largest number of these are in Italian. The English books, many of which are rare and curious, will be dealt with in the next supplement.

NOTE TO §§84-87 (JUNE 24, 1898).

Since the above was printed I have examined the copies in the British Museum, and have made the following notes, which will be enough to show what connection exists between the different editions there.

1684. Traité de Mignature. . . . Troisième Edition, Reveuë, corrigée & augmentée. A Paris, Chez Christophle Balard. . . . 1684.

Small 12mo. ã in 7, A to G in twelves; or pp. [14] 351 (a misprint for 151, but as pp. 83-88 inclusive are dropped, the true number is 146). Table [9].

1688. Traité de Mignature. . . . Derniere Edition corrigée & augmentée de plusieurs Secrets.

à la Haye, 1688.

Small 12mo. * in eight; A to G in twelves; H in 4; or pp. [16] 168 [8].

A few secrets to make Chinese varnish of all colours are added at the end.

1696. Traité de Mignature. . . . Troisiesme Edition, Reveuë, corrigée & augmentée par l'Autheur.

Paris, Christophe Ballard, 1696.

Small 8vo. ã, A to L, in eights; or pp. [16] 166 [10].

This does not contain the receipts about Chinese varnish. It is a reprint of 1684.

1708. Traité de la Peinture en Mignature. . . .

A La Haye, 1708.

12mo. * in eight (including frontispiece); A to M in twelves (with two plates of colours between 154-155); or pp. [16] 269 [11, and 8 of book advertisements].

This includes the *Traité*, but not exactly in the same order, and various chapters have been added. This treatise is therefore fuller and more complete.

1711. Traité de Mignature. . . . Nouvelle edition.

Paris, Christophe Ballard. 1711.

Small 8vo. ã in eight, A to P in fours and eights alternately; or pp. [16] 175 [1]. Reprint of 1684 or 1696.

1769. L'Ecole de la Miniature. . . . Nouvelle edition, revue, corrigée & augmentée.

Paris, J. B. G. Musier. 1769.

Small 8vo. 1 leaf, a, A to L in eights, M2; or pp. [2] vj [10] 179 [1 blank].

The *Traité de Miniature* occupies to p. 103; then comes another tract, *Méthode pour étudier l'Art de la Peinture*, which goes on to p. 179.

1802. L'école de la Mignature. . . . Nouvelle édition, revue, corrigée et augmentée.

Paris, Moutardier. An xi. 1802.

12 mo. 2 leaves, a in six, A to O in alternate eights and fours, P in six; or pp. [4, 12] 178 [2 blank? wanting]. An inferior reprint of the preceding.

It would appear, therefore, that the *École* includes the *Traité* and another small treatise, and that the *Traité de la Peinture en Mignature* is an enlarged and rearranged edition of the earlier work. They are quite entitled, therefore, to be catalogued under distinct headings, but not without sufficient cross-references.

BIBLIOGRAPHICAL NOTES

ON

HISTORIES OF INVENTIONS

AND

BOOKS OF SECRETS

FOURTH SUPPLEMENT

BY

JOHN FERGUSON, LL.D., F.S.A. LOND. AND SCOT.

REGIUS PROFESSOR OF CHEMISTRY IN THE UNIVERSITY OF GLASGOW

GLASGOW

Printed at the University Press by

ROBERT MACLEHOSE & CO.

1900

Read to the Archæological Society of Glasgow, 17th March, 1898.
100 *Copies Reprinted.*

No. IX.

BIBLIOGRAPHICAL NOTES ON HISTORIES OF INVENTIONS AND BOOKS OF SECRETS.

(*FOURTH SUPPLEMENT.*)*

BY

JOHN FERGUSON, LL.D., F.S.A.

(*Read at a Meeting of the Society, held on 17th March, 1898.*)

1. Having in the three preceding supplements disposed of such foreign books on secrets as I have recently seen, I come finally to those in English. They stretch over some two hundred and fifty years, from the middle of the sixteenth down to the early years of the present century. They are fraught with more than one kind of interest, and present attractions, literary, antiquarian, practical and historical, as well as bibliographical, to which last, however, I must mainly confine myself.

2. The larger proportion consists of books which swarmed from the press in London from say 1650 to about 1730. Cheap hand-books of practical instruction were in great request, and the flying stationers (however contradictory that title may sound) carried them about and disposed of them all over the country. Some of the books were popular, and passed through endless editions; but, to judge from the examples I shall put before you, there was a steady market for every book of the kind that was printed. In particular, works on household economy were much demanded; they

* This supplement contains English books down to the middle of the seventeenth century.

embraced a wide range of topics, being intended for women in every rank and condition of life, from the "accomplish'd lady" and youthful maiden to the scullery maid and "widow, etc."

3. The volumes have a strong family resemblance. Chap-books they are, indifferently printed on coarse paper, usually in small 12mo, and bound rudely in brown sheep. The copies one meets with are soiled, worn, and not unfrequently imperfect. On rare occasions, however, one turns up as clean and fresh as if it had just come from Little-Britain, or Duck-Lane, or London-Bridge; and then it is found to be quite attractive, despite its plain exterior. Sometimes, after they have been in the hands of wealthy collectors, they are to be seen masquerading in morocco, but such decoration is incongruous, however complimentary it be to the rarity and curiosity of the volumes so adorned.

4. Of the publishers of these chap-books, it was George Conyers, who, towards the end of the seventeenth century, turned them out in greatest numbers and advertised them freely. By his books he may have been the "Weale" of his day; by their want of dates he is a woe of ours.

5. A considerable number of books and a great number of secrets in other books refer to the toilet, and copious directions were given to ladies for the preservation and enhancement of their personal beauty. It is notable how much attention seems to have been devoted to this topic by the authors of the old receipt books; whereas, so far as I am aware, nothing has been published on it, at least in this country, for a good many years. This makes the older literature all the more curious, even though it be out of date. Special methods and materials have alike fallen into desuetude, and the very phraseology employed is as obsolete as the patches and toupees themselves; still some of the receipts have retained their footing, and there are general principles upon which writers, early and recent, seem to be in perfect accord with one another.

6. As in all my previous papers, what I have to say on the subject will be by the chronological arrangement of authors, so that even late reprints may be found in a comparatively early position in my remarks.

7. Albertus Magnus is the bibliographical embodiment of the inevitable:

one must expect to have to notice his books on every occasion. This time it is a translation of the *Liber Aggregationis*.

The booke | of secretes of Alber- | tus Magnus of the ver- | tues of Herbes, stones | and certayne | Beastes. | ¶ Also a booke of the same | Author of the maruaylous | thinges of the world, and | of certayn effectes cau- | sed of certayne | Beastes. | (.·.)

Small 8vo. A to K (?) in eights. Black letter, 22 lines excluding head-line and catchword. The title is surrounded by a border of four different pieces. As this copy is defective at the end—it wants Kviij at least—and there is no record of place, date, or printer, one can only infer these matters by comparison with other books.

In a MS. note on the cover the type is said to be similar to that of the *Boke of Knowledge*, printed by Robert Wyer; but as I have not been able to make the comparison, I do not know whether this is the case or not. If it issued from Wyer's press, the fact has not been noticed by Mr. Plomer.* I have compared it with two specimens of Wyer's printing which I happen to have, and with one of them it agrees so closely that I should be almost inclined to say that it was really Wyer's work. The book to which I refer is that entitled *The Judgemẽt of all Urynes*, described by Herbert, p. 377.† The two books agree closely in type, in the coarse paper, through which the print shines from each side, and generally in their wretched typographical execution. On the other hand, they differ in the size of the printed page, the Albertus being both shorter and narrower, and in having a headline, which the other wants. I am not at all certain, therefore, that it came from Wyer's press. Judging from its appearance, I should suppose its date is about 1540 1550.

Herbert does not quote a copy of this book as printed by Wyer, but he mentions‡ *Albertus Magnus, the boke of Secretes*, printed by William Seres, which might possibly be the volume I have. Herbert gives no particulars about it, and I know of no copy at present.

---

* "Robert Wyer, Printer and Bookseller." Printed for the Bibliographical Society, London, 1897.

† The copy I have belonged to Herbert; it has his initials and the above reference to his book.

‡ *Typographical Antiquities*, London, 1786, ii., p. 705.

8. [*March*, 1899.

As there will be no other opportunity of mentioning it, I may describe here a copy of a very rare edition of the *De Secretis Mulierum* which came into my possession since the above was written. It begins, f. 1 *recto*:

Scribit ph'S | ph'o24 prin- | ceps. iiij°. | Ethicorum |
Hō eſt opti | mū eo24 que | ſunt in mū | do, *etc.*

It ends f. 41 *recto*, line 8:

Finis h9 Tractatuli venerabil' Alberti magni. |

The *verso* is blank.

It is a small 4to, 41 leaves, 33-34 long lines to the page.

The print is $5^9/_{16}$–$5^{13}/_{16}$ inches by $3^3/_{16}$ inches. It is of a peculiar semi-Gothic shape, with unusual capitals, especially H, M, and R.

It has no date, place, or printer's name; no catchwords, signatures, or leaf numbers.

The initial S at the beginning of the introduction measures $2^3/_8$ by $2^2/_8$ inches, and is a rough woodcut, but vigorous, with a dash of humour. On the *verso* of leaf 1, at the beginning of the text, is a large and elaborate Gothic D; these are the only attempts at decoration in the volume.

Hain, who describes this book fully and accurately under No. *550 from an actual copy, says it is printed *Typis Reyserianis*. Reyser printed a large number of books at Eichstadt in Franconia. Kloss, however, to whom my copy belonged—it has his book-plate, and is No. 63 in his sale catalogue—has added a note in his characteristic handwriting to the effect that the edition was unknown to Panzer (which is correct), and that it was printed at Speier by the brothers Hist or Hijst between 1480 and 1485.

It has been said—I do not know with what accuracy—that the books printed by Reyser were all in folio, which, of course, would exclude the present work. It is to be observed, however, that Hain says only *Typis Reyserianis*, which may be quite compatible with Kloss's ascription to Hijst; that is, Reyser and Hijst may have used similar if not identical founts of type. But this is a matter on which, as I have no facts to go upon, I am unable to express an opinion.

So far as I have observed, there is no copy of this edition in the British Museum Catalogue.

This with the above makes now fifty editions of the two sets of *Secrets* ascribed to Albertus, which I have brought before the Society; but the number of those which exist is not exhausted.]

9. Next to the work of Albertus, which is *hors concours*, so to speak, the earliest I have to show on this occasion is that by Edward Fenton, *The Secret Wonders of Nature*, London, 1569. The copy here was obtained in 1891 at the Hailstone sale, and makes the sixth or seventh I know of at present. The reason for this rarity will be understood by inspection of this copy. The book was a popular one. Not to speak of the amazing tales which fill it, and which would attract readers old and young at all times, it contains a number of woodcuts which would fascinate far more than even the text would do. The result of this popularity has been the almost complete annihilation of the edition through hard usage, and the only survivors I know of, beside the present copy, are: two in the British Museum; one in the Bodleian; one in the University Library at Cambridge; one at Abbotsford; one in Mr. Huth's library; and one or two others in private hands.

On a former occasion,* when I described the book from the Museum copy and gave some specimens of the anecdotes which make it so entertaining, I was in the belief that Fenton was the author. The title-page declares openly that it is by him; and it is only by a statement at the close of his dedication that one ascertains that it is a translation, though he specifies neither the author nor the language of the original. While, therefore, he just saves his own credit for honesty, he has done his best to get the credit—such as it may be—of being the author of the book.

The fact is, however, as I have already pointed out,† it is neither more nor less than a translation of the work of Boaistuau, *Histoires prodigieuses*, which first appeared in 1566. Of this author and certain of his works I gave a short account in my third supplement,‡ so that it is unnecessary to say more about them at present. I may repeat, however, that after having acquired a copy of the 1574 edition of Boaistuau's book, I found that the woodcuts in it, though on a much smaller scale, bore an

* Part IV., p. 306. (Reprint, p. 8.) † Supplement III., § 33.

‡ *Transactions, Archæol. Soc., Glasgow*, III., pp. 381-85.

unmistakable likeness to those in Fenton's. Comparison having thereafter revealed identity in the text, there could be no question as to the source whence Fenton had obtained his "secret wonders of Nature." But he ought to have made it quite clear on the title-page.

10. Reference has been already made* to the fortuitous concourse of paragraphs huddled together in Thomas Lupton's book, *A Thousand Notable Things of Sundrie Sorts.* Although devoid of systematic form, it contains such an array of old-fashioned stories and notions that it continues to be of some use and value to the folk-loreist, and is appealed to as a source of information even at the present day.

11. Among the editions in the British Museum formerly quoted (Part IV., p. 313) is one of 1627. I have a copy of it here:

> London, | Printed by E. All-de for N. Fosbrooke, | by the Affignes of Io: Wright, and | R. Bird. 1627. |

It is a small 4to, in black letter, and consists of sigs. A to Z, Aa to Gg in fours; or, pp. 6 not numbered, containing the title and author's preface; 174 numbered (but a misprint for 214), containing the text; 19 not numbered, containing the table of contents; and a blank page. This, therefore, is a page for page reprint of the undated edition of about 1596,† and of that of 1601, the earliest in the British Museum.‡ There is nothing to be specially observed in this edition.

12. In 1631 another reprint appeared, also formerly mentioned as being in the Museum.§ Of this, too, I can show a copy, and the interest of it lies in its being probably the first issue of the book in 8vo, and the first step in its degradation into the chap-book and falsified forms in which it was published last century.

This edition is still in black letter. The later ones were in roman type. It contains A to Z in eights, or pp. [6] 336 [26].

13. Another edition has just come to hand. It was printed by Edward Allde for Edward White, and was "to be folde at the little North doore of Pauls, at the ſigne of the Gunne," in 1612. It is in 4to, as usual,

* Part IV., p. 313. (Reprint, p. 15.)

† Part V., p. 455. (Reprint, p. 39.)

‡ Part IV., p. 313. (Reprint, p. 15.)

§ *Ibid.*

and black letter. The signatures run from A to Z, Aa to Dd in fours and Ee2, or pp. [6] 214. There is no index; but I suppose that it has been taken out of this copy, not that it was never printed. It is a poorly got-up book.

14. One of my most recent acquisitions is the edition of 1815, of which I gave the title formerly from the copy in the British Museum.* It is a 12mo, uncut, in the original paper boards as it was issued, and it contains sigs.: A to K in twelves, L2; or pp. xxiv., 220.

This edition, it will be remembered, also includes a reprint of the Marquis of Worcester's *Century of Inventions*, to which reference will again be made under the year 1663.

15. Of the *Treasurie of Hidden Secrets*, two editions have been described, dated respectively 1596† and 1600.‡ There is a later one: "At London, Printed by I. W. for Edward White, and are to be fold at his ſhop at the little North doore of Paules, at the ſigne of the Gunne. 1608." It is printed, like the others, in black letter, and is a small 4to with no numbering of leaves or pages, but with signatures A to I in fours. Though it professes to be enlarged "with diuers neceſſary Phiſicke helpes," there is no appreciable difference between it and that of 1596.

16. Still another edition appeared long after, of which I know two copies: one is in the British Museum (1038, i. 35 (8)); the other is in my own collection. It is as follows:

The | Treasury | of | Hidden Secrets, | Commonly called, | The Good-huſwives Cloſet of proviſion, | for the *health of her Houſhold.* | Gathered out of ſundry experiments, lately | practiſed by men of great knowledge: And now newly | inlarged with divers neceſſary Phyſick helps, and knowledge | of the names and diſpoſition of diſeaſes, that moſt commonly | happen to Men and Women. | Not impertinent for every good Huſwife to uſe in her | Houſe, amongſt her own Family. | [Device.]

London. | Printed by *Jane Bell*, and are to be ſold at the Eaſt- | end of *Chriſt-Church*, 1653. |

Small 4to. A to I in fours; no leaf or page numbering. In the Museum copy I4 is wanting; it contains the contents of the chapters after Cap. 56. This too is printed in black letter.

---

* Part IV., p. 314. (Reprint, p. 16.)
† Part V., p. 454. (Reprint, p. 38.)
‡ Part IV., p. 310. (Reprint, p. 12.)

Inserted in the Museum copy is a printed slip: "The powder for sore eyes," containing printed directions for the use of the powder enclosed in it.
The device on the title-page is the same as Plantin's, viz. a hand holding a pair of compasses, with the motto: Labore et Constantia.

17. There is a curious and rare tract by a certain Walter Gedde, or Gidde, on a method of drawing designs upon direct and collateral lines. Its title is as follows:

A Booke of Svndry Dravghtes, Principaly ferving for Glafiers; And not Impertinent for Plafterers, and Gardiners: be fides fundry other profeffions. Wherevnto is annexed the manner how to anniel in Glas: and alfo the true forme of the Fornace, and the fecretes thereof.
London, printed in Shoolane, at the figne of the Faulcon by Walter Dight. 1615.
Small 4to. A to R in fours; or pp. [24] 3-103 [1 blank, 8]. Folding plate. The greater part of the book is taken up with representations of various designs.

The second tract about glass occupies sig. R, and it has a title-page:

The Manner, Howe to Anneile, or Paint in Glas: The true receptes of the cullors, The ordering of the Furnace, and all the fecretes there vnto belonging. 1616.
It contains pictures of the furnace, pot, pan for test pieces, etc.

The first copy of this book which came under my notice was in Pepys' Library, Magdalen College, Cambridge, and at the time I knew of no other. Since then two copies have come to my knowledge—the Ashburnham copy and that which I now exhibit. There is one also in the British Museum (C 21, b (1-2)), where it is specially reserved. It is quite possible that there may be others, and that the book is not so rare after all, but in the meantime I have not had an opportunity to search for them.

18. If one may judge by its price, a book sought after at the present time is that on the arts by John Bate. It is not an absolute rarity, but, on account of its contents and illustrations and the difficulty of getting it in perfect condition, it commands a certain amount of attention.

19. On a former occasion* I was able to describe only the first edition from an actual copy, printed in small quarto at London by Ralph Mabb in 1634, but for the later editions I had to rely upon Lowndes and

* Part VI., p. 12. (Reprint, p. 14.)

Hazlitt. Of these I have now got copies. The second edition, also printed by R. Mabb, came out in 1635. To this is prefixed a portrait of the author, and the title is in black and red. The book is a small quarto, and contains: the frontispiece and pictorial title (two leaves), and signatures A to Qq in fours, sig. I having two extra leaves with diagrams. The following points may also be noticed: the first leaf of Q is misprinted P. N1 is the title-page of book second; T1 of book third; and Hh2 of book fourth.

The portrait, for which the book is prized, is far from complimentary, and represents Bate, who, doubtless, was a decent law-abiding citizen,* as a person for whom no villainy would be too great to perpetrate. G. Gifford, the engraver, has no reason to be proud of the "lively pourtraicture."

Though said to be enlarged in every part, the chief additions will be found in part one, to which in this second edition has been added an appendix.

20. The third edition did not appear till 1654, and it was printed for Andrew Cooke at London. It has no portrait, but it has a number of illustrations over and above those in the previous editions, but all of rather inferior execution. The book is in small quarto, and besides the engraved and printed titles (two leaves) contains B to Gg in fours and Hh in two.

In some way or other this edition has got modernized—though with only twenty years interval—and it is not nearly so attractive as its predecessors. All the editions, more particularly the first and third, are moderately rare, and are not readily procurable in really nice state. It may, therefore, be of some interest to see them all brought together as in the present instance. So far as I know, there were no more than the three editions now specified.

21. Bate, as I pointed out before, was a practical man, and described receipts and apparatus which he had tried and found would work, not like some of the compilers who often gathered nothing else but mere nonsense.

The first book contains a good many ingenious devices worked by

---

* His name, notwithstanding, does not appear in the *Dictionary of National Biography*.

water power. In particular, there is a weather glass, a water clock, force pumps, and other contrivances. The second book deals with pyrotechny; and as it gives minute directions about the materials and various kinds of fireworks, such as rockets, crackers, fire drakes, as well as fixed pieces, it is worth comparing his receipts with those of other writers on the subject.

The third book is occupied with drawing and painting, and an account is given both of mediums and of colours. This part may be compared with Salmon's *Polygraphice*.

The last book, called "Extravagants," is merely a collection of miscellaneous secrets relating chiefly to the metals, and there are a few medical receipts at the end. On the whole the book justifies the demand for it that still exists.

22. As an example of a successful compilation, the Countess of Kent's *Choice Manual of Rare and Select Secrets in Physick and Chirurgery* (first published, I should suppose, about 1651) has been displayed in virtue of its having gone through at least nineteen editions.* One might have expected therefore, to find occasionally a copy of some edition or other for sale. But that is just what very seldom happens; and if one which I have be inspected, the reason of the scarcity, as in the case of Fenton above quoted (§ 8), will be easily comprehended: the books have been simply worn out. When a volume gets into the state exhibited by my copy of the twelfth edition, it is within a very short distance of final destruction. It has been squeezed out of all semblance to a book; it has been blackened by smoke, and fouled by use till it is almost illegible; the owner may have got a fresh copy of a newer edition, and may have had no special sentiment about the old one which had seen so much service, and so it would drift to its end. It has, however, escaped tearing up and burning as waste paper, and it is worth preserving now, as an illustration of the kind of treatment to which useful books are subjected; and how whole editions may disappear merely by bad usage and want of a little care and consideration. Would it not be a pleasant recreation to have a few minutes' talk with the person who maltreated this volume?

---

* Part IV., p. 318. (Reprint, p. 20.)

23. The editions I have recently acquired are the fourth, eleventh, and twelfth.

The fourth is a tiny 24mo, similar to the copy of the nineteenth edition I showed in 1885, and it has a queer little portrait of the Countess. It was printed by Gartrude Dawson, and sold by William Shears at the Bible in St. Pauls Churchyard, in 1654. It contains A to Q in twelves, but the two parts of which the book consists are paged separately: [18], 206, [18], 140. This copy is remarkably fresh and clean.

24. The eleventh edition was printed and sold by the same persons in 1659. It is a 12mo, and is printed in larger type than the other two. It had a portrait, but it has been removed by some Grangerising thief, and its existence, therefore, can only be assumed in the collation:—portrait leaf (?); title; A in six, B to M in twelves; l in six, m in twelve, N to R in twelves, S in eight, T, (T2 was probably blank); or, pp. [2-? 2, 12], 258, [6]; 20, 154. It contains certain additional secrets of medicine in Part I., pp. 231-258, which were introduced into the work in some edition which I have not seen, subsequent to the fourth.

25. The twelfth edition is of the same size and style as the fourth edition; the portrait is reversed, and is not only more debased in execution but is more disreputable in expression. It contains A to R in twelves, S in six, or pp. [18], 233, [7]; [20], 140, including the added secrets to Part I., pp. 207-233 already specified. This edition was also printed by Gartrude Dawson and was sold by William Shears, who, however, had removed to the Blew Bible in Bedford Street, Covent Garden. The date on the book is 2659, an uncommon kind of misprint.

A curious point in the publication of these books is the issue of the eleventh and twelfth editions in the same year, by the same people, but in quite a different form.

26. Since the preceding was written, a copy of a still later edition has fallen into my hands.

A | Choice Manual, | or, | Rare and Select | Secrets | in | Physick | and | Chirurgery. | Collected, and Practifed by | the Right Honourable, the | Countess of Kent, | Late Deceased. | Whereto are added several Experiments | of the Virtues of Gafcon-powder, and | Lapis contra Yarvam, by a Pro- | feffor of Phyfick. |

As alſo moſt Exquiſite ways of Pre- | ſerving, Conſerving, Candying, *&c.* | The 21th. (*sic*) Edition with Additions. |
London, | Printed for Henry Mortlock, at the Phœnix in | St. Paul's Church-yard, 1708. |
12mo. pp. [12] 112 ; [4] 68 [7, 1 blank]. Portrait extra.

The second part has a title-page as follows :

A True | Gentlewomans Delight. | Wherein is contained all manner | of | Cookery :
Together with { Preſerving | Conſerving | Drying and | Candying. | }
Very Neceſſary for all | Ladies, | And | Gentlewomen. | Publiſhed by W. J. Gent. |
London, | Printed for H. Mortlock, at the Phœnix in St. Paul's | Church-yard. 1707.

This copy, too, with one of the Hailstone bookplates, bears evidence of mal-treatment, for it wants a leaf. It has exactly the same portrait as that in the nineteenth edition.

27. When first* I had to speak about the Countess, I consulted Granger† and Walpole.‡ So very different a view, however, of her character (or want of it) is given by Abraham de la Pryme,§ that I have consulted some other authorities, and have found considerable uncertainty and confusion due to a want of agreement among them.

It is unnecessary to go into the details of her life.‖ She is called the second of the three daughters of Gilbert Talbot, Earl of Shrewsbury; she was married to Henry Grey, seventh Earl of Kent; "she was eminent for her virtue and piety," but "her being an author was not the most estimable part of her character."¶ She died at her house in Whitefriars in 1651.

* Part IV., p. 318. (Reprint, p. 20.)

† *Biographical History of England*, 5th edition, London, 1824, Vol. III., p. 209. The list of her portraits seems to be contained in the edition of 1775, Vol. II., p. 374, but I have not seen it.

‡ *Royal and Noble Authors*, by Park, London, 1806, Vol. III., p. 44.

§ *Diary*, edited by Charles Jackson, Surtees Society, 1869-70, Vol. 54, p. 8. I have not seen this authority referred to in any notice of either Selden or the Countess.

‖ See *Dictionary of National Biography*, xxiii., p. 181, under "Elizabeth Grey."

¶ *Memoirs of John Selden*, by Geo. W. Johnson, London, 1835, p. 355.

Selden had been very intimate with the Earl, after whose death in 1639 "he appears to have been domesticated with the widow both in town and country, taking the management of her concerns, and, as Wood says, living with her in a conjugal way: but what he meant by that expression I do not undertake to conjecture."* "There is strong reason to believe," says Johnson, "that he was bound to her by other ties than those of service and friendship." He repeats the "equivocal" words Aikin ascribes to Wood, adding that Aubrey states that he was privately married to the Countess.

Granger says that she was a different person from the Elizabeth "who cohabitated with Mr. Selden, and left him a considerable fortune." There is no equivocating about Granger, but he does not inform us who this other Elizabeth was, and Nichols† says distinctly that Granger is wrong in this supposition. I myself have no doubt that Granger is wrong as to the want of identity.

Lastly, De la Pryme puts the rumour with quite a seventeenth century frankness. Under the year 1686, he says:

> Being reading this day a book entitled "The Countess of Kents receipts," I asked my aunt Prym, who is an ingenious woman, who this Countess was, etc. Shee answered me that when shee, my aunt, lived in London, she lived just over against her, and knew her very well. She sayd that the Countess was a widdow and never had a child in her life: that she was an exceeding good charitable woman, and that she spent twenty thousand pound a year yearly in physick, receipts and experiments, and in charity towards the poor. Shee caused every other day a huge dinner to be got, and all the poor people might come that would, and that which spared they took home with them. My aunt says shee has seen the poor at her tables several times. Sometimes there would have been sixty, sometimes eighty, sometimes more, sometimes less. And shee sent vast quantitys of meat out to those that could not come. She would oft go to the houses of the poor, and visit them and dress their soars with her own hands; and shee distributed a vast deal in money herself yearly to all those that stood in need. Yet for all this (adds De la Pryme) as I have since heard, she lived in common whoredom with the famous Mr. Selden, who (*sic*) she entertained as her gallant.

* *The Lives of John Selden, Esq., and Archbishop Usher*, by John Aikin, M.D., London, 1812, p. 155.

† *Literary Anecdotes*, London, 1814, viii., p. 510.

The editor of the *Diary*, Charles Jackson, in a note to this passage here mentioned, says:

> It is but an act of ordinary justice to the character of the noble lady whom the diarist has named in the text, to mention that the story to which he refers, whether true or false, does not, at all events, or in any way, relate to *her*. The "good Countess of Kent," so called from her deeds of charity and hospitality, was Amabel, the second wife of Henry Grey, tenth Earl of Kent * (who died 1651), daughter of Sir Anthony Benn, Recorder of London, and widow of the Hon. Anthony Fane. She lived to be 92 years of age, surviving her husband forty-seven years, and dying 17 Aug., 1698. But the "Countess of Kent" who was the real subject of the evil report, was an earlier lady, viz., Elizabeth, second dau. and co-heir of Gilbert Talbot, Earl of Shrewsbury, and wife of Henry Grey, eighth [seventh] Earl of Kent. The latter nobleman died in 1639, without issue, when the title passed to his cousin, Anthony Grey, ninth [eighth] Earl, the father of Henry the tenth [ninth] Earl, husband of the "*Good* Countess" aforesaid. Elizabeth Talbot was born in or about 1581, and died 7 Dec., 1651, aged 70. John Selden, who is here (let us hope) so unjustly brought under our notice, was the famous patriot and lawyer. . . . Mr. Selden appears to have been employed as solicitor or legal steward to the Earl of Kent, the husband of Elizabeth (Talbot) above mentioned, with both of whom he was necessarily much associated, and lived for many years in the strictest degree of friendship. John Aubrey, the Wiltshire antiquary, a great collector of the rumours of the day, has not omitted to notice that which De la Pryme had heard as to the Countess and Selden. . . . It is difficult at this day, in the absence of any positive testimony, to believe that he was likely to be a party to any shameful intrigue like that suggested. Selden died 30 Nov., 1654, at the Friary House, in Whitefriars, London, which, amongst her other property, he possessed as devisee of the Countess, who, by her will dated 20 June, 1649, and proved 12th Dec., 1651, appointed him her executor and residuary legatee.

Before considering the questions raised by this note, it will be best to let Aubrey † speak for himself, since he is referred to by several writers.

> He (Selden) was quickly taken notice of for his learning, and was solicitor and steward to the Earle of Kent, whose Countesse being an ingeniose woman, * * * * * * After the Earle's death he maried her. He had a

---

* She was married to the ninth Earl, according to the *Dictionary of National Biography*, xxiii., p. 187.

† Aubrey's MSS. in Wood's *Athenæ Oxonienses*, edited by Philip Bliss, London, 1814, iii., cols. 377, 378. It is to be observed that the asterisks and dots in the first extract are Aubrey's own.

> daughter, if not two, by . . . . . . one was maried to a tradesman in Bristowe. * * * * * *
>
> He never owned the mariage with the Countesse of Kent till after her death, upon some lawe account. He never kept any servant peculiar, but my ladie's were all of his command; he lived with her in Ædibus Carmeliticis (White Fryers) which was before the conflagration, a noble dwelling. He kept a plentifull table and was never without learned company.

Consideration of the statements of Wood, Aubrey, De la Pryme and his aunt, shows that Jackson has got into a confusion of his own making, and that there is no reason whatever for importing the name of the 'Good Countess' into the discussion.

In the first place, the diarist has named nobody in the text but the authoress of the *Receipts*; he does not even insinuate any one else. In the second place, Jackson assumes that because "my aunt Prym" spoke of her charity to the poor, therefore *she* meant the 'Good Countess.' There is no foundation for that, and besides there is no proof that the said Countess spent money on "receipts and experiments," whereas the Countess of Kent's *Receipts* are still with us. In the third place, when it is clearly affirmed that the Countess spoken of by "my aunt Prym" was childless, he overlooks the strongest confirmation of his own statement that Henry, the husband of Countess Elizabeth, died without issue, whereas the 'Good Countess' had a family, so that *she* could not have been meant. In the fourth place, he neglects the obvious implication that the Countess Elizabeth was dead, whereas the 'Good Countess' was still living in 1686. In the fifth place, Aubrey does *not* "notice what De la Pryme had heard"; for he says nothing indicative of a "shameful intrigue" at all, but alludes only to a private marriage. If there is any slur on Selden's character, indicated by Aubrey, it is certainly not in connection with the Countess, but with some person unknown. But Jackson takes no notice of that. In the last place, Jackson would seem almost to believe that the *Receipts* were compiled by the Countess Amabella, which was not the case.

Next, with regard to Wood, as he does not refer to any kind of connection between them, I cannot discover whence comes the phrase with which Aikin credits him, and which Johnson styles 'equivocal.'

Aikin emitted it before the publication of Aubrey's MSS. by Bliss in 1814, and he could hardly have known of De la Pryme's version of the story. One thing seems certain that "my aunt Prym" knew nothing of the report current about the Countess Elizabeth, or, if she did, she could still affirm of her personal knowledge and observation that she was an "exceeding good charitable woman."

De la Pryme evidently believed that the Countess described by his aunt was the same as the Countess Elizabeth who was said to have been intimate with Selden. The aunt was as clearly of opinion that the Countess she was describing was the authoress of the book.

If she meant the Countess Amabel, why did she so emphasise the fact that the person she was thinking of was childless; and why did not De la Pryme correct his aunt's misunderstanding, and remind her that the lady whose book he was reading had been dead for five and thirty years, whereas the Countess Amabel was alive?

It is plain, I think, that De la Pryme got his story about Selden and the Countess from a source different from Aubrey's; "as I have since heard" to use his own words, seems to indicate mere gossip, and without more reliable evidence than that, one inclines to consider it the depreciatory inference which is usually, because easily, deduced from misunderstood or imperfectly known facts, and that both the unauthenticated expression attributed to Wood and the grosser fama of De la Pryme are neither more nor less than the low-minded and vulgar interpretation of an unacknowledged marriage emanating from a shallow cynic. In the most recent notice of Selden,* Aubrey's story is stated to be "probably false," but no reason is given in support of that position; and the other reports —whether true or not—are ignored.

It is no advantage to Selden to gainsay his private marriage, when there was obviously a persistent rumour, good, bad, or indifferent, about him and the Countess, and the most reasonable origin and explanation of it is in what Aubrey says. One can sympathise with Jackson's desire to clear Selden from De la Pryme's aspersions; it can be best done by

---

* *Dictionary of National Biography*, li., p. 220.

accepting Aubrey's statement as true. The existence of the two daughters alluded to by Aubrey is much more difficult to explain away.

Walpole, it may be observed, does not mention the Countess at all in the first edition of his book.* The account of her in the fifth edition was inserted by the editor, Thomas Park, mainly from Granger. He quotes the sixteenth edition, 1670 (which I have not seen), and says that the title-page states that the receipts were "collected and practised" by the Countess, and he then adds the following oddly ideaed note:

> This information, if it were given on any authority, would reduce her ladyship to be considered in the present work as a mere transcriber of receipts for making confections and cordials, unguents and distillations; though it would still leave her the more exalted character, of having contributed with Christian condescension to administer to the comforts or the necessities of others.

Did he too confuse the two Countesses?

As to Mr. Park's remarks, I may say that, in the editions I have seen, the 4th, 11th, 12th, 19th, and 21st, it is stated that they were "collected and practised" by the Countess, so that the 16th edition resembles the others in this phrase. I have not seen the first or second edition, but it is possible that the book was issued shortly after her death in 1651, printed from her manuscript collections, so that even in the first edition she would appear as the Countess of Kent, *late deceased.*

But whether or not she was as black as common rumour has tried to paint her does not concern the bibliographer. All that he has to do with is that she was the compiler of a book which was in constant use for at least fifty years, and, as it was not devoid of intrinsic merit, deserved the popularity it enjoyed.

28. In 1635 N. and I. Okes printed an abstract in English of Dupleix's *Curiosité naturelle*, under the title of *The Resolver, or Curiosities of Nature.*† The absolute ineptitude of this book—as of the original—did not deter the same firm from bringing out a companion volume two years later.

---

* *Royal and Noble Authors*, Strawberry Hill 1758, 2 vols., small 8vo.

† Part IV., p. 316. (Reprinted, p. 18.)

Curiosities: *or* The Cabinet of Nature. Containing *Phylofophical, Naturall,* and *Morall* queftions fully anfwered and Refolved. *Tranflated out of Latin, French, and Italian Authors. by* R. B. *Gent.* Never before Publifhed. *London* Printed by *N.* and *I. Okes.* 1637.

It is a small 12mo, A to N in twelves, pp. [24] 287 [1 blank]. Besides the printed title there is an engraved one: Pliny and Plutarch to the left and right; at the top: Natura seated, holding out bare arms, with celestial and terrestrial globes to the left and right; sun, moon, and stars behind. At the bottom: a landscape with mountains, trees, water; and birds, fishes, quadrupeds, serpents, etc.

29. The author, Robert Basset, in a very "conceited" address to the reader, says that this volume was compiled from a French writer mainly, along with a few others. Some of the sections are taken from Dupleix's *Curiosité naturelle,* but it is possible that it is the same author's *Physique* from which he has borrowed the chief part of the book. I have not been able to lay hands on the *Physique,* and cannot say whether this conjecture can be justified or not.

The contents of this book hardly bring it within range of these notes, except in so far as it includes secrets of nature, which it unquestionably does. The secrets of which explanations are attempted remain as such, some of them, to this day; others have been explained, but by a very different method of investigation and with very different results from those of Robert Basset. There is not a single answer to one of the questions propounded that has any intelligible meaning, much less gives an explanation of the natural phenomenon enquired into. Even the questions are generally those of confusion and ignorance, arising from want of perception of what does and what does not lie within man's capacity of investigation.

This book must not be confused either with Dupleix's *Curiosities of Nature,** or with T. Brown's *Nature's Cabinet Unlock'd,* which will be referred to subsequently under the year 1657.

30. Of Gabriel Plattes' *Discovery of Subterraneal Treasure* a fourth

* Part IV., p. 316. (Reprint, p. 18.)

edition has to be added to those formerly described.* Although without date, it must be subsequent to 1712; for, in a note at the end of the preface, there is an account of an explosion of fire-damp which occurred during that year in a mine in Durham, whereby 72 persons were killed.† The edition is, probably, several years later than that event. Unless by the dedication of the first edition being omitted, there is no difference between this and the earlier editions as to matter, but as to form there is marked deterioration.

The 1639 edition is a small quarto of 60 pages, in large legible type, with 26 lines to the page, well leaded out, comfortable and agreeable to read. The 1679 edition is also a quarto, but it has shrunk to 24 pages, and contains 41 lines to the page, in smaller type. This undated edition is still further contracted to 22 pages of 43 lines each, and the pages are crammed even to the verge of repulsiveness. The first edition under Plattes' supervision is an attractive pamphlet; the present is a catchpenny impression got up by the "Booksellers of London and Westminster" for selling—not for reading—and they have made it as nasty as they could for its cheapness. A similar falling off is not unobservable in successive editions of books at the present day.

31. There comes next by date a gathering of receipt books which I lighted on accidentally in the British Museum, very much as I discovered the volume made up of the Italian pamphlets described in the first Supplement. It shows the difficulty of exhausting the tale of these books, especially as there is no clue usually either to the name of the author, even when it is given on the title-page, or to the title under which the book was published, if it appeared anonymously. One becomes acquainted with them merely by chance. Past experience convinces me that there may be collections still to be discovered and described.

---

* Part I., p. 194. (Reprint, p. 15.) Part II., p. 251. (Reprint, p. 41.) Part III., p. 226. (Reprint, p. 41.)

† Oddly enough the date of another book, White's *Hocus-Pocus*, is fixed by the same catastrophe.

32. I shall now give the titles of this new set of pamphlets in order.

The | Ladies | Cabinet | Opened : | Wherein is found hidden ſeverall Expe- | riments in Preſerving and Conſerving, | Phyſicke, and Surgery, Cookery | and Huſwifery. | [Device.]

London, | Printed by M. P. for Richard Meighen, next to the | Middle Temple in Fleetſtreet. 1639. |

Small 4to. Signatures: A to H in fours; or pp. [2] 59 [3].

The device is a winged skull surmounted by an hour-glass resting on a crossed bone and scythe with the motto: Non Plvs, surrounded by a border of scrolls and flowers.

It is a shabby pamphlet, containing receipts for making preserves, jellies, cordials, medicines, for cooking, for perfumes, rather confusedly arranged.

The Ladies Cabinet Enlarged and Opened, 1654, if it can hardly be called a new edition of the above, certainly contains a large number of the receipts. The present, therefore, seems to be the first form of the collection, which was afterwards extended and rearranged by "M. B.," who signs the preface to the later form of the book. I shall consider the question under the edition of 1667.

The present copy is in the British Museum, 1038, i. 35 (1).

33. The next four works deserve a little attention, as the problem of authorship which they present does not yield a solution all at once.

The tracts may be first taken in order, and the question arising out of them then considered.

34. A Precious | Treasury | of | Tvventy | Rare Secrets. | Moſt neceſſary, Pleaſant, and Pro- | fitable for all ſorts of People. | Published | By La Fovntaine | an Expert Operator.

London, | Printed Anno Dom. 1649. |

Small 4to. Sig. A, 4 leaves.

A1, Title leaf. A2 *recto*, Contents. A2 *verso* to A4 *verso*, the Text.

From the copy in the British Museum (1038. i. 35 (2)).

Among the twenty receipts are these: Writing by black or indigo transfer paper; writing secret letters with alum, and juice of lemon; dyeing hair with silver nitrate; perfuming wine by a lemon or orange stuck full of cloves, hung up in the barrel; appeasing toothache by cotton dipped in essence of cloves; taking out grease spots; making instantaneous ink, and so on.

La Fountaine was what is termed a quack, but part at least of his quackery is recognised now as legitimate, and is licensed regularly under the name of Dental Surgery. He undertakes to do what is done by the dentists of the present day.

At the conclusion of his tract (A4 *recto*) he says:

> I have ſome things extraordinary which I cannot communicate to any but them that have neede of my helpe, and will make uſe of me. I have written here ſomething to ſatiffie the curious but I muſt confeſſe they are but trifles in compariſon of my manuall opperations which I preſume to be as expert in as any man in Europe whatſoever he be, as my approbation of ſome of the beſt Doctors of this age ſhall juſtify if need requires.
>
> I glory not in knowing or doing much, but in doing well that little which I undertake, and the things which I undertake are these:
>
> Firſt, I pull out hollow Teeth or roots, though ever ſo ſhor, (*sic*) with wonderfull dexterity, facility, and eaſe.
>
> (A4 *verso*) | I make even smooth, and plaine Teeth, that are extravagant or ſharper, longer or ruggeder than the reſt.
>
> Hollow Teeth that for ſome reaſon you would not have drawn, as the falling in of the cheek, or an invincible apprehenſion or fear, or the like, I can ſtop them ſo that they ſhall neither ake, putrifie, nor ſmell ill.
>
> I have the beſt ſecret in the World abſolutely and infallibly, to make Teeth moſt perfectly white and clean, though they be ever ſo black or ruſty.
>
> I can ſet in Artificiall Teeth ſo well that none can poſſibly know them from naturall ones.
>
> Finally, I cannot chooſe but convince their ignorance that are againſt drawing hollow Teeth or roots, their common fooliſh objection is, that I do it more for my own profit then the patients eaſe. I ſay they may as wel object the ſame againſt all other Arts, Sciences, or Trades in the world, for it is evident and cleare that the drawing of Teeth from others is as neceſſary for them as the profit is for me, by reaſon that all the Art in the world cannot make a periſhed Tooth ſound again, nor hinder it from infecting the next adjoyning teeth in time. Besides, hollow teeth breed often a ſtinking breath, a canker in the mouth, and other diseaſes in the body, and their intolerable aking is able to make a body fall into a dangerous Feaver; all which inconveniences may be ſhunned by having ſuch teeth and roots neatly drawn. Laſtly, if it be true that it is good to take the broken from the whole, or that one ſcabbed ſheep may ſpoyl a whole flock, or that it is good and neceſſary to repaire imperfections of nature with Art; then all that I have ſaid herein is true, and ſo I reſt fully ſatiſfied.
>
> I live at preſent in St. Martins Lane, at the next Belcony to the Signe of the Anchor neere Charingcroſſe, where they that have need of me, may finde me to give them all reaſonable ſatisfaction, God willing.
>
> Per me La Fountaine Operator.

35. In the same year a similar collection appeared, as follows:

A New | Dispensatory | of Fourty Phyficall | Receipts. | Moft neceffary and Profitable for | all Houfe-Keepers in their Families. | Befides three other pleafant arts fit for | young Gentlemen. | *Published by* SALVATOR WINTER *of* Naples. | *An Expert Operator.*
London, | Printed in the Yeare, | 1649.

Small 4to. A, B, in fours; or pp. [6] 9 [1 blank]. From the British Museum copy (1038, i. 35 (4)).

Winter's preface "To the Reader" is worth quoting:

In my manifold Travels through *Europe*, *Asia* and *Affrica*, I have observed and learned many secrets, yea there is scarce anything profitable for mans body, but I have knowledge thereof. Here I do freely present unto you fourty, most necessary Receipts and very useful in every family, whereof I have oftentimes made good proofe; accept of them thankefully, and as you shall have occasion make use thereof, So Farewell.

Salvator Winter.
Operator.

The receipts are entirely medical and surgical, as for sore eyes, deafness, toothache, palsy, to heal a cut, for the squinzie, for the jaundies, for the swelling of the legs, for piles, for the itch, pleurisy, spitting of blood, scrophulus or King's Evil, for the plague, to break a carbuncle, for worms in children, for the ague, falling sickness, a stitch, sore throat, a scald with fire or hot liquor, to ease the gout, for a tympany, "for an Eare wigg in the Eare," for the green sickness, convulsions in young children, to dye hair black, and three receipts for young gentlemen, which are as follows:

To make a stone which catches fire when spit upon. It consists of lime, saltpetre, tutia, calamine, brimstone, and camphor. To make two waters which turn like milk when mixed, viz. Lead Acetate and oil of Tartar, or potassium carbonate. To make a bullet to hit all marks; take pewter, lead and quicksilver—melt together and cast into bullets. When fired they burst and scatter.

At the very end he states that he pulls out "all manner of Hollow Teeth stumps or roots, with great dexterity, and ease without almost any paine. I make Teeth Smooth and White, set in Artificiall ones, that they shall not be discerned from the naturall ones, yea I doe all to Teeth: (without arrogancie and prejudice to others) as well as any whosoever, and I know none shall goe beyond me in that art." Winter lived at that time "in Bedford-street next doore unto the signe of the Foxe over against the New Exchange."

This may be compared with La Fountaine's advertisement. Were they rivals?

36. A third work, subsequent to the preceding, appeared the same year:

> A Pretious | Treasury: | or a New | Dispensatory. | Contayning 70. approved Phyficall rare Receits. | Moft neceffary and needfull for all forts of people to have | in their Families. | Collected out of the moft approved Authors, both in Phyfick and Chyrurgery: | by *Salvator Winter*, and Signieur *Francifco Dickinfon*, expert Operators. | *in two Bookes.* | [Vignette.]
>
> London, Printed by *Tho. Harper* for *Richard Harper*, and are to be fold | at his fhop in Smithfield, at the Bible and Harpe. 1649. |

Small 4to. A and B in four, C2, or pp. [1-6] 7-20. From the British Museum copy (1038, i. 35 (3)).

The vignette is a quack doctor on a platform handing out doses to the public below. He says: "Me cure all Difeafes." His servant has a small pamphlet and a box of pills or some sort of drug, and says: "Your money Gent." The costumes are of the time.

The first book by Winter consists of receipts for some very common complaints, and is identical with the preceding *New Dispensatory*, with the same address to the reader and the same short statement about teeth at the end.

The second book, claimed by Francisco Dickinson, is in reality a reprint of La Fountaine's *Precious Treasury of Twenty Rare Secrets* of the same date, only the long advertisement about teeth is omitted. This volume, therefore, consists of the two preceding pamphlets printed together, with La Fountaine's name changed to Francisco Dickinson.

37. The fourth collection is in the Museum (1038, i. 35 (5)), and is by La Fountain.

> A Brief | Collection | of many Rare | Secrets. | Moft of which are approved, | and Phyfical, and the reft moft | Pleafant and Recreative. | Very neceffary for all forts of People. | Newly fet forth and publifhed by *La Fountain* | an Expert Operator, living over againft *Effex* | Houfe in the *Strand* near *Temple-bar*. [Device of a burning heart.]
>
> *London*, Printed *Anno Dom.* 1650.
>
> Small 4to. A in four, B in two. No pagination. Very closely printed. The note to the reader is signed Edw. Fountain.

The secrets, which are quite different from those in the above two issues of the *Pretious Treasury*, are medical, cosmetical (to take out wrinkles and spots from the face, etc., to dye hair, to remove hair, to grow a beard), to make perfumes, some chemical experiments, the tree of Diana, to melt a metal, etc.; and once more his dentistry advertisement in full.

In his note to the reader, he speaks of the general acceptance of his little *Book of Twenty Secrets*, which has obliged him to present this which is "farre better"; and from what he says, it appears that this was sold for twopence.

In this *Brief Collection*, B1 *recto*, receipt No. xxxv. shows how to remove hair and prevent it growing again, by laying on the place a piece of the finest gold, heated red hot, and then anointing the place with oyle of roses, or of violets. He adds: "The Author hereof found this Secret, much practised among the Moores, and since used it much himself throughout Europe, to many Ladies and Gentlewomen to their great content, and to his own Fame and Credit." He does not explain that the author he means is not La Fountain but is Alexis!! in whose work it will be found, Part I., English Translation, 1558, f. 80 *recto*, dated 1521.

38. Looking at these results, it is a question if the names La Fountaine, Salvator Winter, and Francisco Dickinson denoted different people, or were aliases assumed by one person for professional purposes. That La Fountaine and Winter were distinct might be inferred from their having consulting rooms at different places, and from their advertisements indicating that they were rival dentists. But that may have been an essential part of the alias. That they were not rivals, however, may be inferred from their tracts being printed together, and one of the dentistry advertisements being dropped. That La Fountaine and Dickinson were identical is made certain by the identity of the tracts assigned to each, and by no objection being raised by La Fountaine in his *Brief Collection* in 1650 to Dickinson's appropriation of them.

It is curious that the first and last tracts by La Fountaine and that by Winter have no printer's name. This may be a mere coincidence, but when taken in connection with their other peculiarities, it seems rather to point to their having been issued by the same individual. But why suppress the printer's name unless there was something questionable in the transaction?

Transformation of name was not unknown to one at least of these reputed writers. In the 1650 tract, though the title-page bears the name of La Fountain, the prefatory note is signed Edward Fountain. I have doubts as to the Neapolitan origin of Mr. Winter, and still more as to the nationality of Signieur Francisco Dickinson, who, I take it, was none other than Edward Fountain himself. The good people of the sixteenth century must have been indifferent, or uncritical, if so transparent a bit of mystification could be passed upon them for any reason whatever.

Anyhow, whether these authors are to be identified or not—and there may be a doubt about La Fountaine and Winter—there can be no question that there was juggling with their respective writings somewhere. It is amusing to get them together, for otherwise this curiosity of authorship could hardly have been detected. It may be observed that these names are not mentioned by Lowndes.

39. One more hand-book of medicine for household use has to be mentioned. Of the author, Daniel Border, I have found no notice. He is not included in the *Dictionary of National Biography*.

ΠΟΛΥΦΑΡΜΑΚΟΣ και ΧΥΜΙΣΤΗΣ: | Or, the Engliſh unparalell'd (*sic*) | Physitian | and | Chyrurgian: | Shewing | *The true Vſe of all manner of Plants and* | *Minerals.* | In which is Explained, *The whole Art and Secresy of* | Phyſick and Chyrurgery. | wherein is contained, |

1. *The cure of any Internal Diſeaſe, which hath taken effect, and* | *brought forth Symptoms, either outwardly or inwardly, in the* | *Head, Stomack, Back, Belly, etc.* |
2. *The making of diverſe rare and excellent Balſumes, Oyls, Plai* | *ſters, and Waters, with the effects thereof.* |
3. *The making and uſe of Purges, Vomits, Drinks, and Cliſters,* | *according to Rules of Phyſick and Aſtrologie.* |
4. *The nature and temperarure* (sic) *of Herbs, Plants, and Vegetables,* | *and the uſe thereof, and the Influence of the Planets as well over* | *Herbs and Plants, as over the Bodies of Men.* |
5. *The way to extract the Spirits and Quintaeſſences of ſeveral ſorts* | *of Herbs and Minerals, and the rare Operations they have.* |
6. *The making of Salt of Wormwood, Balm, or any other Herb.* |
7. *The Cure of any Wound, Contuſion, Puncture, or Vlcer in any part* | *of the Body.* All which have been preſcribed and practiſed by ſe- | veral Learned Doctors and Chyrurgians. |

By D. BORDER, *Practitioner in Phyfick.* |
London, Printed by *B. Alfop,* and are to be fold at his | houfe in Grubftreet, near the Upper Pump. 1651. |
Small 4to. A, *, B to T in fours; or pp. [16] 144.

Black letter, but with a great deal of roman interspersed for headings to chapters, etc.

Though the actual word "Secrets" does not occur in the title, but only the very unusual form "Secresy," it is used by the author several times when referring to different sections of his receipts. It is an illustration of the difficulty of excluding a book of the kind from the category, even if it wants the name. The title gives as full an account of the book as is necessary here. From the copy in the British Museum (1038, i. 35 (6)).

The next supplement will deal with English books subsequent to 1650.

BIBLIOGRAPHICAL NOTES

ON

HISTORIES OF INVENTIONS

AND

BOOKS OF SECRETS

FIFTH SUPPLEMENT

BY

JOHN FERGUSON, LL.D., F.S.A. LOND. AND SCOT.

REGIUS PROFESSOR OF CHEMISTRY IN THE UNIVERSITY OF GLASGOW

GLASGOW

Printed at the University Press by

ROBERT MACLEHOSE & CO. LTD.

1908

Read to the Archæological Society of Glasgow, 21st December, 1905.
100 *Copies Reprinted.*

No. X.

BIBLIOGRAPHICAL NOTES ON HISTORIES OF INVENTIONS AND BOOKS OF SECRETS.

(FIFTH SUPPLEMENT.)

BY

JOHN FERGUSON, LL.D., F.S.A.

(Read at a Meeting of the Society, held on Thursday, 21st December, 1905.)

1. Since my last communication to the Society in 1898 about the older literature which treats of the wonders of the world, of the practical arts, and of the secrets of medicine, I have had some experience of that policy of waiting which is said to be so remunerative to the man who can carry it out. Everything, certainly, has not come to me, and I do not expect that it ever will; nevertheless, visits—they might even be called, not inappropriately, angels' visits—have been paid me by a few very scarce books, which would deserve notice if for no other quality than their scarcity. But I have found, what is perhaps of even more importance, that quite a number of persons are now interesting themselves in the literature of receipts and secrets. I suppose we must regard this phenomenon as further evidence of that wave of science which is advancing on every hand.

2. Set aside illogically by the self-styled practical man, the maintainer of rule-of-thumb methods, as of no further use, never in the eyes of the

literary critic of any significance or importance at all, beneath contempt to the modern "professour of Phisicke and Chyrurgerie," of no esteem even as curiosities, these books had dropped out of mind, out of the booksellers' catalogues, almost out of existence.

One reason perhaps of this neglect has been that, for the understanding of them, technical familiarity with the topics of which they treated was required, so as to recognize what was actual matter of fact in them, accurate to this day, and what was erroneous and had to be amended. The ordinary reader is not likely to possess this familiarity; the literary student certainly is not, and he will find nothing in the subject and no elegancies of style in the books to appeal to him, and the modern technologist and scientific man are too engrossed in the manufactures and science of the moment, to trouble themselves with the supposed rudimentary knowledge and defective applications of older epochs. But, in this section of antiquarian research, it is precisely technical insight that is wanted. No one but the modern metallurgist can describe the metallurgy of the past; it requires the skill of a dyer to say how the old fabrics were tinted; it is the physician alone who can interpret the cures employed in Babylonia and Egypt, recorded on clay tablets and papyrus slips; only the botanist and pharmacist can place a value on the work of Macer, or on the herbals of Dodoens or Gerard, or even of Lovel.

3. Now, some persons having found out that our predecessors had practical arts, were not so ignorant as is supposed, could make beautiful and useful things, and described their methods in books more or less fully and correctly, have shown a desire to procure the books, and this is what has been recently brought home to me. A new class of collectors has arisen with the scientific attainments necessary to comprehend the subjects, with curiosity prompting them to inquire into what was the state of the practical arts in previous times, with enough historical and antiquarian taste and skill to perceive that there must be a connection between that stage of development and our own. Such people, therefore, who possibly never thought of book-collecting before, have discovered in the literature of early science and technology a

source of interest and an attraction which had failed them elsewhere. Anyhow, one result is that, while the books are diminishing rapidly in number, the demand and the competition for them are increasing, and, in consequence, their value in the market is considerably—in certain cases, enormously—enhanced. Prices of books which I know have increased five, ten, even twenty fold what they were but a few years ago.

4. Like many other sections of literature, this one too seems to be drifting into American colleges and libraries—for their appetite is insatiable—and soon there will be little of it to get. Such a collection as has been described in my previous papers could not be formed now without years of waiting and watching, if indeed it could be got together at all.

The little volumes on medicine and the arts, which were published and circulated in such numbers some two centuries ago, have been long since worn out and superseded, so that it is only a chance survivor that comes now and again in one's way, and though the specimens to be described in what follows may appear to be fairly numerous, they amount to but a fraction of what at one time were in circulation. There seems all the more reason, therefore, for putting on record the books of which one can authenticate the existence, before they too disappear.

5. But let there be no mistake. It is not for any beauty or merit in them as books that they are submitted for consideration. They belong to one of the worst periods of book production, notable for bad paper, wretched type, crude illustrations, binding of brown skiver without ornament, without a title, without even a bit of paper to line the covers. Nor is it because they are of interest as rarities and curiosities, which they are undoubtedly. It is really on account of their contents, and for the reason that they are items in the history of the arts, and afford data to the antiquary or historian who wishes to know what were the conveniences of life two or three centuries ago. How did people live? What was their food and how did they prepare it? What were their luxuries, if they had any? What was the ordinary treatment of disease? What did people do who lived in the country and had little chance of getting medical advice, even supposing they could meet the cost of it, and yet

had to face and treat illness in their own families and among their neighbours, and who, in case of mishaps and accidents, had to render what goes at present by the name of "first aid," though without the advantage of the instruction which an "ambulance class" now supplies? How did they deal with the plague, which hung like a dark cloud over Europe and a visitation of which was always anticipated and, as far as possible, provided against? It was to these books they had to go for advice in cases of illness; for directions as to what external applications were considered effective; what diet and drugs were necessary; what plants and herbs were to be employed, when they were to be collected, and how they were to be prepared. The precautions to be taken during the plague, which are detailed in some of the books, throw a lurid light on the unsanitation of the sixteenth and seventeenth centuries; and the wonder is that the plague was ever absent from London, considering the filth of the place and the habits of the people.

Then, as to the practical arts, how did they extract the metals from their ores, alloy them, and use them? What were the colours which the painters employed for the masterpieces of art that remain, while the work of some recent men has become blurred and unintelligible? How did they dye their silks and velvets and fine cloths, and tan and colour leather? How, without scientific understanding of the composition and reactions of their materials and with empirical skill alone, did they produce the enamels, the glass, the imitation gems, which are now the delight of the connoisseur and the envy of the workman?

6. In these, and analogous topics, there is no lack of matter, both general and special, bearing upon the history of civilization, matter as indispensable in its way as the cathedrals and churches upon which so much study and criticism are lavished to the exclusion of other objects. There is a research for the antiquary in a coat of mail no less than in a Norman castle, for, when it comes to what is the essence of all the inquiry, it is not the steel, or the stone and lime, but the human agency involved; and I, for my part, am of opinion that the works of the smith and the dyer, the pharmacist and the physician, merit fully as much attention as those of the mason, even with all his marks thrown into the bargain.

I do not see, therefore, that there can be any question as to the importance and value of these shabby little volumes for illustrating the illnesses, beliefs, superstitions, ignorance, resources, arts and processes, the material details of daily life as it was passed two or three or more hundred years ago, and I welcome another opportunity of introducing a new and interesting group of these volumes to the Society's notice.

7. In the previous paper were included books of receipts in English, dating from 1550 to 1650. It was my intention to continue the investigation by describing those from 1650 to the close of the seventeenth century which I have seen; but as a considerable number of books which belong to the earlier period have come quite recently under my notice, it will be best to dispose of them before going on to those of a later date. Two or three of them have been alluded to before, but the fact that those here are the only copies for sale within my experience, demonstrates their rarity. With a few exceptions these older works happen to deal with medical secrets, more particularly for the use of country folk who had to look after their own curing. One gathers again from some of them that medicine was in the hands of the women of a household, and it is not difficult to see how this came about. The men were otherwise engaged, perhaps in outdoor labour, or in sport, or in handicrafts of all sorts, and could not relinquish their avocations to turn to doctoring or nursing. It was essentially a domestic affair —people who are ill cannot go afield—and the duty of attending to the sick fell naturally to the women at home. Moreover, by having to look after their own health as well as that of the children, they had gradually acquired some actual, if empirical and traditional, experience in the healing art. It was not inappropriate, therefore, for Owen Wood, in compiling his book of medical secrets, to have remembered that it was "for the help of such Ladies and Gentlewomen who of charity labour to do good."

8. The following new list, then, comprises books of natural history, and of receipts in medicine and surgery, pharmacy, husbandry and housewifery, pyrotechny, and practical arts of various kinds. It is arranged, as far as

possible, according to dates, even should that order sometimes involve a separation of books concerned essentially with the same themes.

9. In 1633 a small book appeared under peculiar circumstances. The title runs as follows:

A | Storehouse | of | Physicall and | Philosophicall | Secrets. | Teaching to diſtill all manner of Oyles | from Gummes, Spices, Seedes, Rootes, Hearbs, | and Mineralls, &c. With their ſeverall vertues, | out of ſundry approved Authors. | *The first Part.* |
London, | Printed by Thomas Harper. | M.DC.XXXIII.

Small 4°, pp. [4] 57 [1 containing the colophon; 2 blank]. The second part, p. 33, has no separate title-page.

This tract is appended to the surgical works of John Banester, published by Thomas Harper in 1633, 4°, and, as it is mentioned on the title-page of that collection, it obviously forms an essential part of the volume. So far as the general and special titles inform us, this might have been quite well the first publication of the tract, but examination proves that such is not the case, and that the history of the collection is less simple than might be supposed, as will be seen from the following account.

10. As early as 1575 appeared a little volume on the distillation of oils, etc., of which a description may be given.

The True and | *perfect order to diſtill oyles* | out of al maner of ſpices | ſeedes, rootes, and gummes | with their perfect taſte, ſmel, and | ſauour: where vnto is added | ſome of their vertues | gathered out of | ſundry auc- | thors. As Gualterius, Riffius, Guinthery An- | dernaty, Phillipus, Hermanus, Leo- | nardo, Phirauante, Phallo- | pius, Cardanus. |

Small square 8°, signatures, A to D in fours, or ff. [16]. At the end is this advertisement:

Theſe oyles with a number of others are | to be ſolde at the ſigne of the furneſeies | adioyning to Powles wharfe at the | dwelling houſe of John Heſter, | practiſſioner in the arte of | diſtillacions. | 1575.

No place [London?] or printer's name. Black letter, except the first two and last four lines of the title, the headings of the sections, the initials and proper names in the sections, which are in roman. This little tract is by John Hester. It will be noticed that the spelling of the proper names is peculiar, and it may be also noticed in passing that there should be no comma after Gualterius, Phillipus and Leonardo.

11. This book is mentioned by Cooper,* but it does not occur in any other catalogue that I am acquainted with. On Cooper's authority I included it in my list † of Paracelsus' works in English, but, as I had never seen it, I was unable to say anything about it, except that it must be excessively rare. It is not referred to by Ames, Herbert, or Dibdin; it is omitted by Watt and Lowndes, and, so far as I have been able to ascertain, there is no copy in the British Museum. I have been long disposed, therefore, to doubt the existence of this edition, though Cooper I have always found to be accurate and reliable. Still, the want of a certified or actual example made me hesitate. I do so no longer, for only last Friday, not a week ago, a copy of this slender tract came most unexpectedly into my possession, and the preceding is a description of it, the first ever made from the book itself, I believe. Evidently, something does come to him who waits, but in this instance the delay has been protracted almost beyond hope and expectation. I suppose the present copy is unique in the meantime, the sole representative of the edition. It is just three hundred and thirty years old.

About this edition certain things may be specially observed. 1°, It consists of one part only; 2°, it is not called "The Key of Philosophy," as Cooper's notice of it would imply; 3°, among the authorities from whom the receipts are taken, Paracelsus is not included, and indeed his name occurs only once in the text. The authorities whose names do appear on the title-page are well known, and some of them have been referred to on previous occasions, but in date they are a few years later than Paracelsus, except Riffius, who was a contemporary. These points deserve remark in view of the later editions of the work.

---

* Will. Cooper, *A Catalogue of Chymicall Books*, London, 1675, sigs. *C* 2 *recto*, The Key of Philosophy, 1575, 1580, 1633, 8°, and the second part 1596, 8°; *C* 4 *verso*, Paracelsus, Secrets of Physick and Philosophy, 1633, 12°; *D* 3 *recto*, The Store-house of Physical and Philosophical Secrets, 1633, 4°; Q 4 *recto*, Paracelsus, The Key of Philosophy, 1580, 1633, 8°; R 2 *verso*, The Store-house . . . . 1633, 4°. [*See* "Addenda," § 11.]

† *Bibliographia Paracelsica*, 1877, Part I., p. 27, No. 12; p. 29, Nos. 14, 16; 1890, Part III., p. 32, No. i.; p. 33, No. iii.; p. 38, No. vi.; p. 40, No. ix.; 1892, Part IV., pp. 20-23.

12. Five years elapsed, and then the second edition was published. The title was modified, the text was rearranged and enlarged with new receipts, a second part about minerals, etc., was added, and the authorship was assigned to Paracelsus. The title is as follows:

> The firft part of the Key | of Philofophie. | . . . deuided into | twoo Bookes. | In the firft is fhewed the true and perfect | order to diftill, or draw forthe the Oiles, | of all maner of Gummes, Spices, | Seedes, Rootes, and Hearbes, | with their perfect tafte, | fmell, & vertues. | In the feconde is fhewed the true and perfect | order to prepare, calcine, fublime, and diffolue | all maner of Mineralles, | . . . Firft written in | the Germaine tongue by . . . Theophraftus Parafelfus, and now pu- | blifhed in the Englifh tongue | by Ihon Hefter . . . | . . . | 1580. | At London | Printed by Richard Day, . . . | . . . |
>
> 16°, sigs. *, A to G, in eights; or ff. [64].

This was followed by a third edition, printed at London, in 1596, by Valentine Simmes, in 16°, pp. [16] 111 [1 blank]. It is in the British Museum.*

13. Now, when we come to the year 1633, we find two different editions. One of them is thus entitled, after the foregoing model:

> The Secrets of Phyfick and Philofophy Divided into two Bookes: In the firft is fhewed the true and perfect order to diftill, or draw forth the Oyles of all manner of Gummes, Spices, Seedes, Roots, and Hearbs, . . . In the fecond is fhewed the true and perfect order to prepare, calcine, fublime, and diffolue all manner of Mineralles, . . . Firft written . . . by . . . Theophrastus Parafelfus, and now publifhed . . . by John Hester. . . .
> London, Printed by A. M. for William Lugger, . . . 1633.
>
> 12° pp. [24] 196 [16].

14. The other edition of 1633 is that which is described in § 9. What reason Harper could have had for printing this tract under a different title, without any allusion to Paracelsus, the alleged author, or to John Hester, the nominal translator, or to the fact that the book had appeared three times some forty to fifty years previously, is not at all obvious. He could hardly have meant to ascribe it to John Banester, and, from the way he speaks in the preface, he did not claim the authorship for him-

[* March 1, 1907. I have now got a complete copy of this edition.]

self. For, in his note to the reader, he says: "What experiences the learned author with much coft, and great trauaile, hath digged out of hard ftones, blowne out from hot fire, raked out from foule afhes, that haft thou in this Booke to fee, and with great eafe to ufe at thy command." But he does not name the "learned author," who is Hester, and he does not say that the words just quoted are borrowed from Hester's "epiftle to the Reader."

Curiously enough, Cooper, who quotes the 1575, 1580, and 1596 editions of the *Key of Philofophie*, mentions the 1633, 12°, edition of it under Paracelsus, but puts *The Store-house of Physical and Philosophical Secrets*, 1633, 4°, under *Storehouse*. At the same time, however, he gives cross-references to *Paracelsus* and to the *Key of Philofophie*, showing that he was aware of some connection between the two works, although he does not indicate what it really was, to wit, their identity.

As to which of these two issues of 1633 came out first I have no knowledge. In the preface of the 4° edition the last of July is mentioned, but whether that be of 1633, or not, is not stated. The 12° edition contains no guiding date.*

15. The next book is one of a group of which several have been exhibited to the Society from time to time. The volumes will be found under the names of Andree or Andriessen, *Ettliche Künste*, Petrus Kertzenmacher, *Kunstbüchlin*, and *Rechter gebrauch d'Alchimei*.†

That which I have to show is of interest, as it deals with some of the practical subjects which have just been spoken about. It has the following title:

> A profitable boke | declaring dyuers approoued re- | medies, to take out fpottes and ftaines, in Silkes, | *Veluets, Linnnen* (sic) *and Woollen clothes.* | *With diuers colours how to die Vel-* | vets and Silkes, Linnen and Woollen,

---

* For details about these editions, 1575, 1580, 1596, and 1633, see my *Bibliographia Paracelsica*, above quoted. Sudhoff, *Bibliographia Paracelsica*, 1894, p. 291, 1575, No. 168; p. 320, 1580, No. 184; p. 422, 1596, No. 240; p. 561, 1633, No. 356, 12° edition.

† *Transactions of the Glasgow Archæological Society*, N.S. 1890, I., pp. 454-455 (Reprint V., p. 38).

Fuftian | *and Threade.* Alfo to dreffe Leather, and to co- | lour Felles. How to Gylde, Graue, Sowder, and Ver- | nifhe. And to harden and make fofte | Yron and Steele. | Very neceffarie for all men, fpeciallye for those | *which hath or fhall haue any doinges therein: with* | a perfite table herevnto, to fynde all | thinges readye, not the like reuealde | in Englifh hereto- | fore. | Taken out of Dutche, | and englifhed by *L. M.* | ¶ *Jmprinted at London by* Thomas | Purfoote, and William Pounfonbie. | 1583 |

Small 4°, pp. [2] 78 [6] [2, probably blank, wanting]. Black letter; but the title (except two lines in black-letter and half-a-dozen in italics) and the headings of the sections and paragraphs are in roman.

Another edition was: "Jmprinted at London by Thomas | Purfoot, dwelling in the newe | Rents. 1596." | It is a small 4°, pp. [2] 78 [6, 2 blank] and is in black letter, except the title-page and headings of sections and paragraphs. The title is the same, and the book is a page for page reprint, but it is not quite so well done. The spelling also is modernized in places.

There was a third edition: "Imprinted at London by Thomas | Purfoot, dwelling within the new Rents, in S. Nicholas Shambles. 1605." | It is a small 4°, pp. [2] 78 [6], and it shows a farther decline in paper and printing. [*See* "Addenda," § 15.]

16. These editions were enumerated, and an analysis of their contents given along with that of the other members of the set, in a paper I read to the [Royal] Philosophical Society of Glasgow in 1886.* All three are valuable. The first and third were then described from copies in the British Museum, where they are specially reserved on account of their rarity. Of the second, which was not in the Museum, I happened to have a copy. From that time—that is, for twenty years past—I have seen nothing more of these books till last August, when I acquired the copy now described, dated 1583, which is of the first edition.

As the work purports to be, and is indeed, a translation from the Dutch, I have tried to ascertain the source of the receipts, but without success. Some of those in the second part, about softening iron and steel, etc., correspond to sections in the *Kunstbüchlin*, 1537, and in the first tract of

---

* *Proceedings of the* [*Royal*] *Philosophical Society of Glasgow*, 1888, XIX., pp. 126-159.

Andriessen's *Kunst-Boeck*, 1549; but of the first part, relating to the removal of spots and stains, etc., I have found no Dutch or German original, although there is a general resemblance to the receipts which are contained in several of the collections.

17. The translator, who uses the initials L. M. only, was, it is generally agreed, Leonard Mascall, of whom notices are given by Fuller,* Tanner,† and others, and a summary of them by Thomson Cooper.‡ According to these authorities, he belonged to an old family settled at Plumsted, in Sussex; was the first, says Fuller, "who brought over into England from beyond the seas Carps and Pippins . . . about the fifth year of the reign of King Henry the Eighth, Anno. Dom. 1514," though Cooper and others call this statement erroneous. The "Registrum parochiae de Farnham Royal comit. Buckingh." was written "per me Leonardum Mascall generosum, clericum coquinae de hospitio reverendissimi patris dom. Matthaei Cantuar. archiepiscopi. 25 Jun. A.D. MDLXXIII."; that is, he was clerk of the kitchen in the *hospitium* of Matthew Parker, archbishop of Canterbury. Tanner adds that he died and was buried at Farnham Royal, 10 May, 1589, but Fuller says: "The time of his death is to me unknown."

Donaldson's § account of Mascall is different. He says that nothing is known of his parentage, birth, education, or general employments: he only relates of himself that he was chief farrier to King James, and dedicated his work [on cattle] to Sir Edward Montague Knight. His practical knowledge, adds Donaldson, "relates chiefly to diseases, with a small notice of the animal and its breeding, but such as it is, a large advancement was made by it towards an improved practice." Donaldson also says that Mascall "has always been reckoned a genius in that department of human industry," meaning agriculture.

---

* Thomas Fuller, *The History of the Worthies of England*, London, 1662, II., p. 113.

† Thomas Tanner, *Bibliotheca Britannico-Hibernica*, Londini, 1748, p. 517.

‡ *Dictionary of National Biography*, London, 1893, XXXVI., p. 404.

§ John Donaldson, *Agricultural Biography: containing a Notice of the Life and Writings of the British Authors on Agriculture, from the earliest date in* 1480 *to the present time*. London: printed for the Author. 1854. 8°, p. 10, No. VIII.

The books which appear under Mascall's full name, or his initials, relate to country life and affairs, as to cattle, fishing,* poultry, planting and grafting of trees, all which seem to have been in request, judging by the number of editions of them which appeared, and which are enumerated by Herbert † and by Thomson Cooper.‡ Most of them are in the British Museum. They show that Mascall was a forerunner of Gervase Markham and others.

18. Inspection of the different notices, supposing that they are individually accurate, reveals certain discrepancies which cannot be reconciled with the assumption that there was only one person of the name of Leonard Mascall.§ If Fuller's date, 1514, be correct, it is hard to see how the horticulturist could be King James's farrier. If he died in 1589, how did he publish his book on cattle in 1596? There is nothing in the book to indicate that it was issued by anybody but its author. The epistle "to the Right worshipfull Sir Edward Montegue Knight," the notice "to the Reader." duly signed L. M., and verses "To the Husbandman," also initialled, are all in order and contemporaneous, and there is not the slightest indication that the book was published after the author's death.|| There is evidently opportunity for research into these conflicting dates and statements. [For more about Mascall, *see* "Addenda," §§ 17-20.]

* J. Russell Smith, "A Bibliographical Catalogue of Books on Angling," in Robert Blakey's *Historical Sketches of the Angling Literature of all Nations*, London, John Russell Smith, 1856, p. 320; T. Westwood and J. Satchell, *Bibliotheca Piscatoria*, London, 1883, p. 137. *Athenæum*, 5 July, 1884, p. 9; review of Satchell's reprint of L. M.'s "Booke of Fishing."

† William Herbert, *Typographical Antiquities*, London 1786, pp. 782, 784, 947, 990, 998, 1018, 1182, 1186, 1730.

‡ *Dictionary of National Biography*, as above.

§ This difficulty had been already observed by Edward Solly, *Notes and Queries*, 6th S. IX., p. 178, March 1st, 1884.

|| [Nov. 1906. I have a reprint of this work: The Government of Cattel. Divided into three Books. The first, Treating of Oxen, Kine, and Calves: . . . The second, Discoursing of the Government of Horses; . . . The third, Discoursing of the Order of Sheep, Goats, Hogs, and Dogs; . . . Gathered by Leonard Mascal. Chief farrier to King James. London, Printed for John Stafford, and W. G. and are to be sold at the George-yard near Fleet-bridg. 1662. Small 4°, pp. [8] 307, [2, 1 blank]. A1 is an engraved

19. Two of the treatises assigned to Mascall may be quoted here: one, because it is a rare book and is said to be the first on its subject in English and is interesting besides; the other, because it is an adaptation of a well-known medieval treatise on the Materia Medica.

¶ The Huf | bandlye ordring and | *Gouernmente of* | Poultrie. | Practifed by the Lear- | nedfte, and fuch as haue bene | knowne skilfulleft in that | Arte, and in our | tyme. |

¶ Imprinted at London, | by Thomas Purfoote, for | *Gerarde Dewfe.* | 1581. |

Small 8°, sigs. A in four, B to K in eights, L in four. Black letter; but the title-page, dedication, head lines, titles of the sections, and index, are in roman. The note to the reader is signed L. M., 1581. The dedication is headed as follows: "To Miftriffe Katherine | VVoodford (vvife to maifter Iames | Woodforde Efquier) and cheefe Clarke | of the Kitching to the Queenes Ma- | ieftie Leonarde Maftall (*sic*) vvifheth a profpe- | rous eftate in health long to | continue. |

'Mastall' here is presumably a misprint.

A list of the contents of this book is contained in Donaldson's sketch.

The dedication is quite appropriate as coming from the "clarke of the kitching" of the Primate. There is a copy in the British Museum, 1257. a. 3.

20. The other book is as follows:

Prepofitas his Practife, | A Worke very ne- | ceffary to be vfed for the better | preferuation of the Health | of Man. | Wherein are not onely moft excellent and approued Medi- | cines, Receiptes, and Ointmentes of great vertue, | but alfo moft pretious Waters, againft | many infirmities of the body. | The way how to make euery the faid feuerall Medicines, | Receiptes, and Ointmentes. |

With a Table for the ready finding out of euery the Difeafes, | and the Remedies for the fame. |

Tranflated out of Latin into Englifh by L. M. |

LONDON | Imprinted by Iohn Wolfe for Edward White, dwelling at | the little North doore of Paules, at the Signe | of the Gunne. 1588. |

Small 4°, sigs. A to Q in fours (with the inserted leaf extra); or pp. 111 [17]. Black letter; but lines in the title-page, the head lines, and titles of the sections, are in roman. Vignette of a fleur-de-lis on the title-page. Between A1 (the title) and A2 is inserted a leaf, sig. ¶2, containing an address "To the Reader." There is a copy in the British Museum, 1038. k. 34. (1.).

---

title, or frontispiece; M4, the last leaf, probably blank, is wanting. The book is printed in black letter, but with a good deal of roman interspersed. There are indices and title-pages to each part, included in the pagination. There is nothing in this book to indicate that it is not the first edition. It contains the dedication to Montegue, the address to the Reader, and the verses to the Husbandman, all initialled L. M., but they are not dated.]

L. M. here is again considered to be Leonard Mascall. This is a Materia Medica, which describes the materials and the mode of preparing them for use in medicine. A register of diseases, with references to their appropriate medicines, is given at the end. It is a counterpart to other books, in which the treatment of the diseases themselves is the main theme.

21. Of Præpositus,* called Nicolaus, nothing is definitely known. He has been placed as early as the sixth century and as late as the sixteenth, and, according to some, he received his name as head of the School of Salerno in the twelfth century. The 'Dispensarium ad aromatarios' ascribed to, but probably not by, him, was printed in the fifteenth century, and subsequently passed through numerous editions. A better authenticated work by him is the "Antidotarium parvum."

The connection of L. M.'s tract with the *Dispensarium* is quite close. It is a much contracted translation of the second division of Nicolaus's book, or a collection of prescriptions taken from it, and it is arranged alphabetically in the same way. The chief difference is that L. M. describes the uses and virtues of each prescription, and, at the end, has added a dozen pages of "Approved Medicines, most of them practised by the Translator hereof." There are two alphabetical tables, one of the medicines, the other of diseases, with the appropriate remedies. How does this practice of medicine tally with his farriery, cookery, and arboriculture?

22. At the close of the second supplement (read 1894) † I devoted a paragraph to the consideration of a popular book on medicine: Vicary's *Englishman's Treasure*, published as early as 1586, afterwards edited by Bremer, and I exhibited one or two copies in my possession, and enumerated others among which was that dated 1633. I can now show a recently obtained and remarkably clean copy of that edition, in the original vellum wrapper. It corresponds with the account formerly given; it is in 4°, pp. [8] 264 [8], and is printed in black letter.

---

* [See my *Bibliotheca Chemica*, 1906, ii., p. 223; Schelenz, *Geschichte der Pharmazie*, 1906, p. 193.]

† *Transactions of the Glasgow Archæological Society*, N.S. 1899, III., p. 211; (Reprint, p. 211). [*See* also "Addenda," § 22.]

23. In the second half of the sixteenth century there lived Peter Levens, or Levenes, of whom Anthony à Wood * gives this account:

> Peter Levens, or *Levins*, was born at, or near, *Eske* in *Yorkshire*, became a Student in the University, *an.* 1552, was elected Probationer-Fellow of *Mag.* Coll. into a *Yorkshire* place 18 *Jan.* 1557, being then Bach. of Arts, and on the 19 *Jan.* 1559 was admitted true and perpetual Fellow. In 1560 he left his Fellowship, and one *Thomas Dunne*, M.A., succeeded him, but whether our Author *Levens* proceeded in Arts, or took a degree in Physic, or was licensed to practise that Faculty, it appears not in our Registers. Afterwards he taught a Grammar School, and practised Physic, which is all I know of him, only that he wrote and publish'd these things following,
>
> A Dictionary of *English* and *Latin* Words, *&c.* the *English* going before the *Latin*; necessary for Scholars that want variety of words, and for such as use to write in *English* Metre. *Lond.* 1570. in 18 sh. in qu.
>
> A right profitable Book for all Diseases, called the *Pathway to health*; wherein are most excellent and approved Medicines of great virtue; as also notable Potions and Drinks, and for the distilling of divers Waters, and making of Oils, and other comfortable receipts, *Lond.* 1587. qu. The Author is stiled in the
> *Clar.* title Page *Master of Arts of* Oxon, *and Student in Physic and Chirurgery.*
> 1588. This Book was afterwards several times printed with Corrections; and one Edition came out at *Lond.* 1664. what else *Peter Levens* hath written I cannot yet tell.

In the *Fasti* Wood adds:

> 1556.
>
> July 6. *Peter Levens* of *Magd.* coll. was admitted [B.A.] the same Day. He was afterwards an eminent Physician.

and again:

> 1559.
>
> Feb. In the said month of *Feb.* did supplicate for the said Degree [M.A.] *Pet. Levens* of the same Coll. but whether admitted it appears not; however, in his Book entit. *The path way to Health,* he writes himself M. of A. of *Oxon.*

24. Of Levens' two books, the first, *Manipulus Vocabulorum*, seems to be of extreme rarity. There is a copy of it in the British Museum, there is one in the Bodleian, and an imperfect one is in the possession of Mr. H. B. Wheatley, who edited the work for the Camden Society, 1867, 4°,

* Wood, *Athenæ Oxonienses*, London, 1721, I., No. 269, cols. 237-238. *Fasti*, I., col. 84, 1556; col. 88, 1559. *Id.* ed. Bliss, London, 1820, I., col. 548 (Editions 1587 and 1644, and Bliss adds 1596 and 1608). *Fasti*, London, 1815, I., cols. 149 and 156.

Joseph Foster, *Alumni Oxonienses*, Oxford 1892, Early Series, III., p. 903.

and for the Early English Text Society, 1867, 8°. In drawing up a notice of the author, Mr. Wheatley had to fall back upon the above sketch by Wood, to which he said he was unable to add anything. It has also been followed by Jöcher,* Zedler,† and the Rev. Ronald Bayne,‡ who, however, has put 19 Jan. 1557 as the date of Levens' probationer-fellowship, instead of 18, and simply stated that he was made perpetual fellow in Jan. 1559, instead of 19 Jan.

This is a book of importance for the study of English, as it is a sort of rhyming dictionary and thesaurus, and affords a clue to the pronunciation of the language in the sixteenth century. But this subject does not come within the scope of the present research.

25. What entitles Levens to a place in the present list is his treatise on medicine, or rather his collection of medical receipts. The editions of this book are those already enumerated:§ 1587, 1596, 1608, 1632, 1644, 1654, 1664, but it is possible that this list is not complete, and I can believe that there may have been more than one edition between 1608 and 1632, and others subsequent to 1664. The book had thus a long and apparently successful career and must have proved useful to a number of people. Respecting the editions the following notes may be offered.

26. 1587. This is the earliest edition mentioned by Wood, and it is probably the first. The title is given by him as above. It is not in the British Museum, and I have not seen a copy elsewhere. Possibly it contains a preface or dedication which would throw light on the first publication of the book. According to Wood it is in quarto, and presumably it is printed in black letter. [*See* "Addenda," § 26.]

---

* Jöcher, *Allgemeines Gelehrten-Lexicon*, 1750, II., col. 2403.

† Zedler, *Grosses Universal-Lexicon*, Halle und Leipzig, 1738, XVII., col. 606.

‡ *Dictionary of National Biography*, 1893, XXXIII., pp. 136-137.

§ These, I find, are all quoted by Lowndes (*The Bibliographer's Manual*, ed. Bohn, 1860, Part V., p. 1349), except that of 1664. He mentions the edition of 1654 twice, adding: "with a doubtful portrait of Levens, with an urinal by J. C(hantry)." I have not confirmed this statement, and am at a loss to say whether the repetition of 1654 is a misprint for 1664 or not. If the latter be meant there is no notice by Billings of a portrait in that edition.

27. 1596. There is a copy of the 1596 edition in the Surgeon-General's Library at Washington, and I have taken the account of it from the catalogue: *

A right profitable booke for all diseases, called The pathway to health. Wherein are to be founde most excellent and approved medicines of great vertue, as also notable potions and drinks, and for the distilling of divers precius waters, and making of oyles, and other comfortable receits for the health of the body, never before imprinted. 1 p. l., 114 ff., 3 l. 12°. *London, E. White*, 1596.

This edition and that of 1608 have been added within square brackets by Bliss to his reprint of Anthony à Wood, but Bliss gives no details, and I do not know where he obtained his information, for these editions are not in the Museum and they are not in the Bodleian Catalogue of 1843. I have not seen any notice of the edition of 1608.

28. 1632. I take the account of this edition from a copy in my own possession:

A | Right profitable Booke for | all Difeafes, called, The Path-way | to HEALTH. | Wherein are to bee found moft ex- | cellent and approued Medicines of great | vertue; as alfo notable potions and drinks, and | for the diftilling of diuers precious waters, and | making of Oyles, and other comfortable receits | for the health of the body, neuer before | imprinted | Firft gathered by *Peter Leuens*, Ma- | fter of Art of *Oxford*, and Student in Phy- | ficke and Surgery: and now newly corre- | cted and augmented. |

ECCLES. (*sic*) 38. | *Giue vnto the Phyfitian that vnto him belongeth.* |

LONDON, | Printed by *Iohn Beale* for *Robert Bird*, and are | to be fold at his houfe in S. *Laurence-lane*, at the | figne of the Bible. 1632. |

Small 4°. Sigs. A to Z, Aa to Gg in fours (Gg4 wanting); or, ff. [2] 114 Table [3]. Title within a border.

The Text is in black letter; the title page, address to the reader, head lines, and titles of the paragraphs are in roman.

The B.M. copy (1038. f. 3) is not in good state and also wants Gg4. The prefatory note to the Reader is not signed in this edition as in the later ones.

There is a copy in the Surgeon-General's Library at Washington. †

* *Index-Catalogue*, 2nd Series, 1904, IX., p. 494.

† *Index-Catalogue*, 1887, VIII., p. 89.

29. 1644. This edition is quoted by Wood, but no details are given. I have not seen it.

Those which follow are in the British Museum.

[B.M. E. 1472. (1).]

30. THE | PATH-VVAY | To | Health ; | Wherein are to be found moft | excellent and approved Medicines of | great vertue, as alfo notable Potions and Drinks, | with the Art of diftilling divers precious Waters, | for making of Oyls, and other comfortable | Receits for the health of the Body, | never before printed. | Firft gathered by *Peter Levens*, Mafter of | Arts in *Oxford*, and Student in Phyfick and | Chirurgery, and now newly | Corrected and | augmented. |

ECCLUS. 38. | *Give unto the Phyfician, that unto him be- | longeth,* |

London, Printed for *J. W.* and are to be fold by | *John Andrews*, dwelling at the White Lion in | the Old Bailey. 1654.

Small 8°. Sigs. A in four, B to Z in eights [of which Z8 is blank]; or, pp [8] 331 [1 blank] Table [18] [2 blank]. A1 recto is blank ; verso contains the short title printed up the page lengthways. The title is enclosed in a narrow border of acorns. Roman letter and italics throughout.

In this edition the Epistle to the Reader is signed Peter Levens.

[B.M. 1039. f. 26.]

31. THE | PATH-WAY | TO | HEALTH ; | Wherein are to be found most ex- | cellent and approved Medicins of great | Vertue, as alfo notable Potions and | Drinks, with the Art of Diftilling divers | precious Waters, for making of Oyles, | and other comfortable Receipts, for | the health of the Body, never before | Printed. |

Firft gathered by *Peter Levens*, Mr. | of Arts in *Oxford*, and Student in Phy- | fick and Chirurgery, and now newly | Corrected and augmented. |

Publifhed according to Order. |

ECCLUS. 38. | *Give unto the Phyfician, that unto him belongeth.* |

LONDON, | Printed for *J. W.* and are to bee fold by | *Charls Tyns*, at the three Bibles on | *London-Bridge*, *MDCLXIV.* |

12°. Sigs. A to R in twelves, (but R 12 (? blank) is awanting) or, pp. [8] 380, Table [21], advertisements [3], 2 awanting. A1 recto is blank, verso contains the short title printed lengthways along the page from the foot. The title page is enclosed in a narrow acorn border. Roman and italics.

In this edition the Epistle to the Reader is signed Peter Levens.

This is a reprint of the 1654 edition.

There is a copy in the Surgeon-General's Library at Washington.*

---

* *Index-Catalogue*, 2nd Series, 1904, IX., p. 494.

32. In this book there is no very strict classification of the subject, but it begins with diseases of the head and follows the body downwards. The trouble, whatever it may be, is specified: for him that hath lost his mind; against forgetfulness; for the meigrim in the heade; for to purge the head, and to cleanse it; a gargell for the rume; for all manner of evill aches in the head; for paine in the eares; for the pin and the web, and the fellon in the eye; for wormes in the eyelids; for the tooth-ache; for drynesse of the tongue and for thirst; a sirrop for a sore throat; for the cough; against the letting of the breath; for spetting of blood; for weaknesse of the heart; for the paine of the lungs; for the spleene, a good experiment; for sore sides within and without; a notable plaister for the plurisie; for all manner of ache in the backe; for a great ache in the knee; for a leg be it never so sore; rules for blood-letting; for agues; for burning; for consumption; for the dropsie; for the falling evill; for the Jaundies; to make divers kindes of oyles, oyntments and plaisters, Aurum potabile, antidotes for poisons; for ringworm; salves for wounds; for shingles; Dr. Stephens water; Dr. Willowbies water, and so on. There is no description of the signs or symptoms of the disease; pains in a given place are all the same, and there is no recognition of their being possibly due to different causes, and naturally no inquiry into their origin, but, after the disease is specified, follow receipts for its cure, and, when these are specially recommended, the word 'Proved' is added, as is often done in these books.

33. With the exception of a few mineral substances, such as tutty, calamine, common salt, gold, mercury, copperas, clay, powder of burnt bones, unslaked lime and one or two others, and even these occur but seldom, the remedies employed are entirely of vegetable origin. Decoctions, extracts, juices and powders of plants are administered in wine, or ale, or other vehicle, and the plants are all of a common and familiar kind. For external application the selection was a little wider and included a few mineral and animal substances, the mixtures being sometimes the reverse of attractive, and more likely to cause disease than to cure it.

The receipts concocted from the commonest plants were quite ineffective, and often superstitious, for Levens was not above using periapts and magical words against some troubles. There was no perception of the gravity of some diseases which the receipts were expected to heal or remove in a few days, though the same diseases at the present moment are incurable and their origin is unknown.

This collection of Levens's is but one of the crowd of popular manuals of medical receipts which appeared in the sixteenth and seventeenth centuries. The writers were mere compilers, and, when their books are contemplated, it is obvious how much need there was for drastic criticism of the whole system of diagnosis and treatment and of the materia medica. No real reformation, however, was possible, until the physical sciences themselves had progressed beyond the state they were in at the time.

[33A. *See* "Addenda."]

34. Other books of medical secrets are the following:

AN | HOSPITAL | *for the Difeafed.* | Wherein are to be found moft excel- | lent and approoued Medicines, as well emplai- | fters of fpeciall vertue, as alfo notable Potions or | Drinks, and other comfortable receipts, | both for the reftitution and preferuati- | on of bodily health. |

Very necefsary for this time of common Plague and | mortalitie, and for other times, when | occafion fhall ferve. | *Newly augmented and enlarged.* | Gathered by *T. C.* |

Ecclefiafticus. 38. 4. |

The Lord hath created medicines of the earth, and he | that is wife will not abhorre it. |

AT LONDON, | Printed by Iames Roberts for Edward White, | dwelling at the little North-doore of Paules | Church, at the fign of the Gun. | 1595. |

Small 4°, sigs. A to G in fours, or ff. [28]. Black letter, except the title-page, To the Reader, head lines, and titles of the paragraphs, which are in roman.

The diseases are of the usual kind, and the 'comfortable receipts' are of the usual kind too; but the drawback to the use of the book is that it has no index [or at least the British Museum copy has none (1038 k. 34. (2)] and is not arranged alphabetically, or on any other principle. The receipts are different from those of "O. W." and "A. T." and "A. R." An interesting paragraph contains Dr. Good's rules to be observed by those who use the waters at Bath (sig. G3 recto).

35. A copy of a later reprint in the Hunterian Library (Au. 2. 33. (1.)), has the following title:

AN | HOSPITALL FOR | THE DISEASED. | Wherein are to be found moft excellent | approued Medicines, as well Emplaifters of | fpeciall vertue, as alfo notable

Potions or Drinkes, | and other comfortable Receites, for the refti- | tution and preferuation of bodily health. | Very neceffary for the time of common Plague and | mortality: and for other times when | occafion fhall ferue. |

Newly augmented and enlarged. | Gathered by *T. C.* |

Ecclefiafticus. 38. 4. |

The Lord hath created Medicines of the earth, and he that is | wife will not abhorre it. |

[Vignette of a large fleur-de-lis.]

LONDON: | Printed for *Robert Bird*, and are to be fold at his | fhop in *Cheape-fide* at the figne of the | *Bible.* 1630. |

Small 4°. Sigs. A to G in fours; no leaf or page numbers. Black letter, except the title-page, To the Reader, titles of sections and head lines or running titles, which are in roman.

It shows what little progress was making in medicine when it was still worth while to reprint a book like this so long after its first appearance.

36. In previous parts of these notes* various editions between 1596 and 1815 of Lupton's collection of wonders, culled from the old physicians and naturalists and secret-mongers, have been enumerated. There is still another which may be described:

A Thoufand | Notable Things | of fundry forts, enlarged. | Whereof | Some are VVonderfull, fome | Strange, fome Pleafant | divers Necef- | fary, a great fort Profitable, and many | very Precious | Whereunto is now added, many Excellent New | Conceits never before Printed, very witty, | ufefull, and delightfull. | [the two 6 line Stanzas, beginning: *This Book bewrayes, that fome had rather hide, &c.*]

LONDON | Printed by *Henry Bruges*, for *John Wright*, at | the *Globe* in little *Britain.* 1675. |

8°. Sigs. A to X in eights; pp. [6] 403 (*sic, for* 304), [35, 1 blank].

Among the authorities he quotes the "Natural and artificial Conclusions of the Scholars of Padua," translated by Thomas Hill. I shall have an opportunity hereafter of giving some account of Thomas Hill.

* *Transactions of the Glasgow Archæological Society*, N.S. 1890, I., p. 312 (Reprint, IV., p. 14). *Ibid.*, I., p. 455 (Reprint, V., p. 39). *Ibid.*, 1903, IV., p. 100 (Reprint, Supp. IV., p. 8).

37. Another book of medical receipts presents some points of interest.

> A Rich | *Store-houfe or Treafury* | for the Difeafed. | *Wherein, are many approued* | Medicines for diuers and fundry | Difeafes, which haue been long hidden, | and not come to light before | this time. |
>
> Now fet foorth for the great benefit | and comfort of the poorer fort of people | that are not of abillitie to go | to the Phyfitions. | By A. T. | Rebus aduerfis conftans. |
>
> AT LONDON, | Printed for Thomas Purfoot, | and Raph Blower, Ann. 1596. |
>
> Small 4°, sigs. A2, B, C in fours, or ff. [10, but f. 10 is wanting in the B. M. copy], and D to V2 in fours [wanting H4 and I1], or ff. 66 [wanting 20 and 21]. Black letter. A border encloses the title-page.

The contents are arranged alphabetically, according to the diseases, or parts affected, and are divided into sections or chapters, 1 to 269 numbered, and 270 to 341 not numbered.

38. This manual also proved serviceable and passed through a number of editions, of which the following in the British Museum may be specified. In the ten years succeeding the first publication two editions must have been issued. Then in 1607 another was printed by Raph Blower. The title is the same as before, but there is added: And now fourthly corrected, augmented and inlarged, by | *G. W.* practitioner in Phificke and Chirurgerie. It is in small 4°, and contains ff. [9] 91; or, sigs. A in 2, B to Z, Aa to Cc_2 in fours. The preliminary nine leaves contain the title and dedication to the Lord Mayor, and "Rules and Inftructions to be obferved, in Phificke and Chirurgerie." This part is in roman, but the body of the work is in black letter, except the titles of the sections, etc., which are in roman. The British Museum copy wants R2 and 3, or ff. 55, 56.* G. W., the editor, seems to have made considerable additions, for the chapters have increased in number to 689. While maintaining the alphabetical order, he has made the class titles more prominent. [*See* "Addenda," §§ 37, 38.]

---

*Over this book there is some confusion in the Museum catalogue. In the list of the editions under the heading T. A., there is one (1038. i. 35. (7.)) which is described as imperfect, wanting the title-page and preface, with a query after its date [London, 1600?], and having copious MS. notes by M. Wattes. It contains the text, and the signatures run from $[C_4]$ to Cc_2 in fours.

39. The fifth edition was published at London, by the same Raph Blower, in 1612. It is a small 4°, ff. [14] 176. The title, the introductory Instructions, the headlines and titles to the chapters, and various paragraphs are in roman, the receipts are in black letter. But there is a curious thing in the title: the author and editor have changed places, either inadvertently, or, I suspect, intentionally. Whether or not, the book is said to be:

> First ſet foorth for the | benefit and comfort of the poorer ſort | of people that are not of abillitie to | go to the Phiſitions, by *C. W.* | And now fifthly augmented and inlarged, by *A. T.* | practitioner in Phiſicke and Chirurgerie. |

which is just the reverse of the fact. It is much enlarged, and contains 784 chapters or sections, as compared with 689 in the 1607 edition. It is also rather nicely got up.

40. The transference of authorship, which was quite modestly and unobtrusively attempted in the fifth edition, asserts itself prominently in the seventh. The title remains the same, but the author is announced

---

In the same list there is another copy (1038, k. 34. (5.)), London, R. Blower, 1607, which also is described as imperfect, containing A to C only. In reality it consists of nine leaves, which have the signatures [A_1] A_2, B_{1-4}, C_{1-3}, and contain the title and preliminary matter. It also is said to have MS. notes by M. Wattes.

Now, these two so-called imperfect copies not only supplement each other, but they are actually portions of one and the same volume which have got separated both in cataloguing and in binding. The typography, the signatures, the ornaments and the contents show the essential connection of the fragments. But even putting these aside as insufficient for identification, the mere fact that they both contain MS. notes signed by Mathewe Watte, or Wattes, over and over again, and that on the title-page of the smaller fragment of nine leaves, stands clearly inscribed the date and the owner's signature: Math: Watte. July 29 1610, identifies the two as parts of one volume beyond all question. It is not easy to see how this could possibly have been passed over by any one making such minute descriptions of the two portions as the Museum catalogue contains. Surely the fact of two imperfect copies, both having MS. notes by the same person, Mathewe Wattes, was enough to necessitate examination and comparison before creating even a queried edition of 1600.

It may be remarked here that while the owner clearly signs his name "Math: Watte," on the title-page, throughout the notes which he has written on the margins he uses the form "Wattes." He seems to have kept the book beside him and to have added the notes from time to time. One on fol. 33 recto is dated 1632.

in large letters as *G. W.*, and there is added in small type the absolutely false statement as regards *A. T.*:

And now feuenthly augmented and inlarged by *A. T.* | Practitioner in *Phyficke* and *Chirurgerie.* | LONDON, | Printed by RICHARD BADGER, for *Philemon Stephens* and | *Christopher Meredith*, and are to be fold at the figne of the | golden Lion in *Pauls* Church-yard. 1631. |

Small 4°, sigs. * in four, A to X in eights (X8 wanting in the B. M. copy, 1038. g. 14); or pp. [24] 317 [1, containing the colophon, which is the same as the imprint], [2 wanting, probably blank].

The text of the receipts and cures is in black letter: all the rest is in roman.

41. In the Hunterian Museum (Au. 2. 33. (4.)) there is a copy of the seventh edition of this book which has the date 1630 on the title-page but 1631 in the colophon, X7 verso:

LONDON, | Printed by RICHARD BADGER, | for *Philemon Stephens* and *Christo-* | *pher Meredith*, and are to be fold | at the figne of the golden Lion | in *Pauls* Church-yard. | 1631. |

X8 is wanting, as in the British Museum copy.

This edition contains 795 sections or chapters, as they are called. The classification headings are much more conspicuous than in any of the foregoing. I do not know if another edition followed, but if there be one, I should not be surprised if "A. T." had disappeared altogether and the book was finally claimed by G. W. as his own. He may have thought that by the labour he had expended on it, and the certainly abundant additions he had made to it, he had become to all intents and purposes. the author and owner, and was entitled to all the credit it could bring him. But if he had that end in view he might have attained it in a much simpler fashion by dropping "A. T." at once, and leaving "G. W." merely. The change would never have been remarked, whereas now one cannot miss what he has done. It would seem, therefore, as if in this book and its editions we were witnessing the process of transference or absorption of authorship, a phenomenon of which a complete illustration was given formerly in connection with Auda and Quinti.* In the present case it is attempted or effected by juggling with initials.

---

* *Transactions of the Glasgow Archæological Society*, N.S. 1890, I., p. 322 (read 1885, Reprint, IV., p. 24); N.S. 1896, II., p. 399 (read 1894, Supp. I., p. 36); 1899, N.S. III., p. 403 (Supp. III., p. 46).

42. At this point there ought to be mentioned a work entitled "The Spanish Mandeville of Miracles or Garden of Choice Flowers," which is a translation from the Spanish of Torquemada's "Jardin de flores curiosas," Salamanca, 1570, and made its appearance in 1600 and again in 1618, in the usual small 4to size. It will be considered more fully when I deal hereafter with the original among foreign treatises. The book contains an exposition of wonders of nature in six dialogues, with curious stories interspersed, so that it is not unattractive. It passed through a number of editions in Spanish, and translations of it into various languages were made. These will be described in their proper place.

43. Already more than once books have been quoted on the subject of fire-works, but there are others of some importance which have not yet been referred to. One of these is the work of John Babington, who tells us himself that he was a gunner and student of the mathematics, which is all that is known about him.* He published the book of which a copy can be shown:

> Pyrotechnia | or, | A Discourse of | Artificiall | *Fire-Works* : | In which the true Grounds of that Art are plainly | *and perſpicuouſly laid downe* : | Together with ſundry ſuch Motions, both *Straight* and | *Circular, performed by the helpe of Fire, as are not* | to be found in any other Diſcourſe of this kind, | *Extant in any Language.* | Whereunto is annexed a Short Treatise of *Geo-metrie, contayning certaine* Definitions *and* Problemes, | for the Menſuration of *Superficies* and *Sollids*, with *Tables* for | the Square Root to 25000, and the Cubick Root to 10000 | *Latus, wherein all Roots under thoſe Numbers are* | *extracted onely by Ocular Inſpection.* |
>
> Written by *John Babington* Gunner, and Student | *in the* Mathematicks. |
>
> *London*, | Printed by Thomas Harper for Ralph Mab, | MDCXXXV. |
>
> Small folio, pp. [14] 72; Treatise of Geometric, pp. [4] 80; Tables of square and cube roots, pp. 200.

Besides this printed title-page, there is also one wholly engraved, with a border in compartments displaying various devices for fire-works, while

---

* *Dictionary of National Biography*, 1885. II., p. 314. Js. Granger, *A Biographical History of England*, 1769, I. ii., p. 516; 2nd Ed., 1775, II., p. 338; 5th Ed., 1824, III., p. 162.

at the bottom there is a portrait of the author, "Anno Ætatis Suæ 31."* There are eighteen full page engravings in the text and two separate folding plates of set pieces. The "Treatise of Geometrie" contains numerous diagrams in the text.

It will be observed that this work, like the "Store-house of . . . Secrets" preceding (§ 9), was printed by Thomas Harper, and that the "Pyrotechnia" was printed for Ralph Mab, who published the book by Bate called: "The Mysteryes of Nature and Art," 1634; second edition, 1635, both in 4°. The second division of Bate's book also treats of fireworks, and it is worth while comparing two contemporary works on this subject with each other, and with that of Jean Appier Hanzelet, published in 1630, and already described.† Among the complimentary verses prefixed to Babington's book is a set from the said John Bate, who commends his inventions and skill in the art, so that Babington was obviously esteemed by his contemporaries.

44. The next book again deals with medical secrets, and it affords a fresh instance of the anomalies and difficulties which occur in connection with publications of the seventeenth century. But, in addition, the question of authorship has been raised, and that has led to unconfirmed and misleading statements in such standard authorities as the British Museum catalogue and the *Dictionary of National Biography.*

An account of the books with their peculiarities may first be given and then the questions in dispute may be discussed.

AN | ALPHABETICAL BOOK | *OF* | Phyficall Secrets, | For all thofe Difeafes that are moft | predominant and dangerous (cura- | ble by Art) in the Body | of Man. |

---

* This portrait in a small oval in the engraved title, by J. Droeshout, is scarce, according to Granger. But why? It could not be any scarcer than the book. The *Dictionary of National Biography* mentions the portrait in the Chatsworth copy as if it were exceptional, but, according to the *Catalogue of the Library at Chatsworth*, 1879, I., p. 102, it is exactly the same as that in the present copy. The title in compartments may be compared with that in Bate's "Mysteryes," in White's "A Rich Cabinet," and in the Dutch treatise on fireworks.

† *Transactions of the Glasgow Archæological Society*, 1899, N.S. III., p. 402. (Reprint, Supplement, III., read 1894, p. 45.)

Collected for the benefit, moſt especially | of Houſe-holders in the Country, who are either | farre remote, or elſe not able to entertaine | a learned Phyſician: as likewiſe for the | help of ſuch Ladies and Gentle- | *women, who of charity labour* | *to doe good.* |

Whereunto is annexed a ſmall Treatiſe | of the judgement of Vrines. | By OWEN WOOD. |

Printed at London by JOHN NORTON | for WALTER EDMONDS, and are to be | ſold at his Shop at the ſigne of | the Crowne neer *Ludgate.* | 1639. |

Small 8°, pp. [7, 1 blank] 238 [1, 1 blank].

Width of the page 2½"; height of the printed page, (excluding the headline and catchword) 4½".

To facilitate reference the current letter of the alphabet is printed in the middle of the front margin of each page, instead of being placed at the top as usual.

The licence to print is as follows:

Martii 21. | 1638. | Imprimatur; ſed intra | quatuor menſes ſin mi- | nus hæc licentia ſit irrita. | Jo. Alsop. |

The preface to the book, signed by Alexander Read, Doctor of Physick, is dated 25 March, 1639, and it begins: In all civill and well ordered Common-wealthes, *&c.* Here arises a question: is 1638 in the licence a misprint for 1639, which is possible; or, was the book held back for a whole year; or, was it actually issued in 1638 without Read's preface, and then a year later the unsold copies, supplied with a new-dated title, and Read's preface, issued afresh? I have no means of deciding.

45. A dozen years later the book re-appeared with this title:

[B. M. 1038. c. 18. (1.).]

AN | EPITOMIE | OF MOST | Experienced, Excellent and | Profitable SECRETS | appertaining to | PHYSICKE | AND | CHIRVRGERY. | For all thoſe Diſeaſes that are | moſt predominant and dangerous (cu | rable by Art) in the Body of Man, | as by the TABLE appeares. | Alſo, *the judgement of* Urines. | For the benefit of ſuch diſcreet | Ladies, Gentlewomen and others | which labour to doe good in that Art | Myſtery and Profeſſion. | By *O. W.* Profeſſour in *Phyſicke* and | CHYRURGERY. |

Printed by *T. B.* and are to be ſold in *London* | and *Westminster-Hall.* 1651.

Small 8°. Pp. [31, 1 blank], 258 [1, 1 blank].

Height of the printed page, excluding the headline and catchword, 4½", width, 2$\frac{11}{16}$".

As the pagination jumps from p. 128 to p. 149, the total number is of course 20 too many, and the correct number is 238.

In this edition the receipts are numbered in Roman numerals I. to CLXXXXVI., and in Arabic numerals 197 to 500. The index initials are omitted from the front margins.

46. An absolutely different edition was published with the same date:

> AN | EPITOMIE | OF MOST | Experienced, Excellent | and Profitable SECRETS | appertaining to | PHYSICKE | AND | CHIRURGERY. | For all thofe Difeafes that are | moft predominant and dangerous (cu- | rable by Art) in the Body of Man | as by the TABLE appeares. | Alfo, *the judgement of* Urines. | For the benefit of fuch difcreet | Ladies, Gentlewomen and others | which labour to do good in that Art, | Myftery and Profeffion. | By *O. W.* Profeffour in *Phyficke* and | CHYRURGERY. |
>
> Printed by *T. B.* and are to be fold in *London* | and *Weftminster-Hall.* 1651. |
>
> Small 8°, pp. [32] 239 [1 blank]. In these two issues the licence is omitted and Alexander Read's Epistle to the Reader in commendation of the book is not dated, and has been enlarged by a preliminary portion about O. W. and Dr. Read.
>
> Height of the printed page, excluding the headline and catchword, $4\frac{1}{8}''$; width, $2\frac{5}{16}''$.

The receipts are numbered throughout in Arabic figures, 1 to 500, and the index initials are omitted from the front margins.

47. Comparison of the two issues dated 1651 shows that they are typographically distinct throughout. Besides the different width of the print, which, though slight, produces an appreciable effect on the appearance of the page, there are variations in spelling, in the ornaments at the beginning and end of the sections, in the capitals, and in misprints, which are perhaps more numerous in the Museum copy. These may indicate either that it is the earlier of the two issues and that the misprints were subsequently corrected; or, that it has been pirated from the other and run rapidly through the press. In any case, there is no doubt that the copy in the Museum shows signs of less careful press work. Apart from these typographical variations, however, the contents of the two issues are identical with each other.

48. Comparison of the 1639 edition with those of 1651 exhibits the following differences: 1°, the title, altered by the new printer; 2°, the important passage—apparently by the printer—prefixed to Read's Epistle, and one or two verbal alterations in Read's part; 3°, the insertion of a table of contents, superfluous in a book already in alphabetical order; 4°, the numbering of the sections in the 1651 issues; 5°, the omission of the licence to print and the date of Read's Epistle in the same issues; 6°, the rearrangement, also in these issues, of the first fifteen sections so as to bring together certain of them which refer to the same subject;* 7°, the omission, again in the two later issues, of the index letter on the front margin of each page, as has been already mentioned.

As regards the contents the three editions agree word for word throughout; that is, the issues of 1651 are verbatim reprints of the edition of 1639. If, therefore, the author of the edition of 1639 was Owen Wood, as he calls himself, there is no room for the slightest doubt that the initials O. W. in the 1651 issues, denote the same person and author. In fact anything else is impossible and contradictory.

---

* In the 1639 edition the first fifteen receipts are in the following order; the prefixed numbers denote the sequence in the 1651 issues:

8. An Vnguent for Ache, Bruife, or Spraine.
5. Ache in any Member.
6. Another for the fame.
9. For any Ague falne into the Leggs, and fwolne, the Cure.
10. Afpafmus or Numneffe in Ioynts or elfe-where, to Cure.
11. Argentum vivum to Mortifie.
12. Apoplex to Cure.
13. A Salve for the Apoplex.
14. Aqua Compofita good for head, Memory, and all Difeafes.
15. Another Aqua Compofita.
4. Aches or Lameneffe in Members or Joints.
2. Aches olde; the Cure.
3. Ache in Bones or Ioynts, the Cure.
1. For all Aches in generall.
7. An Oyle that cureth all Aches in Man, or Beaft.

49. Concerning this person I have found only Anthony à Wood's very brief note.* It occurs in his paragraph about Owen Wood, dean of Armagh, who left a benefaction to Jesus College, Oxford. He says:

> One of both his names was a practitioner in physic, and wrote *An Alphabetical Book of Physical Secrets*, as also *A small Treatise of the Judgment of Urines*: both printed at Lond. 1632, and 39, oct.

The 1632 edition has not come under my notice anywhere else, and nothing more seems to be on record about Owen Wood. It will be observed that while Anthony à Wood ascribes the book definitely to Owen Wood, he does not make any reference to Alexander Read at all.

50. The question of authorship has now to be considered. This, as it seems to me, is irrefragably established in favour of Owen Wood by the name and the initials on the title-pages, and the identity of contents. But as this evidence seems to be outweighed, in the opinion of the Museum cataloguer, by the fact that Alexander Read signs the preface, and that that entitles him to be called the author, there may be considered what the prefaces do say, which will show what Alexander Read had to do with the book.

That prefixed to the 1639 edition runs in the following terms:

> To the courteous Reader.
>
> In all civill and well ordered Common-wealthes, even where best Schooles of Learning, and famous Vniversities are, you shall finde published not onely methodicall discourses concerning the curation of infirmities of the body of man; but Treatises also containing parable medicaments, which any one who is endued but with a meane judgement, may apply to himselfe, whereas neither Physician nor Apothecarie can bee had. Great plenty of such Bookes are published in the German tongue: for the Germans delight in uttering of Pamphlets, promising the revelation of secrets in Physick, which prove nothing else but triviall toyes. England in like manner is reasonably ill stored with such stuffe, which emboldeneth no small number of bold Knaves, and impudent Queanes to meddle with the Practise of Physick, to the utter ruine of no small number of rude and improvident persons, who commit themselves

* *Fasti Oxonienses*, 1721, I., col. 126; ed. Bliss, London, 1815, I., col. 227. It was copied by Zedler, *Universal-Lexicon*, 1748, LVIII., col. 1528, and by Jöcher, *Allgemeines Gelehrten-Lexicon*, Leipzig, 1751, IV., col. 2069.

to the Skill and Cure of such unworthy persons. That this is a truth, the manifold complaints which come to the Physicians Colledge when the Fellowes sit doe make good. So that it might seeme requisite rather to suppresse most of these Treatises, then to publish any more of this kinde. But this Booke may bee priviledged to come abroad, if you either respect the method or matter. The Remedies may easily be found out, the Diseases being set downe Alphabetically. The Remedies are not to be rejected; but, seeme to have beene set downe by one who was no Stranger in the Practice of Physick. The Printer having beene enformed of so much, was not willing to keepe the Treatise within his private walls; but to Print some Copies of it, and to present them to the publick view and censure. Hee onely desireth that his good intentions, and purpose to further the publick good, may be taken in good part of those who have not too censuring Spirits. If he shall finde thy favourable acceptance of this Treatise, hee will not spare to publish Bookes of the like Subject come to his hands, if they shall bee thought worthy of the Presse.

25 March 1639,

ALEXANDER READ,

Doctor of Physick.

From this epistle two or three inferences may, I think, be drawn:

1st. Whether the book was previously published or not, the printer got Dr. Alexander Read—a distinguished man of his time—to write a recommendation of it, so as to make this edition acceptable to the public.

2ndly. Apparently Read did not know Wood personally, if at all.

3rdly. Read viewed the author of this book, not as a qualified physician, but as an empiric of a superior type, and with some skill in medicine. He approves of the alphabetical arrangement of the book as convenient for the end in view.

4thly. The whole epistle refers obviously not to Read himself but to the author of the work, whose name was Owen Wood.

51. With one or two slight verbal alterations this epistle of Read's is repeated in the two issues of 1651. But the following sentences are prefixed, which are well worth observing:

Discreet Ladies, Gentlewomen, and others; this small Epitomy of Physicke and Chyrurgerie; being most usefull, excellent and necessary for all People, as well Rich as Poore, it being such a precious Jewell, and gathered by Doctor O. W. and recommended by that worthy and learned Man Doctor Read, that writ so many things of Physicke and Chyrurgery, and it having his approbation and judgement, I could not let it lye in obscurity; and the rather I put it abroad, because that I have knowne some hundreds of late have found the Benefit of it; and by the practise of this and some others, have with Gods blessing, preserved and recovered many a sicke and diseased Body. And as in all civill and well ordered Commonwealthes, &c., &c.,

as in the preface given above.

This preface is also signed by Alexander Read, but there is no date.

From this addition other inferences, confirmatory of the preceding, can be drawn:

1st. The addition has been made by T. B., the printer of the 1651 issues, so far as one can judge, and not by Alex. Read.

2ndly. The writer says distinctly that the "Epitomie" was gathered by Dr. O. W. and recommended by Dr. Read, not that it was composed by Dr. Read; and surely he had better opportunity for knowing than the Museum cataloguer.

3rdly. The writer does not seem to have thought it necessary to give the author's name in full, if he knew it. But he must have known it, for the present edition is a reprint either from that of 1632, in which the author's name was probably in full; or from that of 1639, in which it is certainly in full.

4thly. The writer speaks of Dr. Read in the past tense; "that worthy and learned man Doctor Read, that writ so many things of Physicke and Chyrurgery," is not the way in which one would refer to a person still alive. The difficulties connected with the date of Dr. Read's death will be referred to presently.

These prefaces, therefore, confirm the internal evidence from the identity of the text and the name and initials on the title-pages, that the author of the book was Owen Wood, and not Alexander Read.

52. The entries in the Museum Catalogue have next to be reviewed. They are these:

READ (ALEXANDER), *M.D.*, *F.R.C.P.* *See* WOOD (OWEN) An alphabetical book of physicall secrets, *etc.* [With a commendatory preface by A. Read.] 1639. 8°. 1038. d. 36. (6.)

—— A (*sic*) Epitome (*sic*) of . . . secrets . . . By O. W. [or rather by A. Reade.] 1651. 8°. *See* W., O., *Professour in Physicke, etc.* 1038. c. 18. (1.)

W., O., *Professour in Physicke and Chirurgery.* An epitome (*sic*) of . . . secrets appertaining to physicke and chirurgery . . . By O. W. [or rather by A. Read?] *Lond.* 1651. 8°. 1038. c. 18. (1.)

WOOD (OWEN). An Alphabetical Book of Physicall Secrets, for . . . diseases . . . most predominant in the body of man. [With a commendatory preface by A. Read.] FEW MS. NOTES. *J. Norton for W. Edmonds: London,* 1639. 8°. 1038. d. 36. (6.)

The first and fourth of these entries require no comment. The second, putting aside the misprints, is inaccurate in ascribing the book to A. Reade, and ignoring the possibility and, as a matter of fact, the actuality of the identity of O. W. and Owen Wood, and of "An Alphabetical Book" and "An Epitomie." How the cataloguer missed that it is hard to say; possibly his vision was dimmed by some cataloguing rule; but the two volumes might have been compared and the grounds for passing over Owen Wood's claim to the authorship and for ascribing it to Dr. Read should have been considered. Some doubt seems ultimately to have arisen, for the authorship of Dr. Read is queried in the later entry under W., O.; but, again, if there could be any doubt in the matter at all, the question should have been fully examined and the allotment to Read finally cancelled.

What further strikes one in these entries is the fact that while there is a reference from Read to Wood (Owen) and to W., O., there is none from Wood (Owen) to W., O., or from W., O., to Wood (Owen), which would seem much more requisite and obvious.

53. Dr. Alexander Read himself may now be introduced. He compiled a book in the same style as the "Epitomie" of Wood, which ought not to be forgotten here.

MOST | EXCELLENT | And approved | *Medicines & Remedies* | For moſt Diſeaſes and Maladies | Incident to Man's Body, | Lately compiled and extracted out of | the Originals of the moſt Famous and | beſt Experienced Phyſicians both in | *England* and other Countries, | By *A. R.* Doctor in Phyſick deceaſed. | And ſince reviſed by an able Practitioner in | the ſame Science, and now Publiſhed for | the univerſall good and benefi (*sic*) of | this Common-wealth. |

Vita ſine Valetudine Odium eſt, non Vita. |

Eccleſ. (*sic*) 38. 1, 2, 4. |

Honour the Phyſician *with that honour that is* | *due unto him, becauſe of* Neceſſity: *for the* | LORD *hath created him. For of the moſt* | HIGH *cometh* healing. |

The LORD *hath created* Medicines *of the Earth,* | *and he that is* Wise *will not abhor them.* |

LONDON, Printed by *J. C.* for *George Latham* | *Junior*, and are to be fold at the figne of | the Bifhops head in St. *Pauls* Church- | yeard 1652. |

Small 8°. Pp. [16] 144. Small border round the title.

"The Epistle to the Reader," which is a protest against fatalism and an exhortation to use every means known to preserve oneself in health and to combat disease, is signed with the initials T. A., who has used the same motto from Ecclesiasticus as T. C. did in 1595.

54. This Dr. Alexander Read, then, was dead by 1652, and it may not be out of place here to point out certain errors and contradictions which have found their way into the current biographies of him.

The first confusion was introduced by Granger,* who calls him a Scotch Physician and says:

DR. (ALEXANDER) READE. A small head by Gaywood; in the title to the "Secrets of Art and Nature," fol. 1660.

This is entirely wrong; but the biographical notice which he adds is correct enough, so far as it goes.

Granger's error was copied by Henry Bromley,† whose account runs thus:

ALEXANDER READE, M.D. Scotic. A small oval, in the title page to his "Secrets of Art and Nature," 1660, *fol.* . . . Gaywood.

There is scant room here for more mistakes. He was not a Scottish M.D.; "The Secrets of Art and Nature," 1660, are not "his," but belong to Hans Jakob Wecker. The portrait there given is not of Dr. Alexander Read, but of Dr. R. Read, the translator of Wecker's book, who wrote and signed a preface to it, and published it nineteen years after the accepted date of Alexander's death.

---

* *A Biographical History of England*, London, 1769, I. ii., p. 482. 5th ed. 1824, III., p. 120.

† *A Catalogue of Engraved British Portraits from Egbert the Great to the Present Time*, London, 1793, p. 102.

Lowndes is no better, for to Alexander Read he assigns "Secrets of Art and Nature," 1660, folio, and adds: "With a portrait of the author in the title-page by Gaywood." Yet he gives a reference to WECKER, John, under which name, however, there is no allusion to Read, or to the book being a translation, but there is mention of "a frontispiece containing portrait by Gaywood," of whom is not specified.

Robert Watt * gives a list of Read's anatomical and surgical works, and includes among them:

> The Secrets of Physicke and Chirurgery. Lond. 1651. 12mo.—Most excellent Medicines and Remedies for most Diseases. Lond. 1652. 12mo.—Secrets of Art and Nature. 1660. fol.

This last treatise is quoted also under 'WECKERUS, Johannes Jacobus,' as being the only one of his in English, but there is no reference to the translator, so that one cannot tell whether Watt thought them the same or not.

Under 'WOOD, Owen,' he quotes: "An Alphabetical Book of Physical Secrets against most Diseases. Lond. 1639. 8vo.," again without reference to Read, or to the later reprints of 1651. Allibone † simply condenses Watt's list of Read's works and copies the entry about Owen Wood. Weckerus is not mentioned. But under 'Reid, Alexander,' he has the following: "Physician to Charles I., of the same family as Thomas Reid, the philosopher, pub. some forgotten works on medicine and surgery," unaware, apparently, that he had already given a list of them under Read. He makes no reference to a portrait. As to the works being forgotten, surely it is the business of the author of a dictionary of English Literature to rescue them from that very oblivion he seems to think they deserve.

55. The most recent and fullest account of Read or Reid is that given in the *Dictionary of National Biography*, but unfortunately in it are repeated the statements of Granger, Lowndes, and others, which are so inaccurate. One or two points in this article may be considered.

---

* *Bibliotheca Britannica*, Edinburgh, 1824, Authors, II., Read, 794 *a*, *b*; Weckerus, 956 *b*; Wood, 980 *y*.

† *A Critical Dictionary of English Literature*, Philadelphia, 1870, II., p. 1751; and London, 1871, III., p. 2821, for Wood, and 1870, II., p. 1763, for Reid.

The dates assigned to Read are 1586?–1641, and it is said that "he died in October 1641, his will being proved 24 Oct. 1641." This looks definite enough, and it is something to start from, but it does not tally so well with other statements as one could desire. It seems curious that so much as eleven years after his death, in 1652, that event should still be referred to; but possibly, "deceased," the word employed, was equivalent to the modern expression "late." It corresponds, however, with the expression in the preface of Owen Wood's "Epitomie," 1651, already quoted, where he is called the Dr. Read, "who writ so many things of Physicke and Chirurgery."

As to his birth it is placed, in the article, *about* 1586. That date, however, does not agree at all with what he himself says. If he died in 1641, and was born about 1586, he would be about 55 years old at his death. But in his address to the Reader, dated 5th August, 1634, prefixed to his Chirurgicall Lectures, delivered in "Chirurgeans Hall" in 1632, 1633, and 1634, and published at London in 1635, he remarks that he had practised "Physicke and Chirurgerie now 42 yeares." If that be right, and he was born in 1586, he must have begun practice about the year 1590-92, when he was no more than four years of age! Supposing, however, that he was 21 when he began to practise that would place his birth about 1571-73, and would make him about 63 in 1632-34, and a man of 70 at his death. If he were still alive in 1655, he would be 85, which corresponds still better with the phrase he uses about the running out of the sands, than his death at the age of 55.

There is another difficulty connected with the date 1641, which I am unable at this moment to remove from want of the necessary facts. It is this. In the 3rd edition of Read's *Manual of Anatomie*, 1642, the following passage occurs at the beginning of the address to the Reader:

> Courteous Reader, Now I prefent to Thee the third edition of the Manual of Anatomy, which fhall be the laft that fhall be publifhed in my lifetime, which is not far from its period. The Hour-glaffe hafteneth, and but few fands remain unrun.

The first edition of this book was published in 1634, and this address is not contained in it. It is wanting also in the 4th edition, 1650, but it is present in the 5th, 1655, with the change of "third" to "fifth" so as to make the address fit the edition, irrespective of the author's being alive or dead. It is wanting also in the 1659 edition. To settle this point one would need to ascertain the date of the 2nd edition, and whether or not it contains the passage just quoted. If the date is before 1641, and the passage is present, then no difficulty arises in connection with his death in 1641, but if the passage be not contained, and if it do not appear till the 3rd edition in 1642, there are two dates which at present are conflicting. So far I have not been able to examine the 2nd edition.*

It may be remarked that in the various editions of this work, subsequent to 1641, there is nothing whatever to indicate that the author was dead. On the contrary they all appear to have been revised and edited by himself.

56. Now, still assuming 1641 as the year of his death, how is it possible for him to have written "An Epitomie of Secrets" in 1651, or "The

* [January 1907. I have found the second edition in the British Museum [548. b. 5.]:

The manuall of the Anatomy or dissection of the body of man. . . . Enlarged and now methodically digested into 6. Books. By Alexander Read, Doctor of Physick, a Fellow of the Physitians College of London, and a brother of the Worshipfull Company of the Barber-Chirurgeons. London, Printed by J. H. for F. Constable, and are to be sold at his shop under Saint Martins Church near Ludgate. 1638.

12°. Pp. [12 including the engraved title], 574, [1. 1 blank]. 5 engraved plates. The engraved title represents five doctors in birettas at a table on which is lying a dissected body. The centre figure with the fur-trimmed gown, lace cuffs and piped collar, has a pointed beard, and, with a wand in his hand, is apparently giving a demonstration.

This is said to be a portrait of Read, and the others are presumably barber-chirurgeons. Represented as attached to the front of the table by two large nails is a human skin with the face and hands, and on it is printed the fly-title and the date 1638. The dedication to King Charles is in Latin, and is dated: Londini 4 Calend. Octob. Anni ab exhibito in carne Messia, supra millesimum sexcentesimum tricesimi septimi. It is signed: Alexander Reidus Scoto-Britannus. There is no other preface, or address to the Reader, so that the passage above quoted about the running out of the sands does not appear in this second edition. Consequently the difficulty remains of reconciling Read's death in 1641 with the third edition of his Anatomy printed in 1642, containing a preface obviously by himself, but without any indication that he had died in the meantime.] [*See* "Addenda," § 55.]

Secrets of Art and Nature" in 1660? If he wrote "An Epitomie," the fact of his death would surely have been intimated on the title-page, just as much as on his own book in 1652; and, if he were the author, why should the initials be O. W. and not A. R., as in his own book?

For ascribing "An Epitomie" to Alexander Read—even doubtfully, as in one of the Museum entries—there is not the smallest reason, and everything is definitely against it. What was said by Read in his preface shows that Owen Wood was a distinct person. What was said in the 1651 preface, ten years after Read's alleged decease, shows still more clearly that the writer believed in O. W. as the author, and in Read's commendation of him, and in their being different persons.

Why, then, should a book, acknowledged without a suggestion of doubt or difficulty in one place in the Museum catalogue to be by Owen Wood, when Read was alive, have a reprint of it assigned to the same Read ten years after he was dead, in spite of Owen Wood's initials on the title-page and in the preface, and in disregard of the identity of contents which the most superficial comparison would have established?

To "save face" in this confusion, it might be suggested that Read wrote under the pseudonym of O. W., or of Owen Wood, and could thus speak safely of the "Epitomie" as "a precious jewell and gathered by Dr. O. W." There is no evidence in support of this, and Anthony à Wood's plain statement contradicts such a supposition. It is nowhere stated in the Museum Catalogue that Wood was a pseudonym of Read's, for the identity of Wood's "Alphabetical List" with O. W.'s "Epitomie" has been ignored.

But, besides, judging from his other writings, Read was not the man to conceal himself under the name or initials of Owen Wood or of anyone else. He was too well aware and assured of his own merit as a practitioner, teacher and writer,* to have declined the authorship of this or of any book if it had been his, and if the "Epitomie" had been his and appeared

---

* See the address to the Reader in *The Chirurgicall Lectures of Tumors and Vlcers*, London, 1635, 4°; 3rd edition, 1669; and in *The Manuall of the Anatomy . . . of the Body of Man*, 1655, 12°.

after his death, there was no more reason for concealing his connection with it than with that of his own book of 1652.

57. There remains one more error to be cleared out of the way, and that is the ascription of the translation of Wecker's *Secrets of Art and Nature* to Alexander Read. This error is of old standing and is most persistent. It began with Granger, who is answerable for so much vicious taste and the ruination of so many English books, was made worse by Bromley and so has come down to the present day.

The translation of Wecker's *Secrets* was first published in 1660, and then with the date altered to 1661. The translator's name is Dr. R. Read, *not* Dr. A. Read, and the engraved title-page by Ric. Gaywood contains portraits of Alexis and Albertus Magnus, Dr. Harvey and Lord Bacon, Dr. Reade (*sic*), Dr. John Wecker and Ramund Lullius, most of which are imaginary.

There are certain possibilities here.

One is that Dr. R. Read is really Alexander, who, instead of his initial A, used the last letter, R, of his surname. But this implies again an attempt at concealment on the part of Alexander, an assumption which has been already disposed of. There was no occasion for concealment, and no reason can be assigned for a pseudogram here, except the necessity of explaining away the R by those who would have us believe that it is actually or ought to be A.

If this could be confirmed, then the portrait of Dr. R. Read might pass for that of Dr A. Read, but there is no proof of that, and Granger merely makes a statement without once thinking that the portrait was that of "R. R," the translator, and not of Dr. A. Read, who was in 1652 described as "deceased." But, having once gone wrong he continued in error.

Bromley, as has been already shown, has gone farther astray, and it is to be regretted that, before he was followed as an authority in the *Dictionary of National Biography*, Wecker's book with Gaywood's engraved title had not been inspected. Prefixed to the 2nd edition, 1638, 3rd edition, 1642, and that of 1659 of Read's *Manuall of Anatomie* is the engraved title already described, representing five surgeons or physicians sitting at a table, and the centre figure is said to be that of the author. If that be correct, it is a very different portrait from what was drawn by Gaywood, as can be easily seen by comparing them.

Next, on turning to the address 'To the Reader' in the translation, R. Read, after some general remarks on the aims of various persons in life, and on the contents of the book, says:

> One thing more I have to impart unto thee, and that is, the great pains, coft and charges that have been expended in Publifhing this Book: and thus far I think my felf obliged modeftly, to fay, that he had regard to the world and his credit, more than his peculiar profit.
>
> Laftly, I fhall humbly intreat, that if there be any thing that through the Publifher or Printers fault hath been omitted, to pafs a gentle cenfure, and herein you will oblige
>
> Your reall Friend
>
> R. R.

This is obviously contemporary with the publication, and R. Read must have been living in 1660, and could not have been A. Read said to be deceased in 1641, and certainly dead in 1652.

58. The points may be summarised:

Owen Wood and Dr. Alexander Read are two different persons.

Owen Wood is the Author of the "Alphabetical List," afterwards called "An Epitomie."

Alexander Read wrote a preface to the "Alphabetical List," afterwards printed in "An Epitomie" also. He did not write the "List," and therefore could not have written the "Epitomie."

Owen Wood's book, with the title "An Epitomie of . . . profitable Secrets," 1651, small 8vo, is not to be confused with the translation of Wecker's book: "Secrets of Art and Nature," 1660, folio.

Dr. R. Read, the translator of Wecker's book, is a different person from Dr. A. Read, who had nothing to do with that book.

The portrait in the engraved title-page of Wecker's book by Gaywood is not that of Dr. A. Read, but of the translator, Dr. R. Read.

59. On previous occasions reference has been made to books of receipts for ladies, two of the best known being Partridge's "Treasurie of

hidden Secrets, commonly called the good huswives Closet of Provision," first printed about the middle of the sixteenth century,* and Hugh Plat's "Delightes for Ladies" printed in 1602.† The next that I know of, and the first of a series of similar books which appeared in the seventeenth century, has been already described,‡ but for convenience of comparison the description may be repeated from the British Museum copy (1038. i. 35. (1.)):

THE | LADIES | CABINET | OPENED: | Wherein is found hidden feverall Expe- | riments in Preferving and Conferving, | Phyficke, and Surgery, Cookery | and Hufwifery. | *LONDON*, | Printed by M. P. for *Richard Meighen*, next to the | Middle Temple in Fleetftreet. 1639. |

Small 4°. Sigs. A to H in fours; or pp. [2] 59 [3]. Vignette of a winged skull resting on a scythe and bone crossed and surmounted by a sandglass, with the motto "NON PLVS" below the skull; the whole enclosed in a scroll and flower border.

This first edition, "notwithstanding the disorderly and confused jumbling together of things of different kinds," as "M. B." puts it, was very well received and, a second being called for, the said "M. B.," who may or may not have been the original author, or who may be simply M. Bedell, the publisher of this new edition, "resolved" (as he says in his preface addressing the ladies) "to smooth your way a little, by bringing each particular to its proper head, or (since its called A Cabinet) laying each Jewel in his peculiar box; and so having fitted it for readier use, to have sent it abroad again to salute your gentle hands, the second time." "But," he continues, "hearing in the mean time of certain rare *Experiments*, and choice extractions of Oyls, Waters, &c. the practice of a Noble hand, and of approved Abilities (to testifie how ready I am to further ingenuous undertakings in this kind) I have with much pains,

* *Transactions of the Glasgow Archæological Society*, N.S., 1890, I., pp. 309-10 (Reprint, IV., pp. 11-12); 454 (Reprint, V., p. 38); 1903, IV., p. 101 (Reprint, Supp. IV., p. 9). [*See* "Addenda," § 59.]

† *Ibid.*, 1890, I., pp. 459-60 (Reprint, V., pp. 43-44.).

‡ *Transactions of the Glasgow Archæological Society*, 1903, N.S., IV., p. 114 (Reprint, Supp. IV., p. 22).

and some charges, sought after, and at length happily purchased them for you. Al which, with the addition of many other Secrets of several kinds, (and I hope, of valuable concernment) I have so incorporated together, if I may so say, and methodically digested, that they may be the more easily and profitably improved."

60. This next, therefore, appears to be the second edition, but there was a longer interval between the two than usually exists. The title is modified as follows:

THE | LADIES CABINET | *ENLARGED* and | *OPENED*: | *Containing* | Many Rare Secrets, and Rich Orna- | ments of feveral kindes, and | different ufes. | Comprized | *Vnder three general Heads.* |

Viz. of { 1. Preferving, Conferving, Candying, &c. |
2. Phyfick and Chirurgery. |
3. Cookery and Houfwifery. |

Whereunto is added, | Sundry Experiments, and choice Ex- | tractions of Waters, Oyls, &c. | Collected and practifed, | *By* | the late Right Honorable and | Learned Chymift, | The Lord RUTHUEN, | *With* | A particular Table to each Part. |

London, Printed by T. M. for *M. M. G. Bedell,* | and *T. Collins,* at the middle Temple- | Gate, Fleet-ftreet. 1654. | [November 1, *1653*, really. Compare *The Ladies Companion.*]

Small 12°. Signatures: 4 leaves, A to I in twelves, K in 8, L in 6, or, pp. [8], 1-42 [43-46] [47] 48-168 [169-180] 181-227 [228-232] [12 pp. of book advertisements.]

The preliminary 8 pp. contain the title and the rather euphuistic preface just quoted—signed M. B.

Title red and black. The pages within square brackets contain the table of contents of each part.

There is a copy in the British Museum, E. 1528. (1.).

This second edition is considerably enlarged, new receipts having been added. In the first edition, though there is no division into heads, and there may be some faulty classification, the receipts follow each other practically in the order given here: conserving, physic, and cookery. The only difference is that there has been some rearrangement, and receipts which were a little out of place in the first edition have been brought under their proper class in this one. [*See* "Addenda," § 60.]

61. The third edition I have not seen. The following is an account of the fourth, from the copy in the British Museum, 1037. a. 35:

THE | LADIES CABINET | *ENLARGED* and | *OPENED*: | *Containing*, | Many Rare SECRETS, and Rich | ORNAMENTS of feveral kinds, | and different Ufes. | Comprifed | *Vnder three General Heads.* |

Viz. of { 1. Preferving, Conferving, Candying, &c. |
2. Phyfick and Chirurgery. |
3. Cookery and Houfwifery. |

Whereunto is added, | Sundry Experiments, and Choice Ex- | tractions of Waters, Oyls, &c. | Collected and practifed, | *By* | The late Right Honourable and | Learned Chymift, | The Lord RUTHUEN. | *The fourth Edit. with Additions;* | AND | A particular Table to each Part. |

LONDON, Printed for *G. Bedel*, and *T. Collins*, | at the *Middle Temple Ga* (sic) in *Fleet-Street* 1667. |

12mo. A in four, B to L in twelves, M in eight; or, pp. [8] 251 [4, 1 blank]. Title in red and black.

There are a few alterations and additions in this, as compared with the preceding. 'M. B.'s address is again prefixed; next comes a note to the following effect:

The Stationers to the Reader.

Besides the gracefulness and beauty of this beyond the former Impression of the Ladies Cabinet enlarged, in respect of the manner of its coming forth; there are moreover many profitable and delightful Directions added. Which as they are of severall kinds, according to the threefold division of the Book; So are they carefully inserted in their proper Parts and peculiar places, sutable to its former Method. By the accession whereof we are again imboldned to send it forth into the World, and with greater confidence, commend it to your further perusall. Aug. 22. 1657.

The reference here to a "former Impression" may denote the third edition, which must have appeared in that case between 1654 and 1657, if the above date be correct.

On the other hand, this note—if the date be correct—may have been prefixed originally to the third edition, and may imply that it was superior to the second of 1654. But if the date be a misprint for 1667, then the third edition may have come out any time during these thirteen years, 1654-1667.

Whether or not there was an edition after 1667, I have not ascertained.

62. Along with Babington's book on fireworks may now be taken that of Nathaniel Nye, which appeared in 1647. Another treatise in Dutch, 1613, and a third in French, 1747, will be described under their dates in another paper.

Directions for making fireworks are also included in certain of the receipt books written by J. White, V. Etten and others to be afterwards mentioned. There was therefore no shortcoming of instruction in this branch of art, and one is driven to the belief that pyrotechnic displays must have been popular in the seventeenth and eighteenth centuries.

Nye's book (British Museum, 8827. c. 20.) is as follows:

THE ART OF | GUNNERY. | Wherein is defcribed the true way | to make all forts of Gunpowder, Gun-match, the | Art of fhooting in great and fmall Ordnance: Ex- | cellent ways to take Heights, Depths, Diftances, | acceffible, or inacceffible, either fingle or divers | diftances at one operation: to draw the Map or Plot | of any City, Town, Caftle, or other fortified place. | To make divers forts of Artificiall Fire-works, both for | War and Recreation, alfo to cure all fuch Wounds that | are curable, which may chance to happen by | Gunpowder or Fire-works. |

This Treatife is compofed for the help of all fuch | *Gunners and others, that have charge of Artillery,* | and are not well verfed in *Arithmetick* and | *Geometry*: all the rules and directions | in this Book, being framed both with | and without the help of | *Arithmetick.* | By *Nathaniel Nye* Mathematician, Mafter | Gunner of the City of WORCESTER. |

LONDON, | Printed for *William Leak,* at the figne of the | Crown in *Fleetftreet,* between the two | Temple Gates. 1647. |

8°. Pp. [22] 88; 102 [2 blank] sigs. A in eight, [but A 1, containing the portrait, is wanting] † 4; B to F in eights, G in four; a to f in eights, g in four (but g 4 is blank). Folding plates at pp. 69; second part, 20, 36, 50; 7 woodcuts in the text in the first and second parts; six engravings in the text in the second part.

On d 5 recto, p. 57 of the second part, is the following title-page:

A | TREATISE | OF | Artificiall FIRE-WORKS | for Warre and Recreation. | Conteining a defcription to make fundry | Kinds of FIRE-WORKS both for ufe | and pleafure, with leffe labour and | coft then any hath heretofore | been publifhed. | All of them being experimented and | practized by the Author: as alfo compared | to the beft, which are taught by | any other Author what- | foever. | By *Nathaniel Nye.* |

London, Printed for *William Leak,* at the | figne of the *Crown* in *Fleetftreet,* 1647. |

He makes the same claim in his address to the Reader: A 5 *verso*: "And whatfoever thou findeft in my Fire-works, I do protest to thee, that I have made, and ftill do make practice of them my felf; having by experience found them the beft of all others that ever I have read of: or that are taught by Bate, Babington, Norton, Tartaglia, or Malthus." These firework authors are rather touchy people and apt to go off briskly on the question of originality and of the merit of their processes and designs. Perhaps their characters were influenced by their studies and their experience of the instability of their materials.

63. This book of Nye's was reissued the following year with the date changed to 1648. The copy in the British Museum, 8827, b. 47, has the portrait, but wants the last blank leaf, and the folding plate at p. 50 is defective.

The portrait is an etching by Hollar, and is entitled: The true Effigies of Nathaniell Nye, | Mathematician, | and it is dated: *Ætatis ſuæ 20* | *A.* 1644. It represents him as a rather sulky boy, with long fair hair, in a slashed doublet and a large all-round white collar or tippet trimmed with lace, to his shoulders. The portrait appears to be rare; I suppose the Grangeriser conveys it away.

Nye was the author also of a Prognostication for 1643, so that he was apparently a precocious youth, with a turn for the mathematics.

Lowndes* mentions an edition of the Art of Gunnery, dated 1644, and Norgate† one dated 1670, regarding which I have no information.

The prognostication contains "a description of the foure quarters of the yeare, a description of all the Constellations of Stars, both Northern and Southern, and divers other things very meet to be known." Among these are a Table to know the beginning and end of the terms, eclipses, the position and motions of the planets on the first of each month, the right ascension, etc., of 30 principal fixed stars, some meteorological notes, new astronomical discoveries, tables for the lengthening and shortening of the day and of the dawn and twilight. It contains no calendar.

---

*Lowndes, *The Bibliographer's Manual*, London, 1861, Part VI., p. 1712.

†*Dictionary of National Biography*, 1895, XLI., p. 279. James Granger, *A Biographical History of England*, 1769, I. ii., p. 516; 2nd Ed., 1775, II., p. 338; 5th Ed., 1824, III., p. 162.

64. Gervase Markham,* the author of the tracts next to be described, was born in the latter half of the sixteenth century and died in 1637. He was a soldier, a scholar and linguist, a poet and a literary man, who lived by doing work for the booksellers. He wrote a great deal and is said both to have compiled several books on the same subject and to have issued the remainder copies of a book under a different title. That, however, was a common custom of the time and did not seem to be misunderstood, and some examples are referred to in this very paper. Leaving his poetical and literary writings out of account, the works which are under notice at present are those which deal with practical occupations, such as farriery, agriculture, domestic economy, sport, archery, angling, and so on. His books are full of curious and interesting matter relative to a style of living which has long passed away, and they bear evidence of personal experience of the descriptions and directions given. Markham was enthusiastic on certain subjects, especially horses and dogs, and about the former wrote several treatises.

The first editions of his books appeared early in the seventeenth century, and enjoyed a great vogue, as can be judged from the present collection, which purports to be in its seventh edition. The treatises seem to have appeared originally in a separate form, some at least did, and were afterwards collected, as here, under a general title. Each treatise, however, is provided with a separate title, which, even if long and somewhat tedious, is better given in full, since it affords a tolerably complete view of the contents, and saves the trouble of any description of the subject.

65. There follow, therefore, first, the general title, and then the individual titles in order:

A | WAY TO GET | WEALTH : | Containing ſixe Principall Vocations, or Callings, | in which every good Husband, or Houſe-wife may | lawfully imploy themſelves. | AS, |

I. *The natures, ordering, curing, breeding, choice, uſe, and feeding, of | all ſorts of Cattell, and Fowle, fit for the ſervice of man: As al- | ſo the riding and dieting Horſes, either for Warre or Pleaſur.* |

---

* *Dictionary of National Biography*, 1893, XXXVI., p. 166 (article by Clements R. Markham).

II. *The knowledge, ufe, and laudable practife of all the recreations | meet for a Gentleman.* |

III. *The Office of a Houfewife, in Phifick, Surgery, extraction of Oiles, | Banquets, Cookery, ordering of feafts, preferving of Wine, con- | ceited Secrets, distillations, Perfumes, ordering of Wooll, Hemp, | Flax, dying, ufe of Dayries, Malting, Brewing, Baking, and the | profit of Oates.* |

IV. *The enrichment of the Weald in Kent.* |

V. *The Husbanding, & enriching of all forts of barren grounds, ma- | king them equal with the most fruitfull: with the prefervati- | on of Swine, and a computation of men, and Cattels labours, &c.*

VI. *The making of Orchards, Planting and Grafting, the office of | Gardening, & the ornaments, with the beft Husbanding of Bees.* |

The firft five bookes gathered by *G. M.* The laft by Mafter | *W. L.* for the benefit of great Brittain |

The feventh time corrected, and augmented by the Author. |

LONDON. | Printed by *B. A.* for *John Harifon*, and are to be fold at his fhop | in Pauls-Church yard, 1648 |

4°. Title-leaf.

This collection was often republished, for in addition to the present edition the following are to be found in the British Museum Catalogue: 6th, 1638-31-38; 9th, 1657; 10th, 1660; 12th, 1668; 15th, 1695-83.

Another copy of this work is made up of the six treatises with the following dates:

I. Cheape and good Husbandry, 1631;
II. Country Contentments, 1633;
III. The English House-Wife, 1637;
IV. The Inrichment of the Weald of Kent, 1636;
V. Farewell to Husbandry, 1638;
VI. A new Orchard and Garden, 1638.

It will be noticed that some of these editions are not in the British Museum.

66. Treatise I.

Cheape and good | HUSBANDRY | FOR | The well-Ordering of all Beafts and Fowles, and | for the generall Cure of their Difeafes. | Containing the Natures, Breeding, Choice, Ufe, Feeding, | and Curing of the difeafes of all manner of Cattell as *Horfe*, | *Oxe, Cow, Sheepe, Goats, Swine*, and tame *Conies*. | Shewing further, the whole Art of Riding great Horfes, with the | breaking, and ordering of them, and the dieting of the Running, | Hunting, and Ambling Horfe, and the manner how | to use them in their travaile. | *Alfo approved Rules for the Cramming, and Fatting of all forts* (sic) *of* | *Poultry, and Fowles, both tame, and wild, &c. And diverfe good,* | *and well approved Medicines, for the Cure of all the difeafes* | *in Hawkes, of what kind foever.* | Together with the ufe, and profit of Bees, the making of Fifh- | ponds, and the taking of all forts of Fifh | Gathered together for the generall good and profit of this whole | Realm, by exact & affured experience from Englifh practifes, both cer- | tain, eafie, and cheap, differing from all former and forraign ex- | periments, which either agreed not with our Clime, or | were too hard to come by, or over-coftly, and | to little purpofe: all which herein are avoyded. | Newlye corrected and inlarged with many ex- | cellent additions, as may appeare | by this marke, ☞ | *The feventh Edition.* |

Printed at London by *Bernard Alfop,* for *Iohn Harrifon,* and are | to be fold at his fhop in Pauls Church-yard. 1648. |

4°. Pp. [26] 188 [2 blank].

Separate editions of this tract were published. The 1st, 1614; 2nd, 1616; 3rd, 1623; 5th, 1631; 9th, 1657; 13th, 1676, are in the British Museum.

67. Treatise II.

Country Contentments: | OR, THE | HVSBANDMANS | RECREATIONS. | CONTAYNING THE WHOLSOME | Experiences in which any man ought to Recre | ate himfelfe, after the toyle of more ferious bufineffe. | As namely, Hunting, Hawking, Courfing with | Greyhounds, and the lawes of the Leafe, Shooting in | Longbow or Crosbow, Bowling, Tennis, Baloone. The | whole Art of Angling, and the ufe of the Fighting Cock. | *By* G. M. | *The fixth Edition.* | Newly Corrected, Enlarged, and adorned with many excellent | Additions, as may appeare by this marke, ☞ | [Device.]

LONDON, | Printed by *William Wilfon,* for *Iohn Harifon,* in | St. Pauls-Church-yard. 1649. |

4°. Pp. [8] 118 [2 blank]. The device is a rabbit nibbling a stalk of wheat watched by a man—overhead the sun surrounded by a cloud; the whole in a scroll frame with the words DVM SPERO FERO and the initials I. H.

The editions in the British Museum are of 1611; 1615; 1623; 4th, 1631; 11th, 1675.

68. Treatise III.

THE | ENGLISH | HOUSE-WIFE, | CONTAINING | The inward and outward Vertues which ought to be in a | compleat Woman. | As her skill in Phyfick, Surgery, Cookery, Extraction of Oyles, | Banquetting ftuffe, Ordering of great Feafts, preferving of all forts of | Wines, conceited Secrets, Diftillations, Perfumes, ordering of | Wooll, Hemp, Flax, making Cloth, and Dying, the know- | ledge of Dayries, Office of Malting, of Oates, their | excellent ufes in a Family, of Brewing, Ba- | king, and all other things belonging | to an Houfhold. | A Work generally approved, and now the fifth time much aug- | mented, purged, and made moft profitable and neceffary for all men, | and the generall good of this KINGDOME. | By *G. M.* | [small scroll ornament.]

LONDON. | Printed by B. ALSOP for JOHN HARISON, and are to be sold at his | Shop in *Pauls* Church-yard, 1649. |

4°. Pp. [2 blank, 10] 252. Woodcuts in the text.

In its contents this book resembles many others published in the seventeenth century for the use of housewives and ladies. The edition of 1611, and the 11th edition, 1675, are in the Museum. The fourth edition appeared separately: London. | Printed by Nicholas Okes for Iohn Harison, and are to | be fold at his fhop at the figne of the golden | Vnicorne in Pater-nofter-row. 1631. It is in 4°, and contains A6 (A1 is a blank leaf), B to Q in eights, R in six, or, pp. [2 blank, 10] 252. It is in Roman letter and italics, with ornamental initials, and Harison's device of a rabbit and the legend: Dum spero fero, on the title-page.

69. Treatise IV.

THE | INRICHMENT | Of the Weald of | KENT: | OR, | A Direction to the Husbandman, | for the true Ordering, Manuring, & Inriching | of all the Grounds within the Wealds of KENT, | and SUSSEX; and may generally ferve for all | the Grounds in ENGLAND of | that nature: As, |

1. *Shewing the nature of all Wealdifh Grounds, comparing it with* | *the foyle of the Shires at large.* |

2. *Declaring what the Marle is, and the feverall forts thereof,* | *and where it is ufually found.*

3. *The profitable ufe of Marle, and other rich Manuring, as well* | *in each fort of Arable Land, as alfo for the increafe of Corne and* | *Pafture through the Kingdome.* |

Painfully gathered for the good of this Iland, by a man | of great Eminence and Worth, but revifed, inlarged, and | corrected with the confent, and by conference | with the firft Author. | By GERVASE MARKHAM. |

LONDON, | Printed by *Eliz. Purflow,* for *John Harifon,* and are to be | fold at his Shop at the figne of the Holy-Lambe at | the East-end of *Pauls.* 1649. |

4°. Pp. [4] 24.

Editions of 1625 and 1631 are in the Museum, and possibly others are known.

70. Treatise V.

Markhams farewell to | HVSBANDRY: | OR, | THE ENRICHING OF ALL | Sorts of Barren and Sterile Grounds in our | Kingdome, to be as fruitfull in all manner of | Graine, Pulfe and Graffe, as the beft grounds | whatfoever: |

Together with the annoyances, and prefervation of | all Graine and Seed, from one yeare to many yeares. |

As alfo, a husbandly computation of Men and Cattels dayly | Labours, their Expences, Charges, and utmoft Profits. |

Now newly the fourth time, revifed, corrected, and amended, | together with many new Additions, and cheape Experiments: |

For the bettering of Arable, Pafture, and Woody Grounds: Of | making good all Grounds againe, fpoiled with overflowing of falt | water by Sea-breaches; as alfo, the enriching of the Hop- | garden, and many other things never publifhed before. | [Device.]

LONDON, | Printed by *William Wilfon*, for *Iohn Harifon*, in | St. Pauls-Church-yard. 1649. |

4°. Pp. [12] 158. Woodcuts in the text. The device is the same as that in Treatise II.

In the British Museum are editions of 1620; 1625; 4th, 1638; 10th, 1676.

71. Treatise VI.

A new ORCHARD, and | GARDEN; | OR, | The beft way for Planting, Grafting, and to | make any ground good, for a rich Orchard: Particularly | in the North, and generally for the whole Kingdom of *England*, | as in nature, reafon, fituation, and all probability, may and doth appear. |

With the Country-houfwifes Garden for Herbs of common ufe: | their Vertues, Seafons, Profits, Ornaments, variety of Knots, Models for | Trees, and Plots for the beft ordering of Grounds and Walks. | AS ALSO, |

The Husbandry of Bees, with their feveral Ufes and Annoyances. | *All being the experience of Fourty and eight yeers labour, and now the fecond* | *time corrected and much enlarged, by* WILLIAM LAWSON. |

Whereunto is newly added the Art of Propagating Plants; with | the true ordering of all manner of Fruits, in their gathering, | carrying home, and prefervation. | [Vignette.]

London, Printed by *W. Wilfon*, for *John Harifon*, and are to be fold | at his Shop in *Pauls* Church-yard, 1648. |

4°. Pp. [8] 134. Woodcuts in the text. The vignette (repeated on p. 33) represents three men engaged in grafting, with the mottoes: "Skill and pains, bring fruitful gains"; and "Nemo sibi natus."

This treatise consists of various tracts.

Pp. 1-74, An Orchard.

Pp. 75-108, The Country House-Wives Garden, with a title-page:

THE | COVNTRY HOVSE-WIVES | GARDEN, | Containing Rules for herbs, and Seeds, | of common ufe, with their times and feafons | when to fet and fow them. | Together, | With the Husbandry of Bees, publi- | fhed with fecrets very neceffary for every *Houfe-* | *wife*: as alfo divers new Knots for Gardens. | The Contents fee at large in the laft Page. |

Genef. 2. 29. |

I have given unto you every Herb, and every tree, that ſhall be to you for | *meate.* |

[Harison's Device, as before.]

LONDON | Printed by *William Wilfon*, for *John Harrifon*, and are to be | fold at his Shop in Pauls-Church-yard. 1648. |

Pp. 109-123 [124 blank]. The Art of propagating plants, by Simon Harward.

Pp. 125-134, The Husbandmans Fruitful Orchard. P. 134, Colophon: [Harison's device]; *LONDON.* | Printed by *William Wilfon* for IOHN HARISON, | in *St. Pauls Church-yard.* 1648. |

Editions of the "Orchard" dated 1625; 14th, 1683, are in the Museum, and also one of 1623 of the "Country House-Wives Garden."

72. In order of date the next book is one on which some remarks were made formerly* as to its authors and as to its connection with other pamphlets of similar character. Its title begins:

A Pretious Treasury: or a New Dispensatory . . .

It was printed at London by Thomas Harper for Richard Harper, 1649. It forms a pamphlet, 4°, pp. [1-6] 7-20, and the authors are Salvator Winter and Francisco Dickinson. Having already described it from the British Museum copy, I should not have referred to it here again except for two reasons: first, the printer, it will be observed, is Thomas Harper, whose name has just been before us in connection with the writings of Babington, Banester and Bate. Harper, like Conyer some sixty or seventy years later, seems to have been a printer of scientific and medical books, and it is odd how on this occasion chance has brought four of his productions together. Secondly, it is a very scarce tract, because, being so slight and

* *Transactions of the Glasgow Archæological Society*, N.S., 1903, IV., p. 117 (Reprint, Supp. IV., p. 25).

unimportant, it would be easily overlooked and destroyed. It is indeed little more than an advertisement circular. Few copies have escaped, but one has chanced to fall into my hands; the only one I have ever seen for sale. It is a singular document and affords a glimpse of the ways of the time which only such a tract could do.

73. In a former part* a notice was given of a book issued at London in 1656, called *Queen Elizabeths Closset of Physical Secrets.* This also was a reissue of a book already published under a different name. I have a copy of the earlier issue, and its title runs in the following fashion:

> A | Rich Closet | of Physical | Secrets, | Collected by the Elaborate paines | of four feverall Students of Phyfick, | And digefted together; | *Viz.* | *The Child-bearers Cabinet.* | *A Prefervative againft the Plague and fmall Pox.* | *Phyficall Experiments prefented to our late Queen* | Elizabeth's *own hands.* | With | Certain approved Medicines, taken out of a | Manufcript, found at the diffolution of one of our | Englifh Abbies, and fupplied with fome of his own | Experiments, by a late Englifh Doctor. |
>
> London, | Printed by Gartrude Dawfon, and are to be sold by *William* | Nealand, at the Crown in *Duck-Lane.* 1652. |
>
> Small 4°. Pp. [8] 71 [1 blank]; A Treatise concerning the Plague and the Pox, [6] 65 [1 blank]; Some other few additionall observations, pp. 97-146 [Tables, 14].

There is no doubt regarding the identity of the two editions, for, apart from the title-page of the Treatise on the Plague in both issues bearing the imprint: "Gartrude Dawson, 1652," the irregularity in the pagination, caused by dropping the numbers between 65 and 97, is also the same in both. That omission originated, apparently, in a simple inversion of the figure 6, after which the error was carried on to the end of the book, making it appear 30 pages longer than it is. The most important difference, however, between the two issues is Elstrak's portrait of Queen Elizabeth, with which the edition of 1656 is embellished. This portrait does not appear to have existed in that of 1652.

---

* *Transactions of the Glasgow Archæological Society*, N.S., 1888, I., p. 319 (Reprint, Part IV., p. 21).

ADDENDA.

13TH JANUARY, 1908.

IT will obviate the necessity of recurring to parts of the subject treated of in the preceding paper, if there be now added information accumulated since it was printed.

§ 11. The 1575 edition of *The true and perfect order to diſtill oyles* was mentioned eighty years before Will. Cooper, by Andrew Maunsell in *The Seconde parte of the Catalogue of English printed Bookes*, London, 1595, small folio, p. 8. Cooper, therefore, may have simply taken his notice of the book from Maunsell, but, considering Cooper's opportunities and collector's enthusiasm, it is probable that he possessed a copy of it, or at least had seen one.

§ 15. Maunsell (*Catalogue*, 1595, Part II., p. 23) has the following entry:

> Remedies to take out Spottes and Staines, in Silkes, Veluets, Linnen and Wollen clothes. Alſo how to Die Silks, Linnen, and Wollen, etc. Alſo to dreſſe Leather, and to colour Fells, Howe to Gylde, Grane, Sowder, and Verniſhe, and to harden and make ſoft Iron and Steele, tranſlated by L. M. Prin. by Tho. Purfoote. 1588. 4°.

If this date be correct, and not a misprint for 1583, then this is an edition of which I have seen no copy, but there is no reason why there should not be an edition of this date. Will. Cooper does not refer to the book at all, which is a curious omission for him to have made. The 1588 edition, and one of 1600, "London, Printed by John Wolfe, and are to be sold by Edwarde White, etc., 4to, with woodcuts," are mentioned by W. Carew Hazlitt, *Hand-Book to the Popular . . . Literature of Great Britain*, 1867, p. 382. He adds that the Harleian Catalogue contains an edition of 1606.

§§ 17, 18. It may be convenient for reference to give the names and editions of the rest of the works attributed to Mascall which are in the British Museum.

A Booke of fishing with Hooke and Line. . . . Another of sundrie Engines and Trappes to take Polecats, Buzards, Rattes. . . . 1590. 4°; and London, 1884. 4°.

W. C. Hazlitt (*Bibliographical Collections and Notes*, 1876, I., p. 282) quotes an edition of 1600, 4to.

A booke of the art and maner, howe to plant and graffe all sortes of trees. . . . London, [1572], 1575, [1580?], 1582, 1592, 1596, 1652, 1656, all in 4°.

Maunsell (*Catalogue*, 1595, Part II., p. 10) gives the title of the 1580 edition as follows:

Art of Graffing and Planting all forts of trees, how to fet Stones and fowe Pepins, to make wild trees to Graffe on, as alfo remedies, and Medicines, with other new practifes. &c. by Leonard *Mafcall.* Pr. for Jo. Wight. 1580. 4°.

Of the editions, Richard Weston, in "A Chronological Catalogue of English Authors on Agriculture, Botany, Gardening, etc.," added to *Tracts on Practical Agriculture and Gardening*, London, 1773, 8vo, p. 10, quotes that of 1572, and (p. 27) that of 1652.

W. C. Hazlitt (*Bibliographical Collections and Notes*, Second Series, London, 1882, p. 387) quotes an edition of 1590, besides some of those in the Museum.

Mascall's remaining work is that on Cattle, in three books. In the British Museum are editions of the following dates: 1596, 1600, 1605, 1620, 1633, 1662-61, all in 4°, and 1680, 8°, called 'The Countryman's Jewel.'

Weston (*Chronological Catalogue*, p. 16) mentions under 1596: On Horses, the second book of his treatise on cattle, black letter, in 4to; but he quotes (p. 20) *Government of Cattle*, in three parts, black letter, in 4to, 1627, and (p. 27) the same book, black letter, in 4to, 1653, which editions are not in the Museum. He also quotes (p. 43) *The Countryman's Jewel*, in three books, in 8°, 1680.

In the note ‖ to § 18, I have described the edition of 1662 (the title to the second book on Horses is dated 1661), and I may now describe the edition of 1610, which is not in the Museum and does not seem to be known otherwise.

THE | FIRST BOOKE | OF CATTELL. | *Wherein is ſhewed, the gouernement of* | Oxen, Kine, and Calues, and how to vſe | Buls and other cattell to the yoke, and fell. | With diuers approued remedies, to helpe moſt di- | ſeaſes among cattell: moſt neceſſary for all, eſpeci- | ally for huſbandmen, hauing the gouerne- | ment of any ſuch cattell. | *Gathered and ſet forth by Leonard Maſcall.* |

Thou heardman, keeper of thy beaſtes,
When any beaſt is ſickly:
Search herein, and thou ſhalt finde,
[Vignette.] Of proued remedies quickly.

LONDON, | Printed by *Nicholas Okes* for *John Hariſon*, and are to | be ſold in Pater Noſter Row, at the ſigne | of the white Grayhound. | 1610. |

Small 4°. A to T in eights, V in four, of which A1 is blank; or pp. [2 blank, 6] 1-301 [1], but the numbering is very irregular, and R6 has been omitted. Black letter, except the title, the Epistle, Verses to the Husbandman, the head lines, titles to the sections, rubrics, and alphabetical tables, in roman. The Vignette is a caduceus intertwined with two cornucopiæ, held by two hands out of clouds, and over all Pegasus.

The book is in three parts, with separate title-pages and indices, which however are included in the signatures and pagination.

The second title (on G5) may be abstracted:

THE | SECOND BOOKE | INTREATING OF | THE GOVERNMENT | of Horſes, with the approued | remedies againſt moſt | *Diſeaſes.* | . . . | Gathered by LEONARD MASCAL. |

Although the learned haue reueald
The helpes for horſe great ſtore,
Yet practiſers therein againe,
Haue found for them much more.

[Vignette as in the first title.]

AT LONDON | Printed for IOHN HARISON, | 1610.

The third title is as follows (O1):

THE | THIRD BOOKE, | Intreating of the ordering of Sheepe, | *Goates, Hogges, and Dogs: With* | ſuch remedies to helpe moſt diſ- | eaſes as may chaunce | to them. | . . . | *Gathered by Leonard Maſcall.* | [Device.]

LONDON | Printed for *Iohn Hariſon*, and are to be | fould in Pater-Noſter ROWE, at the | ſigne of the white Gray hound. | 1609. |

The device is a blurred medley of birds, fruit, faces, and a shield. Harison published Markham's books. It is to be hoped that Mascall's farriery was better than his poetry.

§ 19. Mascall's treatise on Poultry, 1581, is not quoted by Maunsell, but it is included by Richard Weston in his "Chronological Catalogue," 1773, p. 12. Weston, however, did not know the author's name.

§ 20. This book, *Prepofitas his Practife*, is quoted by Maunsell, Part II., p. 21, but he gives the initials of the translator inaccurately as T. M., instead of L. M.

§ 22. Maunsell (p. 8) quotes an edition of Vicary's *The Englishman's Treasure*, for John Perin, London, 1587, 4°. Herbert also refers to it (*Typ. Ant.*, 1790, III., p. 1343).

§ 26. The edition of 1587 of Levins' book is thus given by Maunsell (p. 20):

> The Pathway to health, wherein are moft excellent and aproued Medicines of great vertue, as alfo notable potions, and drinkes, and for the diftilling of diuers waters, and making of Oyles and other comfortable receytes, by Peter Leuins. Pri. for Edw. White. 1587. 4°.

33*a*. Between §§ 33 and 34 should be inserted the collection following, with which I became first acquainted through Maunsell's *Catalogue*, 1595, part second, p. 27:

> The widowes treafure, plentifully furnifhed with fundrie precious approued fecretes, in Phyficke and Chirurgerie. Prin. for Edw. White. 1591. 8°.

This, presumably the first, edition is unfortunately not in the British Museum, but there are others which are enough to confirm what I have often remarked upon, the extraordinary demand there was for collections of receipts and secrets. No matter how defective such a book was in matter and in manner, it was certain to sell, and a new edition to be called for every four or five years; then possibly after a period of stagnation or oblivion, it might be brought out anew and meet with fresh acceptance.

Another point, also, is the apparent importance attaching to liquors, extracts, oils and what not, prepared by distillation. In the present paper alone there are numerous books occupied in whole or in part with details of the process and its results.

The editions of this work in the Museum are the following:

[B.M. 1037. e. 1. (2.)]

THE | WIDDOWES | *Treaſure.* | Plentifully furniſhed with ſundry | precious and approued ſecrets in Phi- | ſicke and Chirurgery, for the | health and pleaſure of | Mankinde. | Heereunto are adioyned, ſundry prit- | tie practiſes and concluſions of Coo- | *kerie, with many profitable and wholſome* | *Medicines, for ſundrie diſeaſes in* | *Cattell.* |

AT LONDON, | Printed by I. Roberts for Edward | VVhite. 1595. |

Small 8°. A to F in eights. No pagination. Black letter, except the title, the address, the running title, and headings of the receipts, in roman.

The address to the reader is such an excellent example of how to avoid giving a clear explanation about any matter, and is couched in such quaint language, that, at the cost of prolixity, I reproduce it in full:

To the Curteous Reader.

This Pamphlet being written (not many yeeres past as it should seeme) at the earnest request and sute of a Gentlewoman in the Countrie for her priuate vse, which by these singuler practises hath obtained such fame, that her name shal be remembered for euer to the posteritie. The originall Copie (by great chaunce) was lent me by an especiall friend of mine, in the perusing whereof I found it so furnished with such rare experiments and prittie conceites, as the like are not to my knowledge extant in the English tongue. I thought it my dutie for the health and pleasure of many my friends, to set them forth to the view of the world, to be practised, especially of those that take delight in such matter, or hath any occasion through their defects to vse them for their healthes sake. This I dare presume to reporte of the worke that there are heerein included very many secrets, that I knowe by the Widdowes owne practise, to be most singuler & approued, and though perhappes they are not orderly set downe as many of better skill might haue done: I desire you to consider that this was of the collection of such a one as neuer once thought to haue made them common, but onely to his good friend M. R.

This caueat she also giueth by the way, to reade them aduisedly & practise with discretion, and as you finde cōmend: and so with all humilitie concluding, I dedicate vnto you this her Treasure and cheife Iewell, desiring you to be a defence aswell for her as for me, from all such as shall in your hearing scorne her, or me for disclosing such profitable experiments.

Farwell.

From the wording of this address the writer of it is apparently the responsible editor, but not the author, of the book. Nor is it clear if the Gentlewoman was identical with the Widdow, or if the Widdow wrote it for her, or if the Widdow and Gentlewoman only put the receipts in operation with marked success and distinction. M. R. also is an unknown personage, but the collection was made by some one for this his good friend. The author, therefore, must have been a man. But, if so, what of the caveat 'she,' seemingly the authoress, also giveth by the way? Was M. R. the Widdow or the Gentlewoman? If not, the book had a double origin: on one hand it was compiled for the Gentlewoman at her request and sute and was used by her and by the Widdow as well; and, on the other hand, the author, when he was compiling it, thought of nobody but M. R. Who's who? One may wish that this address had been less involved.

But, alas for the Gentlewoman's promised posthumous fame and the remembrance of her name! The posteritie has but a poor memory. No one has heard either of her or her name; the Pamphlet is unknown which was to perpetuate it; the rare experiments and prittie conceites, curious as they are, have long since sunk into oblivion, and not even a better arrangement of them could have saved them from their doom; no need is there now to defend either the Widdow or her editor, for the scorner has passed away too, his chair is empty and his envy is buried in the dust. Vain are the hopes the daughters of men have built on such works.

The secrets and receipts are not only miscellaneous, but are set down without any order or connection, and as there is no table or index one would require absolute familiarity with the contents before consulting the book. The compiler, whether the widow or another, probably just jotted down any receipt that struck her as useful, and when the collection came to be printed no attempt at arrangement was made. Accordingly the most unlikely 'secrets' come together. The first of them are for the most part medical and refer to cures for all sorts of ailments. Then, in the very middle of these, come receipts 'To keepe venison fresh a long time,' 'to keep pears,' 'against Drunkennes,' 'to make Linnen cloth or Yarne white,' 'a good drink for the Pestilence,' 'to make a perfect black Ink to write or limme withall,' 'a very good Greene,' and so on quite indiscriminately. At the end are some cookery receipts and then others for horses and cattle, and the collection concludes with a receipt 'Approued against the plague.'

The book, notwithstanding the laudation in the address, does not differ from similar collections of receipts already described.

[B. M. C 54. a. 14.]

THE | WIDDOWES | *Treafure.* | Plentifully furnifhed with fundry | precious and approued fecrets in Phi- | ficke and Chirurgery, for the | health and pleafure of | Mankinde. | Heereunto are adioyned, fundry prit- | tie practifes and conclufions of Cookerie, | *vvith many profitable and wholefome Medicines* | *for fundry difeafes in Cattle.* |

AT LONDON, | Printed by I. Roberts for Edward | White. 1599. |

Small 8°. A to F in eights. No pagination. Black letter, except the title, 'To the Courteous Reader,' head lines and headings of the sections, in roman.

This is a page for page reprint of the edition of 1595, but it is not line for line, and there are numerous alterations in spelling.

THE | WIDOWES | TRESVRE, | Plentifully furnifhed with | fundry precious and approved | Secrets in Phyfick and Chirur- | gery, for the health and pleafure | of Mankind. | Hereunto are adjoyned fundry pretty | Practifes and conclufions of Cookery, | with many profitable and wholefome | Medicines for fundry Difeafes | in Cattell. |

[Vignette of a fleur-de-lis.]

LONDON, | Printed by *R. B.* for *Robert Bird*, and are to | be fold at his Shop in S. *Laurence-lane* | at the Signe of the Bible. 1639. |

Small 8°. A to F in eights. No pagination. Black letter, except the title, To the Courteous Reader, head lines and titles of the receipts, in roman.

This is practically a page for page reprint of the 1595 edition, except sheet F, which has run over a little in the 1639 edition. The spelling also is modernised somewhat. The address to the reader is the same in all the editions.

In the British Museum (1038. f. 43.) is another edition of "The Widowes Treasure," which unfortunately wants A1, the title-leaf. It is dated in the catalogue "1600?", but that seems to be too early. It is a small 8vo, A to G in eights, no pagination. Black letter, but the address To the Curteous Reader, head lines and titles to the receipts are in italics. The black letter seems more compressed than that in the other editions, and the initial letters of the paragraphs are roman and not gothic as in the 1595 and 1639 editions.

Cvij *r*. Marmalade of Quinces is on Di *r* in the editions of 1595 and 1639.

Cvij *v*. A whaie for Itche or breaking out is on Di *v* in the same.

§§ 37, 38. Since these sections were printed I have acquired a copy of the 1596 edition of A. T.'s book: *A Rich Store-houſe or Treaſury for the Diſeaſed.* All things considered it is in good condition, and it has the advantage over the Museum copy of being perfect. But I have got besides an edition of 1601, which is not in the British Museum, and which may, therefore, be more fully considered, in view of the universal catalogue of English books that is to be compiled some day:

A | RICH STOREHOVSE, | OR | Treaſurie for the Diſeaſed. | WHEREIN, ARE MANY APPRO- | ued Medicines for diuers and ſundry | Diſeaſes, which haue bin long hidden, and not | come to light before this time. | FIRST SET FOORTH FOR THE | great Benefit and Comfort of the | poorer ſort of People that are not of abilli- | tie to go to the Phiſitions, by A. T. | And now newly Corrected, augmented and inlarged, by | G. W. Practicioner in Phiſicke and Chirurgerie. | [Device.]

¶ AT LONDON, | Printed by Ralph Blower. 1601.

Small 4to. A2, B to Y in fours, Z2; or ff. [10] 78. Black letter, but the title, Epistle Dedicatorie, head-lines, titles of sections and chapters, and technical words are in roman. The device is a clenched right hand with the motto: EX AVARITIA BELLVM. There are numerous florid capitals, and typographical ornaments between the main sections.

As the first edition appeared in 1596, and the fourth in 1607, the question arises whether this of 1601 is the second edition or the third. Judging by the date and allowing for the usual interval of about four or five years for an edition, this might be either the second or third, though more likely the third, for the reason that the second was possibly or probably issued by the author himself immediately after the first, after which the work may have fallen into the hands of G. W. This view seems to be confirmed by the rate at which additions were made successively to the book, as is shown by the following table:

1596.	1st Edition,	341	Chapters.
1601.		604	„
1607.	4th „	689	„
1612.	5th „	784	„
1631.	11th „	795	„

The problem, then, which can be solved only by the discovery of the other edition between the first and the fourth, is, whether it is a mere reissue of the first and contains only 341 chapters, or whether it was enlarged, either by the author or by G. W., and contains fewer chapters or more than 604. Speculation on the subject is futile.

It will be noticed that in the sixteen years between 1596 and 1612 there were five editions, and in the nineteen years from 1612 to 1631, there were six, which illustrates the steady demand there was for the book.

§ 55, Note *. The discrepancy between Reid's death in 1641, and the preface of his book in 1642, could be got over possibly by assumptions like these: 1°. That the address to the reader had been already printed off before the author's death and had to stand as it was. 2°. That, if the author died before the printing was begun, the printer, or publisher, or editor, did not presume, or did not wish, or did not trouble to alter the terms of the passage referred to. But whether such explanation, or some other, be sufficient or not, the actual explanation is probably now beyond our power.

§ 59. Maunsell in *The Seconde parte of the Catalogue of English printed Bookes*, London, 1595, p. 20, mentions: Jo. Partridge, Treafurie of commodious conceits and hidden secrets. Rich. Jones. 1591. 8°. This edition I have not seen, but it fills up the blank between the editions of 1586 and 1596.

§ 60. I have now got a copy of the 1684 edition of "The Ladies Cabinet Enlarged and Opened," which tallies with the description given.

BIBLIOGRAPHICAL NOTES

ON

HISTORIES OF INVENTIONS

AND

BOOKS OF SECRETS

SIXTH SUPPLEMENT

BY

JOHN FERGUSON, LL.D., F.S.A. LOND. AND SCOT.

REGIUS PROFESSOR OF CHEMISTRY IN THE UNIVERSITY OF GLASGOW

GLASGOW

Printed at the University Press by

ROBERT MACLEHOSE & CO. LTD.

1910

Read to the Archæological Society of Glasgow, 20th February, 1908.
100 *Copies Reprinted.*

No. I.

BIBLIOGRAPHICAL NOTES ON HISTORIES OF INVENTIONS AND BOOKS OF SECRETS.

(*SIXTH SUPPLEMENT.*)

BY

JOHN FERGUSON, LL.D., F.S.A. LOND. AND SCOT.

(*Read at a Meeting of the Society, held on Thursday, 19th December,* 1907.)

1. The present paper on receipt books in English, from 1651 to about 1700, was to have been read a couple of years ago, but for it, in the interests of chronology, another was substituted, dealing with English books prior to 1650, which had not been previously described. As these have now been recorded, the postponed paper may be resumed.

It is worth while to notice the difference in style and subject in the two periods. The present books are less varied in theme than those preceding them, and they consist mainly of receipts for the purposes of every-day life: for cleaning, dyeing, cooking, preserving, and so on. In fact, a considerable number of them were intended for the ladies and housewives, and for servants. There are only a few about medicine and natural history.

2. The books themselves were printed for the pedlars and hawkers, who sold them about the country. Consequent upon this was a change of form. They were no longer in the small quarto size which was so common in the

sixteenth century, and to some extent also in the seventeenth, but were in small octavo or duodecimo, and were as cheaply produced as was possible at the time. Notwithstanding their external imperfections they are of interest from their contents, as well as from peculiarities in their publication and production. At all events they are different from the books which have been recently under review, and they thus possess the attraction of novelty.

3. Among the authors of the latter half of the seventeenth century whose compilations found a large and speedy sale there was John White, who has distinguished himself from others of the same name by adopting the title: "a lover of Artificial Conclusions," or, "a lover of Art and Ingenuity." This designation, as has been shown in my paper on Thomas Hill, was borrowed from the collection which Hill translated from the Italian. Of White himself I have not discovered any notice. Granger has omitted the rude portrait of him, which I reproduce from the fourth edition of ***Arts Treasury***, hereafter described (§ 19), as all the contribution I can make to his biography. He appears, however, as the author of four or five collections, though it is not impossible that some of the contemporary anonymous literature of like character, published by Conyers, was his handiwork as well. There is, however, only internal evidence in support of such a view, and that is barely enough. I shall, therefore, credit him only with the books which he himself has acknowledged, and describe the others separately. Unfortunately, a number of the editions of his books are undated, although they belong obviously to the middle or end of the seventeenth century. The want of dates in Conyers' publications is an irreparable defect.

4. As similar titles are borne by the collections, these had better be enumerated to avoid confusion. They are:

(A.) A rich Cabinet, with Variety of Inventions;
(B.) A rich Cabinet of Modern Curiosities;
(C.) Hocus Pocus, or a rich Cabinet of Legerdemain Curiosities;
(D.) Arts Treasury of Rareties and Curious Inventions;
(E.) Arts Masterpiece.

The earliest edition of any of these is of 1651, so that the volume of that date may be made the starting point for observations on White's writings.

5. (A. 1.) The title runs as follows:

A RICH | CABINET, | WITH | Variety of Inventions; | Unlock'd and opened, for the recreation of | Ingenious Spirits at their vacant houres. | Being Receits and Conceits of ſeverall Natures, | and fit for thoſe who are lovers of Naturall and | *Artificiall Concluſions.* | Whereunto is added variety of Recreative | Fire-works, both for *Land, Aire,* and *Water.* | ALSO | Fire-works of Service, for Sea and Shore, | very fitting for theſe Warlike times or Action. | Collected by *J. W.* a lover of | *Artificiall Concluſions.* | [Device.]

Printed for *William Gilbertſon* at the ſigne of the Bible | without Newgate in *Giltſpur-ſtreet.* 1651. |

Small square 8vo. A to L in eights, M 2; no pagination. A 1 (blank, or with an engraved frontispiece, or portrait, or licence?) is wanting. Woodcuts in the text. The Epistle is signed: *An Artiſts Friend,* | JOHN WHITE.

This is the first edition. It is in the British Museum (E. 1295. 2.).

6. (A. 2.) It was followed in 1653 by the second edition. The title-page is repeated and the book is simply reprinted, but at the end there is an addition of fourteen pages on fire-works.

What follows gives the needful details of this edition.

A RICH | CABINET, | WITH | Variety of Inventions; | Unlocked and opened, for the recreation of | Ingenious Spirits at their vacant hours, | Being Receits and Conceits of ſeverall Natures, | and fit for thoſe who are lovers of Naturall and | *Artificiall Concluſions.* | Whereunto is added, variety of Recreative | Fire-works, both for *Land, Aire,* and *Water.* | ALSO | Fire-works of Service, for Sea and Shore, very | fitting for theſe Warlike times of Action. | Collected by *J. W.* a lover of Artificial Concluſions. | The ſecond Edition with ſome Additions. | [Device.]

LONDON | Printed for *William Gilbertſon* at the Sign of the Bible | without New-gate in *Giltſpur-ſtreet.* 1653, |

Small 8°. A to M in eights; or pp. [2, blank] [14] 174 [2]. This edition did not have an engraved title page, for the first leaf in A is blank in my copy but is wanting in that of the British Museum (7955. a. 15.). The copies tally in other respects. There are woodcuts in the text.

7. The receipts are divided into sets or sections, not very sharply defined. The first set is of such a miscellaneous character that every one of the "Receits" would have to be quoted to convey a notion of what they are

all about. The second set consists of a few physic receipts. The third is devoted to fire-works, partly for amusement, partly for war both by land and sea; the fourth, directions for drawing and painting; the fifth, some legerdemain tricks (there are some in the first division as well), which were afterwards repeated in *Hocus Pocus*; the sixth and seventh, exercises in "Arithmetick"; and the last, experiments in geometry.

8. The section in White's book about fire-works can be compared with the writings of Hanzelet, Babington, Nye, Bate, etc., already reviewed, and with others to be afterwards described. These writers, however, are quite independent of one another; the directions are different, the illustrations, though analogous, are distinct, and the devices do not coincide. As the directions are given by authors who had experience in the making of fire-works, we have in these books a full insight into this branch of technology in the seventeenth century. More on the history of the subject can be gathered from the treatises of Biringuccio,* Agricola, Birelli, Tartaglia,† Alberti,‡ Siemienowicz, and others.

9. The third and fourth editions, and two issues of the fifth edition are next to be considered. They are enlarged in size and printed in a different style. The third edition has the following title-page:

(A. 3.)

A RICH | CABINET, | WITH | Variety of Inventions: | Unlock'd and open'd, for the Recreation of | Ingenious Sirpits (*sic*) at their vacant hours. | Being Receits and Conceits of feverall Na- | tures, and fit for thofe who are lovers of Na- | tural and Artificial Conclufions. | AS ALSO | Variety of Recreative Fire-works, both for *Land*, | *Air*, and *VVater*. And Fire-works of Ser- | vice, for Sea and Shore. | Whereunto is added divers Experiments in *Drawing*, | *Painting*, *Arithmetick*, *Geometry*, *Aftronomy*, and | other parts of the Mathematicks. | Collected by *J. W.* a lover of Artificial Conclufions. | The Third Edition, with many Additions. |

---

* Biringuccio, *La Pirotechnia*, Venetia, 1550, 4°.

† Niccolò Tartaglia, *Three Bookes of Colloquies concerning the Arte of Shooting in great and small Pieces of Artillerie . . . translated . . . by Cyprian Lucar*, . . . London, 1588, small folio. Lucar's *Appendix* contains additional matter relating to gunpowder.

‡ Giuseppe Antonio Alberti, *La Pirotechnia*, Venezia, 1749, will be included hereafter among foreign technical books.

LONDON, | Printed for *VVilliam Gilbertſon* at the Sign of the | Bible without New-gate in *Giltſpur-ſtreet*. | 1658. |

Small 8vo. 8 leaves without signature or pagination; B to N in eights; or pp. [16] 191 [1 blank]. Numerous woodcuts in the text.

The first eight leaves contain: 1, Verses, "The Authour to his Book"; 2, Engraved title, by T. Cross, dated 1658, containing, in small compartments, reproductions of the most important devices in the book; 3, Printed title, as above; 4, White's Epistle; 5 to 8, Table of Contents. The text occupies the remainder of the volume.*

10. (A. 4.) The fourth edition is a reprint of this, but, although nearly agreeing page for page, differs typographically from it throughout. The title runs thus:

A RICH | CABINET, | WITH | Variety of Inventions: | Unlock'd and open'd, for the Recreation of | Ingenious Spirits at their vacant hours. | Being Receits and Conceits of ſeverall Na- | tures, and fit for thoſe who are lovers of Na- | tural and Artificial Concluſions. | AS ALSO | Variety of Recreative Fire-works both for *Land*, | *Air*, and *Water*. And Fire-works of Ser- | vice, for Sea and Shore. | Whereunto is added divers Experiments in *Drawing*, | *Painting*, *Arithmetick*, *Geometry*, *Aſtronomy*, and | other parts of the Mathematicks. | Collected by *J. W.* a lover of Artificial Concluſions. | The Fourth Edition, with many Additions. |

LONDON, | Printed for *William Whitwood* at the ſign of the | Golden Lion in *Duck-Lane* near *Smith-field*. | 1668. |

Small 8vo. 8 leaves without signature or pagination; B to N in eights, or, pp. [16] 190 [2]. Engraved title and numerous woodcuts in the text.

The first eight leaves correspond exactly with those in the third edition, except that the date in the engraved title has been altered to 1668. By a little compression at the end the text has been got into pp. 190, and the last two pages are occupied with the names of some books sold by Whitwood.†

11. (A. 5.) In the fifth edition, the title, though altered in spelling, corresponds in arrangement with the preceding down to "other parts of the Mathematicks," and then there follows an addition about bell-ringing. The title may be given in full:

---

* This edition is quoted by W. Carew Hazlitt, *Third and Final Series of Bibliographical Collections and Notes*, 1887, p. 263.

† This edition also is quoted by W. C. Hazlitt, *Supplements to the Third and Final Series of Bibliographical Collections and Notes*, 1889, p. 177, and there is a copy in the British Museum, 1036. c. 5.

A RICH | CABINET, | WITH | Variety of Inventions: | Unlock'd and open'd, for the Recreation of | Ingenious Spirits at their vacant hours. | Being Receipts and Conceits of ſeveral Na- | tures, and fit for thoſe who are lovers of Na- | tural and Artificial Concluſions. | AS ALSO | Variety of Recreative Fire-works both for *Land*, | *Air*, and *Water*. And Fire-works of Ser- | vice for Sea and Shore. | Whereunto is added divers Experiments in *Drawing*, | *Painting*, *Arithmetick*, *Geometry*, *Aſtronomy*, and | other parts of the Mathematicks. | Likewiſe Directions for Ringing the moſt uſual | Peals, that belong to that Art. |

Collected by *J. W.* a lover of Artificial Concluſions. |

The Fifth Edition, with many Additions. |

LONDON, | Printed for *William Whitwood* at the ſign of the | Golden Bell in *Duck-Lane* near *Smith-field*. | 1677. |

Small 8vo. 8 leaves without mark; B to P in eights, Q four; or, pp. [16] 190; 41; [1]. Woodcuts in the text. The engraved title is the same as in the preceding edition, but the date has been altered to 1677, though not so as to obliterate entirely the earlier figures. The preliminary matter and the text are the same as before; the appendix on bell-ringing is paged separately. The last page contains the advertisement of two books. The most notable difference is the partial modernizing of the spelling throughout, and it may be also observed that Whitwood had changed his sign. This is the only edition quoted by Lowndes,* who gives a price: 'Bright, 6s.'

12. The second version of the fifth edition is somewhat of a puzzle. This is the title:

A RICH | CABINET, | WITH | Variety of Inventions, | Unlock'd and open'd, for the Recrea- | tion of Ingenious Spirits. | Being Receits and Conceits of ſeveral | Natures, and fit for thoſe who are lovers of | Natural and Artificial Concluſions. | AS ALSO | Variety of Recreative Fire-works both | for *Land*, *Air*, and *Water*. And Fire-works | of Service, for Sea and Shore. |

Whereunto are added divers Experi- | ments in *Drawing*, *Painting*, *Arithmetick*, *Geo-* | *metry*, *Aſtronomy*, and other parts of the Ma- | thematicks. Together with ſeveral Curious | Receits of Great uſe, collected out of *Alexis*, | *Mizaldus*, *Wecker*, &c. |

By *John White* a lover of Artificial Concluſions. |

The Fifth Edition, with many Additions. |

LONDON, | Printed for *William Whitwood*, next the *Crown* | *Tavern* in *Duck-Lane*. 1684. |

16°. A to I in eights; or, pp. [16] 128. Engraved title extra. The printed title is surrounded by a double line.

A1, The Author to his Book. A2, The title page as above. A3, The Epistle, signed: An Artiſts Friend, | JOHN WHITE. A4, The Table. A8, Books sold by William Whitwood. B to I8, The Text of the book.

---

* *The Bibliographer's Manual*, ed. Bohn, 1864, V., p. 2900.

This is a chap-book, quite inferior to the fifth edition of 1677. The engraved short title is as follows:

A Rich | Cabinet | with | Variety of Inventions | in several Arts | and Sciences. | By J. W. | London | Printed for Will. Whitwood | 1684 |

It is surrounded by the drawings in compartments of the devices and apparatus described in the book. The date has obviously been changed, but the original figures have been so defaced as not to be decipherable. This book further exhibits an anomaly of which there are other examples, for although quite altered in typography and in date from the edition of 1677, it too is called the fifth. If the publisher had been different, one could have understood an error arising and have even condoned it, but why Whitwood should have made the misstatement it is difficult to see. Even though he meant that it was merely a reissue, he, by calling it the fifth edition, introduced confusion and gave rise gratuitously to doubts and difficulties in enumerating the editions. It is a mere accident which has enabled me to see both issues and indicate their peculiarities. The name of the author appears here in full for the first time, though it is given at the end of the epistle in every edition I have examined from the first onward.

13. (A. 6.) In the sixth edition of the *Rich Cabinet* (B.M. 1036. c. 6.) the poorly got-up chap-book style of the 1684 edition is continued, and this edition also is much inferior to some of its predecessors.

The description may be given:

A RICH | CABINET, | WITH | Variety of Inventions, | Unlock'd and open'd, for the Recrea- | tion of Ingenious Spirits. | Being Receits and Conceits of feveral | Natures, and fit for thofe who are lovers of | Natural and Artificial Conclufions. | AS ALSO | Variety of Recreative Fire-works both | for *Land*, *Air*, and *Water*. And Fire-works | of Service, for Sea and Shore. | Whereunto are added divers Experi- | ments in *Drawing*, *Painting*, *Arithmetick*, *Geome-* | *try*, *Aftronomy*, and other parts of the Mathematicks. | Together with feveral Curious Receits of Great | ufe, collected out of *Alexis*, *Mizaldus*, *Wecker*, &c. | By *John White* a lover of Artificial Conclufions. | The Sixth Edition, with many Additions. | *LONDON*, | Printed for *William Whitwood*, next door to the | *Bible* in *Duck-Lane*. 1689. |

Small 8°. A to I in eights; or, pp. [12, 2] 128. The printed title is surrounded by a double line.

A 1, wanting, possibly contained "The Author to his Book," or the engraved title. A 2, the title page as above. A 3, the Epistle, signed: An Artist's Friend, | JOHN WHITE. A 4, the Table. A 8, Books lately printed for W. Whitwood. This list is different from that in the 1684 edition. B to I 8, the text. Woodcuts in the text.

The title-page, as will be observed, is nearly a reprint of the 1684 fifth edition. This sixth edition, however, is not the remainder of the 1684 fifth edition, but is a page for page and almost line for line reprint, though there are typographical differences which show that the text has been set up afresh throughout. The matter however is identical in the two editions.

The first three sections correspond with those in the second edition, but with several omissions in the first and second parts; the third, on Fire-works, has only two or three omissions. In the 1684 fifth, and in this sixth edition are included: Experiments performed by Legerdemain, Experiments in Arithmetick, Pleasant Questions in Arithmetick, Experiments in Geometry, and some miscellaneous Experiments. All these except the last appear in the fourth and 1677 fifth editions, but Necessary directions for Drawing and Painting and the Ringing of Changes in the fourth and 1677 fifth editions are omitted.

14. (A. 7.) In the British Museum (1037. e. 28.) there is still another edition of later date, and in chap-book style, but it wants the title page. The fragment contains A5 to H4 in sixes, or pp. 5-92, which include receipts, fire-works, the experiments on painting, the legerdemain tricks, and the arithmetic and geometry experiments, but it is defective at the end.

Thus, in less than forty years, seven or eight editions at least of this little book appeared, which shows that it was able to hold its own in spite of a good deal of competition.

15. (B.) The second collection made by White, with a somewhat similar title, must not be confounded with the preceding work. Examination of the Contents shows that they are occupied with quite different topics. The title of this volume is as follows:

A
Rich Cabinet
OF
Modern Curiofities.
CONTAINING,

1. Many Natural and Artificial Conclufions.
2. Variety of Fire Works, of great Service for the Sea and Land.
3. Divers Experiments in Drawing, Painting, Arithmetick, Geometry, Aftronomy, Dialing, and other parts of the Mathematicks.
4. The Art of *Legerdemain*, Sports and Paftimes.
5. To teach Children their A.B.C. in play.
6. An Excellent way to teach one to Read fpeedily.
7. To catch all forts of Birds, Fifh, and Fowl,
8. A Secret way of Writing Love Letters, *&c.*
9. Curious Receipts to caft Flowers in Wax, and to make Wax-Work.
10. To preferve all forts of Fruit a whole Year.
11. To Bake Bread without a hard Cruft, and make it whiter, and Keep much longer then ufual.
12. To make Curious Glittering Plaifterings, for Cealings, or for Walls.
13. Some Choice Phyfical Receipts, with the fignification of the Colours of Urine, and when Sickness is at hand.

With a Multitude of other Curiofities, the whole Adorned with above forty Curious Cuts. By *John White*, a lover of Artificial conclufions.

Printed for *G. Conyers*, at the *Gold Ring*, in *Little Brittain*.

Small 12mo. A and B mixed, 6 leaves, C to M in fours and twos, alternately; or, pp. 72.

It is a rudely printed chap-book, with equally rough wood-cuts. It was reprinted with some modifications under the title *Hocus Pocus*. Whether or not it preceded the 1688 edition of *Arts Treasury* (§ 18) is a question. From the look of the volume I should say not; but, anyhow, there must have been an edition of the *Hocus Pocus* before the issue of the fourth edition of *Arts Treasury*, as the inscription to the portrait shows. With regard to the title, it must be remembered that the persuasive and illusive expression "modern" is at all times a purely relative and defective one, and there is no quality so temporary, so evanescent, as that denoted by it.

16. (C.) While engaged with White's works I had the good fortune to acquire a copy of the aforesaid *Hocus Pocus*, which enables me to give a more complete view of his contributions to the literature of receipts and secrets, than possible otherwise.

Hocus Pocus: | OR, A | RICH CABINET | OF | **Legerdemain Curiosities,** | *Natural and Artificial Conclufions.* | SHEWING | 1. How to cleave Money. 2. To make Sport | with Cats, Ducks, or Poultry. 3. To hang two Knives | on the brim of a Glafs. 4. To wafh your Hands in mel- | ted Lead without damage. 5. To make a Sixpence feem | to fall thro' a Table. 6. Teach Children to Read by Dice. | 7. Divers wonderful Things done by the Loadftone. 8. To | catch *Kites, Crows, Magpies, &c.* alive. 9. To catch a | Pick-pocket. 10. To name a Pack of *Cards*, and not fee | 'em. 11. To write *Love-Letters* fecretly. 12. Experi- | ments in *Drawing, Painting, Geometry, Aftronomy*, &c. | 13. To make variety of *Fireworks*. 14. To keep *Fowl*, | *Venifon*, or any *Flefh* fweet a month. 15. To make a | Drink when you cannot relish other Liquors. 16. To | fox *Fish* and *Fowl*. 17. To make one Candle outlaft | three. 18. To preferve Fruit all the Year. 19. To | make excellent *Plaiftering for Ceilings or Walls.* | With many other Natural and Artificial Con- | clufions, affording great variety of innocent | Sport and Paftime. | *Adorn'd with above* 40 *curious Cuts.* | By J. WHITE, a Lover of Art and Ingenuity. |

Sold at the *Ring* in *Little-Britain*, pr. 1*s.* as alfo | *Hill*'s Legerdemain, and Arts Treafury. *Price each* 1*s.* | A thoufand notable Things, The way to Save Wealth, | and The way to Get Wealth, 1*s.* 6*d.* each: Sports and | Paftime, 6*d.* |

12mo. A6, B to F in sixes: or pp. 72. Frontispiece is A1.

I do not know when this work first appeared or if there were more than one edition, but this one was certainly subsequent to 1712. For, on the last page, there is a reference to the blowing up of a coal mine in Durham in 1712, by which 72 persons were killed; apparently the catastrophe which is mentioned in Plattes' book.* It is curious that the date of two quite unrelated books should be limited in one direction by the same event.

17. As to the books mentioned in the advertisement, "Arts Treasury" is by White, "A Thousand Notable Things" is possibly Lupton's, and "The Way to save Wealth" and "to get Wealth" are possibly by Markham.

---

* *Transactions of the Glasgow Archæological Society*, N.S., 1903, IV., p. 113 (Reprint, 1900, Supplement IV., § 30).

On the *verso* of the title of *Hocus Pocus* appear the following notes:

"To all Lovers of
Artificial Conclufions.

I *Have here open'd to your View a* Rich Cabinet *of* Curiofities, *both profitable and diverting, of which I think I may fay, the like was never extant in the* Englifh Tongue. *For which reafon thefe few Receipts (feveral whereof are my own) I dedicate freely to thy ufe, not doubting but there are things contain'd in this fmall Volume that will give fatisfaction to the Ingenious.*

JOHN WHITE.

Some more Books of Art and Ingenuity, fold by G. C. *at the* Ring *in* Little-Britain.

THE French Perfumer, price 1*s.* Arts *Mafterpiece* 1*s.* *Myftery* of Vintners, 1*s.* *Butler* on Bees, 1*s.* Young Sportfmans Inftructor, 6*d.* Ladies Companion 1*s.* Compleat Vermin-Killer, 6*d.* ftitcht. A new Book of Knowledge, by W. Winftanly, 1*s.* The Experienced Fowler, 6*d.* Compleat Fifher, 6*d.* A Family Jewel, 1*s.* The Traders fure Guide, 3*s.* *Tryons* Letters, 2*s.*"

Some of these will be noticed subsequently.

18. (D. 1.) The fourth collection which bears White's name has the following title:

ARTS TREASURY:
OR, A
Profitable and pleafing Invitation to the Lovers of
INGENUITY.
Contained in many extraordinary Experiments,
Rareties and curious Inventions.
In two PARTS.

PART I. Containing the Myftery of Dying Cloths, Silks Stuffs, Hair, Feather, Bone, Horn, Ivory, Leather, *&c.* The manner of preparing Colours, Directions to take out Spots, Stains, Pitch, Tar, Wax, Rofins, *&c.* out of Silks, Stuffs, Woollen, Linen, *&c.* To recover faded Colours and preserve Apparel *&c.* To fcowr Silver and Gold Lace, Plate, *&c.* Preparing Cement for glafs, China, *&c.* Diapering Linen, making Perfumes. The Art of Drawing, Limning, Painting, preparing and laying on Colours, Etching, Engraving, Cutting, Gilding, Enammelling, refrefhing Pictures, hangings, Carpets, preparing Flock-work-cloth, directions to paint thereon, and to make fundry Colours, prepare Gums, Alum-Water. The Art of thickening Linen, Wafhing or Colouring Maps, or printed Pictures, with fundry other things, *&c.*

PART II. Containing the Quality, Generation and Product of Metals and Minerals, Natural and Artificial; directions to harden or foften them for ufe. The Art of Souldering, Burnifhing, and Gilding Metals. Inftructions for making Inks of all kinds and Colours, preparing Gold or Silver for Writing, and all manner of Sealing-wax, Wafers, *&c.* Gold and Silver to know the finenefs, and to detect counterfeit Coins. How to write fundry ways, that it fhall not appear without the help of Fire, Water, *&c.* taking blots and writing off Paper or Parchment. The Myftery of beautifying, writing with Gold or Silver, Colouring, Dreffing, Cleanfing, and Perfuming Gloves, Ribands, *&c.* Imboffing in Plate, Wax-work, Imagery, Dyalling, Gauging, Wafhing Point-lace, Tiffany, Sarcenet, Luteftring, and any coloured Silks, *&c.* Weighing Ships, extinguifhing Fire, blowing up Houfes or Veffels under Water, *&c.* *London*: Printed for *W. Whitwood*, next the Bible in *Duck-Lane*; and Mrs. *Feltham* in *Weftminfter-Hall.* 1688.

(Price Bound One fhilling.)

Small 12°. A in six. B to G in twelves, H six; or pp. [12] 154.

A1, License. A2, Title. A3, the Preface, ends A4 *verso* and is signed: John White. A5 and 6, book advertisements. B to H6, the Text.

This apparently is the first publication of White's *Arts Treasury*, but, if it be, the later editions were much curtailed, and the principal title also was condensed, as can be seen by comparing them. In the Second Part, Chap. III. on the Assay of coins, Chap. IV. on Inks, Chap. VI. on Imbossing, Casting, Chap. VII. on Vinegar &c., Chap. VIII. on Inks Chap. IX. on Cosmetics, Chap. X. Miscellaneous Receipts, Chap. XI. Miscellaneous Experiments, are all omitted, and in the later edition there is inserted a chapter on Leather.*

* This is the only edition quoted by Lowndes, *The Bibliographer's Manual*, ed. Bohn, 1864, V. p. 2900. A delightful travesty of the title is there given: Arts' Treasury; or a profitable pleasing Invitation to the Lover's Ingenuity. Lond. 1688, 12mo.

If this be so, and so far as I have compared the editions I see no reason to doubt it, it is certainly unusual for an author to reduce his work by nearly one half.

19. (D. 2.) The second and third editions have hitherto escaped me, but I have got a copy of the fourth edition, which bears this modified title:

ART'S
Treafury of Rareties,
AND
Curious Inventions.
In Two Parts.

PART I. Containing the Myftery of Dying Cloths, Silks, Stuffs, Hair, Feathers, Bone, Horn, Ivory, Leather, *&c.* The manner of preparing Colours, Directions to take out Spots, Stains, Pitch, Tar, Wax, Rofin, *&c.* out of Silks, Stuffs, Woollen, Linnen, *&c.* To recover faded Colours and preferve Apparel. To fcowre Silver and Gold Lace, Plate, *&c.* Preparing Cement for Glafs, China, *&c.* Diapering Linnen, making Perfumes. The Art of Drawing, Limning, Painting, Japanning and laying on Colours; Etching, Engraving, Cutting, Guilding, Enammelling. Refrefhing Pictures, Hangings, Carpets, preparing Flock-work-cloth, directions to Paint thereon, and to make fundry Colours, prepare Gums, Allum-water. The Art of thickening Linnen, Washing or Colouring Maps, or Printed Pictures. Tanning Leather, *&c.*

PART II. Containing the Quality, Generation and Product of Metals, Natural and Artificial: Directions to harden or foften them for ufe; to Sodder, Burnifh, and Guild Metals. To make Inks of all Kinds and Colours. To prepare Gold or Silver for Writing, and all manner of Sealing-wax, Wafers, *&c.* Gold and Silver, to know the Finenefs, and to detect counterfeit Coins. How to take Blots and Writing off Paper or Parchment. The Myftery of Beautifying, Colouring, Dreffing, Cleanfing and Perfuming Gloves, Ribands, *&c.* Imboffing in Plate. Wax-work, Imagery, Dyalling, Gauging, Wafhing Point-Lace, Tiffany, Sarcenet, Luteftring, and coloured Silks, *&c.* weighing Ships, Brewing Fine Ale, making Wines, Legerdemain, and divers other Curiofities.

The Fourth Edition.
London, Printed for *G. Conyers*, at the *Ring* in *Little-Britain.*
Price 1*s.*

Small 12mo. A to F in twelves, G six; or pp. [4] 152.

This is a chap-book, and the first leaf contains a woodcut portrait of the author, which I have reproduced for this paper. Underneath it are the words: "JOHN WHITE, *Author of* | ART'S Treafury, *and* Hocus | Pocus; *or* a Rich Cabinet *of* | Legerdemain Curiofities." | This same portrait, but with mottoes and a couplet instead of the title, is prefixed to *The Way to Save Wealth*, to be mentioned under date 1697. As the portrait in *Art's Treasury* is not inserted, but is printed on the first leaf of sig. A, and thus forms an integral part of the book, it is plain that the Legerdemain book must have been published before this fourth edition; and if 1712 be its date, then this book must be later. I am not certain, however, that 1712 *is* the date of the first edition of *Hocus Pocus.* But as there may have been an edition of *Hocus Pocus* prior to 1712, this fourth edition too may have appeared before that year. Bibliographers are much indebted to Conyers for pleasing date-puzzles on which to exercise their "art and ingenuity." At the best they come to no certainty, but only to "artificial conclusions."

20. (D. 3*a*.) In the first part of these *Notes* * the fifth edition of this work was mentioned. I find now that there is another edition also called the fifth, but differing entirely in typography. Here is another problem for the "lover of Artificial Conclusions," and it is parallel with that already set in §§ 11, 12, under White's *Rich Cabinet*, where also it is the fifth edition which is duplicated. Possibly one is a reprint of the other, only, if so, why did Conyers not call the later one the sixth edition? But it is also possible that one is a pirated edition upon which Conyers' name has been put, in which case the pirate would expect to profit by the sale. But once more, for the information of collectors, as well as to facilitate the labours of the compilers of the future Universal Catalogue of English books, I shall give transcripts of the title-pages of the two fifth editions, seeing that they are both undated. Perhaps some one may find other varieties besides these.

---

* *Transactions of the Glasgow Archæological Society*, Glasgow, 1883, II., p. 196 (Reprint, 1896, p. 20).

See p. 14.

ART'S
Treafury of RARITIES:
AND
Curious Inventions.
In Two PARTS.

PART I. Containing the Myftery of dying Cloth, Silk, Stuffs, Hair, Feathers, Bone, Horn-work, Leather, *&c.* and to prepare and ufe them. To prepare and colour Skins of Leather, or gild them with Gold, Silver, or Lacquering, *&c.* To dye Briftles, Hair, Marble-Paper; to recover faded Colours in Cloth or Silk, and to take out Spots or Stains of Tar, Rofin, Greafe, Oil, *&c.* and preferve them from Moth and Worms. To fcower Silver and Gold Lace and Plate. To cement broken Glafs and China, and to make Perfumes, *&c.* The Art of Drawing, Limning, Painting in Oil and Japanning, and tanning Leather, Eching, Graving, Writing, Gilding, Enamelling. To make fundry Colours, prepare Gums and Allom-Water; to thicken Linen, colour Maps and Pictures.

PART II. Containing the Generation of Metals, Natural and Artificial, and to Soder, Gild, and to harden and foften them. To cleanfe and perfume Gloves, wafhing and ftarching Lawns, Sarfenets, Tiffany, and Luteftrings.

With divers other CURIOSITIES.
The Fifth Edition.
London: Printed for *G. Conyers*, at the Ring in *Little-Britain*. Price One Shilling.

12mo. A to G in sixes; or, pp. 84. The print is $4\frac{1}{8}'' \times 2\frac{1}{8}''$. "The Fifth Edition" is in roman letter.

21. (D. 3*b*.) The other edition is as follows:

ART'S
Treafury of RARITIES:
AND
Curious Inventions.
In Two PARTS.

PART I. Containing the Myftery of Dying Cloth, Silk, Stuffs, Hair, Feathers, Bone, Horn-work, Leather, *&c.* and to prepare and ufe them. To prepare and colour Skins of Leather, or Guild them with Gold,

Silver, or Lacquering, *&c.* To dye Briftles, Hair, Marble Paper; to recover faded Colours in Cloth or Silk, and to take out Spots or Stains of Tar, Rofin, Greafe, Oyl, *&c.* and preferve them from Moth and Worms. To fcower Silver and Gold Lace and Plate: To Cement broken Glafs and China, and to make Perfumes, *&c.* The Art of Drawing, Limning, Painting in Oyl and Japanning, and Tanning Leather, Eching, Graving, Writing, Gilding, Enamelling. To make fundry Colours, prepare Gums and Allom-water; to thicken Linnen, colour Maps and Pictures.

PART II. Containing the Generation of Metals, Natural and Artificial, and to Solder, Gild, and to harden and foften them. To cleanfe and perfume Gloves wafhing and ftarching Lawns, Sarfenets, Tiffany, and Luteftrings, *With divers other* CURIOSITIES.

The Fifth Edition.

London: Printed for *G. conyers*, at the *Ring* in *Little-Britain.* Price 1*s.*

12mo. A to G in sixes; or, pp. 84. The print is $5\frac{3}{16}'' \times 2\frac{11}{16}''$. "The Fifth Edition" is in black letter. A copy is also in the British Museum, 7943. aa. 17.

22. (D. 4.) Of this book I know two Glasgow reprints; there may have been more. They are as follows; I quote the titles in full:

ART'S
Treafury of RARITIES:
AND
Curious Inventions.
In TWO PARTS.

PART I. Containing the Myftery of dying Cloth, Silk, Stuffs, Hair, Feathers, Bone, Horn-work, Leather, &c. and to prepare and ufe them. To prepare and colour Skins of Leather, or gild them with Gold, Silver, or Lacquering, &c. To dye Briftles, Hair, Marble Paper; to recover faded Colours in Cloth or Silk, and to take out Spots or Stains of Tar, Rofin, Greafe, Oil, &c. and preferve them from Moths and Worms. To fcour Silver and Gold Lace and Plate. To cement broken Glafs and China, and to make Perfumes, &c. The Art of Drawing, Limning, Painting in Oil and Japanning, and tanning Leather, Etching, Graving, Writing, Gilding, Enamelling. To make fundry Colours, prepare Gums and Allom-Water; to thicken Linen, colour Maps and Pictures.

PART II. Containing the Generation of Metals, Natural and Artificial, and to Soder, Gild, and to harden and foften them. To cleanfe and perfume Gloves, wafhing and ftarching Lawns, Sarcenets, Tiffany, and Luteftrings.

TO WHICH IS ADDED,

A variety of Artificial Fire-Works, with divers other Curiofities.

GLASGOW:

Printed for ROBERT SMITH, and fold at his Shop in the Salt-mercat, at the fign of the Gilt-Bible. 1761.

18mo. 4 ff. A to G in sixes, H two; or pp. [8], 3-89 [1 blank]. The first leaf is a frontispiece containing a rude woodcut of St. George and the Dragon; the second is the title, the third has the Address to the Reader and the Contents, and the fourth a list of books sold by Robert Smith, among which is "Art's Masterpiece," referred to below (§ 27).

23. (D. 5.) The other has the following title:*

ART's

TREASURY of RARITIES:

AND

CURIOUS INVENTIONS.

In TWO PARTS.

PART I. Containing the Myftery of Dying Cloth, Silk, Stuffs, Hair, Feathers, Bone, Horn work, Leather, &c. and to prepare and ufe them. To prepare and colour Skins of Leather, or Gild them with Gold, Silver or Lacquering, &c. To Dye Bristles, Hair, Marble Paper; to recover faded Colours in Cloth or Silk, and to take out Spots or Stains of Tar, Rofin, Greafe, Oyl, &c. and preserve them from Moth and Worms. To fcower Silver and Gold Lace and Plate. To Cement broken Glafs and China, and to make Perfumes, &c. The Art of Drawing, Limning, Painting in Oyl and Japanning, and Tanning Leather, Eching, Graving, Writing, Gilding, Enamelling. To make fundry Colours, prepare Gums and Allom-water; to thicken Linen, colour Maps and Pictures.

PART II. Containing the Generation of Metals, Natural and Artificial, and to Solder, Gild, and to harden and foften

* I am indebted to William Young, Esq., R.S.W., of this Society, for the opportunity of collating this volume. There is a copy in the British Museum, 7953. a. 52, and I also have acquired one.

them. To Cleanſe and Perfume Gloves, Waſhing and Starching Lawns, Sarſenets, Tiffany, and Luteſtrings.
With divers other CURIOSITIES.
THE SIXTH EDITION.
GLASGOW:
Printed for JOHN TAIT, Bookſeller, Saltmercat,
M,DCC,LXXIII.

12mo. A to D in twelves, or pp. 96.

The contents on the title-page are the same as in Conyers' fifth edition. Tait's edition is a reprint of Conyers', but in better style.

24. An interesting thing connected with these books is the number of references given incidentally to other books of receipts, both general and special. In the present work,* for example, there is a reference, p. 65, to Lemery's *Curiosa Arcana*; p. 69 to Nurb's *Husbandry*, Plat's *Treasure*, *The Compleat Collier*, *New or Modern Curiosities*; p. 74, to *Art's Masterpiece*, White's *Rich Cabinet*, and Hill's *Legerdemain*, and there are other references to the books *How to get Wealth*, and *How to save wealth*. Some of these which I have seen will be mentioned in their proper place, but others have escaped me.

25. (E.) To the preceding series must now be added *Art's Master-Piece*, a small treatise, which, though anonymous, is White's, for he distinctly claims it. In his *Hocus Pocus*, p. 44, at the beginning of the chapter entitled "Experiments on Painting, &c.," he says: "Having treated of these Matters in *Art's Masterpiece* and *Art's Treasury*, a little I have to add." As it is coupled in this remark with a book which bears White's name, there can be no question as to the authorship, and the propriety of placing it here.

The third edition of this collection of practical art receipts—"with many curiosities and rare Secrets," and additions by C. K., printed by G. Conyers—was referred to by me in 1888.†

26. Again a copy has presented itself, this time of the fifth edition, in the same chap-book style, with a frontispiece of a portrait-painter at work,

*See the fifth edition, § 21, above.

† *Transactions of the Glasgow Archæological Society*, N.S., 1896, II., p. 21 (Reprint, 1890, Part VI., p. 23).

executed in a manner in keeping with the book. This later edition is more closely printed, for it contains pp. [1-6] 7-144. It is undated, but it belongs to the end of the seventeenth or beginning of the eighteenth century.

27. Of this book also there were at least two Glasgow issues. One of these was printed apparently for Robert Smith, as it is mentioned in the list of books on sale by him in 1761, included in his edition of *Art's Treasury of Rarities* (§ 22). I have not seen it.

28. The other, of which there is a copy here, was published in 1707. It has the following title:

ART's Mafter-Piece.
OR,
A Companion for the Ingenious
of either Sex.

CONTAINING,

I. The Art of Limning and Painting in Oil, *&c.* In Particulars, *Viz.* Drawing and Painting Faces, Bodies, Garments, Landskip, Preparing and laying on Colours; Alfo Colouring Mezzatinto Prints, Gilding on Wood, Mettals, and Leather. II. The neweft Experiment in Japanning, to imitate the Indian way, Plain and in Speckles, Rock-work, Figures, *&c.* Receipts for making the feveral forts of Varnifhes, Colours, *&c.* III. To make Artificial Tortoifefhell, to Dye or Stain Ivory, Horn, Bone, Briftle, Feathers, and fundry forts of Woods for Cabinets IV. The Myftery of Dying Silks, Stuffs, Woollen and Linnen Cloth. V. To take Spots, Stains, Pitch, Tare and Iron Moulds out of Silks, Stuffs, Linnen, and Woollen, and to recover faded Silks, Linnen, *&c.* VI. The Art of Perfuming, and Beautifying. VII. Divers Phyfical and Chirurgical Receipts. VIII. To make *London* Powder Ink, other Powder Inks, and the fhining Japan Ink: with many other notable Things. By C. K.

Glafgow, Printed for *J. D.* and fold by *John Crawford* at his Shop in the Head of the *Tron-gate* on the North Side. 1707.

12mo. A to G in sixes, G (*bis*) two; or, pp. [1-5] 6-88.

A1, Title, *verso* blank. A2 *recto*, The Epistle to the Reader, signed C. K. ends *verso*, or, pp. 3-4. A3 *recto*, Text begins, and ends G (*bis*) 2 *verso*, pp. [5]-88.

This book must not be confused with *Art's Compleat Masterpiece* to be hereafter described.

It will be observed also that, doubtless through ignorance on the part of the publisher, C. K. is made to appear as if he were the author, which is certainly a mistake.

John White's name is not included in the *Dictionary of National Biography*, which must be my excuse for the amount of the preceding detail.

29. Besides these, which are unquestionably all by John White, there are other two works of a similar character, of which also he may have been the author. In the collection entitled *The Way to save Wealth*, to be mentioned hereafter, the frontispiece is the same portrait of White as that prefixed to *Art's Treasury* (§ 19). This by itself does not prove anything as regards authorship, but it may indicate either that White acted as editor, or that he actually wrote part of the collection. One or two of the tracts might quite well have been composed by him. The other book is entitled *Beauties Treasury*, the author of which calls himself "J. W. . . . Philo. Chym. & Med." As far as the contents are concerned they might have been drawn up by White; the initials correspond, and the date is not impossible; but "Philo. Chym. & Med." is such a departure from White's favourite designations, that I am doubtful about the identification of the two J. W.s. The problem of authorship must be left for future investigation, and the description of this *Treasury* postponed to the year to which it belongs, namely 1705.

30. On several occasions I have enumerated editions of Gabriel Plattes' work on mines,* first published in 1639. The edition now to be mentioned

---

* *Transactions of the Glasgow Archæological Society*, 1883, II., pp. 194, 251 (Reprints, 1896, I., p. 17; II., p. 26); N.S., 1890, I., p. 226 (Reprint, 1885, III., p. 41); N.S., 1903, IV., p. 112 (Reprint, 1900, Supp. IV., p. 20). For references to Plattes, see my *Bibliotheca Chemica*, 1906, II., p. 207.

was an issue fourteen years later of copies of the first edition of 1639, with a new title page. There is no other difference. Poor Plattes' book, therefore, was a failure, and he himself got no encouragement to prosecute all the discovery and invention of which he was brimful. He was allowed to die of starvation on the streets of London.

The title of the re-issue in question is as follows:

A | DISCOVERY | OF | SUBTERRANEAL | TREASURE: | *VIZ.* | Of all manner of *Mines* and *Minerals*, | from the *Gold* to the *Coal*; with plain Di- | rections and Rules for the finding of them | in all Kingdoms and Countries. | AND ALSO, | The *Art* of Melting, Refining, and | Affaying of them is plainly declared, fo that | every man that is indifferently capacious, may with | fmall charge prefently try the value of fuch *Oares* | as fhall be found either by Rule, or by Accident. | Alfo a perfect way to try what colour any Berry, | Leaf, Flower, Stalk, Root, Fruit, Seed, Bark, or | Wood will give: With a perfect way to make | Colours that they fhall not Stain, nor | Fade like ordinary Colours. |

LONDON, | Printed for *J. E.* and are to be fold by *Humphrey Mofeley* | at the *Prince*'s *Arms* in Saint *Paul*'s Church-yard. 1653. |

Small 4°. Title-leaf; B to K2 in fours; or, pp. [2, 8] 60.

This is an able little book, but the details are chemical and are hardly suited for exposition in this paper. His account of some of the operations are remarkably clear and scientific, considering the time when he lived.

This edition of 1653, along with those of 1639 and 1684, is quoted by Lowndes,* who has a note upon Plattes.

31. In the fourth Supplement † to these notes, I devoted some space to the Countess of Kent and to questions arising in connection with her, and described several editions of her book. The earliest I had seen then was the fourth, dated 1654, and the latest was the nineteenth, dated 1687.

A month ago (January 24th, 1908) I secured a copy of the second edition, dated 1653. As the Countess died on December 7th, 1651, she is described in this edition, as in all others that I have seen, as *late deceased.*

---

* *The Bibliographer's Manual*, ed. Bohn, 1865, IV., p. 1880.

† *Transactions of the Glasgow Archæological Society*, N.S., 1903, IV., pp. 104-111 (Reprint, 1900, Supp. IV., §§ 22-27).

There is no evidence that the *Manuall* was printed in her lifetime, and no edition of 1652 has come under my notice anywhere, though one might have looked for its appearance in that year. On the other hand, there are three editions dated 1653, of which one has no number, while the other two are both called the second edition. I have not seen the third edition, but between those dated 1653, and the fourth dated 1654, not much room is left for a third, and I cannot say whether it was published in 1653 or in 1654. In any case the book had a run of popularity much more rapid than was usual for such works; perhaps the reputation of the Countess for skill and charity helped the sale; perhaps the mere novelty of it attracted the up-to-date gang of the time. The volumes, besides, were small and dainty and attractive to buy. The second editions, like the fourth, are tiny narrow 24mos, but my copy has, for some reason, been inlaid to octavo size, and bound with a similarly inlaid copy of the *Ladies Cabinet*, already described.* The inlaying was doubtless intended to make the books more easily handled and preserved, as well as to admit of notes and possible additions, but it was wrong to bind them together.

32. Examination of the three editions of 1653 inclines me to believe that one of them is the first edition of all. My reasons are: the different style of the book from those which followed; the want of a number on the title-page; the different printer; the differences in the text.

In the belief that it is really the first edition, until some other issue can be shown to have a better claim to that distinction, I may give a description of the copy in the British Museum (C. 48. b. 18.).

> A | *CHOICE MANVAL* | OF | Rare and Select | SECRETS. | IN | PHYSICK | AND | CHVRURGERY; | Collected and practifed by the Right | Honourable, the Counteffe of | *Kent* late deceafed. | As alfo moft Exquifite ways of Pre- | ferving, Conferving, Candying, &c. | Publifhed by *W. J.* GENT. |
>
> *LONDON,* | Printed by *R. Norton,* 1653. |
>
> 12°. A four leaves, B to E in twelves, F four leaves, but F4 is blank; or pp. [8] 102 [2 blank]. The title is enclosed by a single line. The second tract has the following title:

* *Transactions of the Glasgow Archæological Society*, 1908, N.S., V., pp. 164-167 (Reprint, 1908, Supp. V., §§ 59-61).

A | *True Gentlewomans* | DELIGHT. | Wherein is contained all | manner of COOKERY : |

Together with {
Preferving,
Conferving,
Drying
and
Candying.

Very neceffary for all Ladies | and Gentlewomen. | Publifhed by *W. J.* GENT. |

LONDON, | Printed by *R. Norton,* 1653. |

A four leaves, B in eight, C, D in twelves, E four leaves; or, pp. [8] 96. The title is enclosed in a single line. The print is 5″ × 2$\frac{7}{12}$″. Both G. D.'s editions are 3$\frac{8}{12}$″ × 1$\frac{8}{12}$″. There is no portrait. It has Henry T. Buckle's bookplate.

The first tract is dedicated to Letitia Popham, wife of Colonel Alex. Popham, and is signed W. J. The second tract is dedicated to Mrs. Anne Pile, and is also signed W. J., *i.e.* W. Jar. In the first part, all after the Countess of Kent's powder, is omitted in this edition. The second part agrees in the three editions of 1653. Norton's edition is quoted by Hazlitt.*

33. The two second editions of 1653 are next to be considered. Lowndes † mentions an edition of 1653, with a portrait of the Countess, a small oval, but, of course, does not specify which it is. He is hardly to be blamed, for no one who had not compared copies would have thought of the possibility of two quite distinct issues of the same date.

A | *Choice Manuall,* | OR | RARE and SELECT | SECRETS | IN | PHYSICK | AND | CHYRURGERY: | Collected, and practifed | by the Right Honourable, the | Counteffe of *Kent,* late | deceafed. |

Whereto are added feve- | ral Experiments of the Virtues | of Gafcon pouder, and *Lapis con-* | *tra Yarvam,* by a Profeffor | of Phyfick. |

As also most Exquifite waies | of Preferving, Conferving, | Candying, *&c.* |

The Second EDITION. |

LONDON, | Printed by *G. D.* and are to be | fold by *William Shears,* at the Sign | of the Bible in St. *Pauls* | Churchyard, 1653. |

24°. A to Q in twelves; or, pp. [17, 1 blank, including the portrait] 1-206 (i.e. A to K5). Then K6 is the title of the second part:

---

* *Collections and Notes,* 1876, p. 245.

† *The Bibliographer's Manual,* ed. Bohn, 1865, III., p. 1266.

A | *True Gentlewomans* | DELIGHT. | Wherein is contain- | ed all manner of | COOKERY : |

Together with { *Preſerving, Conſerving, Drying and Candying.*

Very neceſſary for all | Ladies and Gentle- | women. | Published by *W. I.* GENT. |

LONDON, | Printed by *G. D.* and are | to be ſold by *William Shears,* | at the Sign of the Bible in | St. *Pauls* Church-yard, 1653. |

This tract goes on to the end, Q12, and is separately paged: [18] 140. The first title is surrounded by a narrow border of fleurs-de-lis within two lines, while the second title is enclosed in a narrow border of the letters X and O, placed alternately, within two lines. G. D. is Gartrude Dawson. The portrait is not badly executed, but it is not complimentary to the Countess. It was reproduced in the fourth edition. This edition is in the British Museum, 7383. a. 24., and I have also a copy of it.

34. The other issue is in the Museum, 7953. a. 67. (1-2.), and a description of it follows:

A | *Choice Manual,* | OR | RARE and SELECT | SECRETS | IN | PHYSICK | AND | CHYRURGERY: | Collected, and practiſed | by the Right Honourable, the | Counteſſe of *Kent,* late | deceaſed. |

Whereto are added ſeve- | ral Experiments of the Virtues | of Gaſcon pouder, and *Lapis con-* | *tra Yarvam,* by a Profeſſor | of Phyſick. |

As alſo moſt Exquiſite waies | of Preſerving, Conſerving, | Candying, &c. |

The Second EDITION. |

LONDON, | Printed by *G. D.* and are to be | ſold by *William Shears,* at the Sign | of the Bible in St. *Pauls* | Church-yard, 1653. |

24°. 2 leaves: B in six, C to K in twelves, L in eight; or, pp. [4, 11, 1 blank], 206.

The first leaf contains a portrait of the Countess, like that in the other 1653 edition, but a better impression. There are the address to the reader by W. Jar, the contents and the text of Physick Secrets. This is followed by the title page of the second part:

A | *True Gentlewomans* | DELIGHT. | Wherein is contain- | ed all manner of | COOKERY : |

Together with { *Preſerving, Conſerving, Drying, and Candying.*

Very neceſſary for all | Ladies and Gentle- | women. | Published by *W. I.* GENT. |

LONDON, | Printed by *G. D.* and are | to be fold by *William Shears,* | at the Sign of the Bible in | St. *Pauls* Church-yard, 1653. |

4 leaves, the second and fourth of which are marked A3; *B* in six; B in four, C to G in twelves, H in six; or, pp. [8, 12] 140.

The title-leaf has no mark. The next, marked A3, contains Jar's address to Lætitia Popham, wife of Colonel Popham. In the other 1653 edition this address is prefixed to the first treatise, to which it seems properly to belong. Next is a leaf, without mark, containing Jar's brief note to the reader. The *verso* is blank. Then a fourth leaf, also marked A3, containing Jar's address to Anne Pile, eldest daughter of Sir Francis Pile, Bart., deceased. *B* in six leaves has the table of contents, after which is the text. The general title-page is surrounded by fleurs-de-lis between two lines; that of "A True Gentlewomans Delight" by X and O placed alternately between two lines.

These two editions of 1653 are identical as regards contents, but while agreeing page for page and almost line for line, differ typographically in many details. In one of them in the second part there seems to have been a shortcoming of italic capital T and lower case k, for roman letters are used instead.

Hazlitt quotes "A True Gentlewomans Delight" of this edition but not "A Choice Manuall," and adds: "This seems to be a distinct edition from that registered in *Coll. and Notes,* 1st Series, p. 245," which is Norton's edition.* In this he is right.

35. To make certain that the two 1653 editions are essentially different, though resembling each other so closely that the differences might be readily overlooked, certain details may be noticed. The alteration of signatures could have been effected possibly without making any change in the text, and that would have been quite obvious on inspection. But there are differences in the text, proving that these portions at any rate must have been reset.

	My own copy (B.M. 7383. a. 24.)	B.M. 7953. a. 67.
p. 4. last lines.	the next morning fet them a ftilling in a glaffe Limbeck.	the next morning fet them a ftilling in a glaffe Limbeck.
p. 5. l. 10.	and profiteth the ftomack, it preferveth a good colour, the vifage, memory and youth, it deftroyeth the Palfie.	and profiteth the ftomack, it preferveth a good colour, the vifage, memorie, and youth, it deftroyeth the Palfie.

* *Third and Final Series of Bibliographical Collections and Notes on Early English Literature,* London, 1887, p. 311.

	My own copy (B.M. 7383, a. 24.)	B.M. 7953. a. 67.
p. 7. l. 6.	and the patient undoubtedly ſhall find great eaſe, and void many ſtones, and much gravel with little pain.	and the patient undoubtedly ſhall find great eaſe, and void many ſtones, and much gravel with little pain.
p. 15. last lines.	when it is in your body keep it half an hour, or three quarters of an hour, or an hour if you can.	when it is in your body keep it half an hour, or three quarters of an hour, or an hour if you can.
p. 17. l. 18.	three a clock in the afternoon, being bloud-warm.	the other part at three a clock in the afternoon, being bloud-warm.
p. 35. last lines.	you ſhall find all the ſpots and ſtains gone.	you ſhall find all the ſpots and ſtaines gone.
p. 41. last lines	mix them well together, and put it into a gally pot, and take thereof firſt in the morning, and laſt in the even-	of red Roſe-water, and mixe them well together, and put it into a gallie pot, and take thereof first in the morn-
p. 69. l. 6.	draw out the Iron, and after heal the wound as it requireth.	you ſhall draw out the Iron, and after heal the wound as it requireth.
p. 173. l. 22.	thick, and annoint the head therewith.	reaſonable thick, and annoint the head therewith.
p. 204. l. 16.	mingle it with ſome *Aqua Theriacalis.*	mingle it with ſome *Aqua Theriacalis.*
p. 206. last lines.	need the Patient to obſerve any preciſe diet.	neither need the Patient to obſerve any preciſe diet.

Differences, like the illustrations now given, will be found on almost every page of the *Choice Manuall*, and from a consideration of the better style of printing of the copy, B.M. 7953. a. 67., I should conclude that it is the earlier of the issues dated 1653. In the other issue the tendency is to print more closely, though nothing is gained by so doing.

Turning to *A True Gentlewomans Delight*, similar differences present themselves, and a few examples may be selected:

p. 39. l. 16.	glaſſes, and let it ſtand till it be cold, and then it is ready to eat.	glaſſes, and let it ſtand till it be cold, and then it is ready to eat.
p. 63. l. 7.	take them up and put them into your glaſſe, and when they be cold cover them.	take them up, and put them into your glaſſe, and when they be cold cover them.
p. 101. l. 7.	ſtew a little while, then pour it on your Tongue, and ſerve it.	dine, let it ſtew a little while, then pour it on your Tongue, and ſerve it.

p. 114. l. 1.	*To ſcald Milk after the VVestern faſhion.*	*To ſcald Milk after the Weſtern faſhion.*
p. 132. l. 12.	Take the former ingredients, and	Take the former ingredience, and

36. The fourth edition appeared in 1654, and has been already described.* Lowndes † mentions an edition of 1656, with a portrait of the Countess, in an oval of foliage, by Chantry, but does not tell us the number of it.

The eleventh edition was described in the fourth Supplement,‡ and an account of it is also given by Hazlitt.§ He has the misprint *Tarvum* for *Yarvam*, and he mentions the portrait. He adds in a note:

> "Annexed to this copy is *A True Gentlewomans Delight*, with separate title and signatures, but apparently belonging properly to another book, as the sheets are from l to T: l, 6 leaves: m, 12 leaves: N-R in twelves: S, 8 leaves: T, 1 leaf. This portion was also *published* by W. J.; but no author is named. It went through divers impressions; the present is dated 1659, and very probably at the time of issue copies were bound up with Lady Kent's work."

This note is not quite clear. It seems to imply that the presence of the tract is exceptional, and that, as in this case the two tracts were printed at the same time, they were bound and sent out together. But, in every edition of the book which I have seen, the two tracts have always been together and printed uniformly in the same style, and, though in some instances the signatures are unconnected, in others they run consecutively through the two tracts, thus showing them to be integral parts of the volume. Whether or not the second tract is by the Countess, or by W. J., or by some unknown author, may be debated, but there seems to be no ground for supposing that the tracts were issued independently.

---

* *Transactions of the Glasgow Archæological Society*, N.S., 1903, IV., p. 105 (Reprint, 1900, Supp. IV., § 23).

† *The Bibliographer's Manual*, ed. Bohn, 1865, III., p. 1266.

‡ *Transactions*, N.S., 1903, IV., p. 105 (Reprint, 1900, Supp. IV., § 24).

§ *Collections and Notes*, 1876, p. 245.

37. The twelfth edition, curiously enough, was published in 1659 also, but in a different form. It was described by me formerly,* and it is mentioned by Lowndes.†

38. Hazlitt ‡ furnishes the following note of the thirteenth edition:

A Choice Manuall, or Rare and Select Secrets: Collected, and practised by the Right Honourable, the Countess of Kent, late deceased. . . . The Thirteen[th] Edition. London. Printed, by Gartrude Dawson, and are to be sold by William Sheares, . . . 1661. Small 12°. A, 9 leaves: B-S, 6 leaves.

On Lii occurs *A True Gentlewomans Delight*, with a new title.

39. The fourteenth edition is in the British Museum, 1037. a. 9.

A | *Choice Manuall*, | OR | RARE and SELECT | SECRETS | IN | PHYSICK | AND | CHYRURGERY: | Collected, and practifed | by the Right Honourable, the | Counteffe of *Kent*, late | deceafed. | Whereto are added feve- | ral Experiments of the Virtue | of Gafcon pouder, and *Lapis con-* | *tra Yarvam*, by a Profeffor | of Phyfick. | As alfo moft Exquifite waies | of Preferving, Conferving, | Candy-ing, *&c.* | The Fourteen Edition. |

LONDON, | Printed, by *Gartrude Dawfon*, and | are to be fold by *Margaret Shears* | at the Sign of the Blew Bible | in *Bedford* Street in *Coven* | *Garden*. 1663. |

24°. A to R in twelves, S in six; or, pp. [17, 1 blank] 233 [1 blank] [5, 1 blank]; [19, 1 blank] 140.

A1 contains a portrait, mounted, and before A3 is inserted another portrait, which I have not seen elsewhere, and which differs from all the others in giving a more flattering likeness of the Countess. The title is surrounded by a border of fleurs-de-lis between two lines.

On Lii is the title of the second tract:

A | *True Gentlewomans* | DELIGHT. | Wherein is contain- | ed all manner of | COOKERY: |

Together with { PRESERVING, CONSERVING, DRYING and CANDYING.

Very neceffary for all Ladies | and Gentlewomen. |
Publifhed by *W. G.* (sic) GENT. |

LONDON, | Printed by *Gartrude Dawfon*, | Living in White Crofs-ftreet | in the Bowling-Alley. | 1663. |

This title is enclosed in a similar border.

---

* *Transactions*, N.S., 1903, IV., p. 105 (Reprint, 1900, Supp. IV., § 25).

† Lowndes, *The Bibliographer's Manual*, ed. Bohn, 1865, III., p. 1266. He quotes a price: 5s., with a portrait.

‡ Hazlitt, *Third and Final Series of Bibliographical Collections and Notes on Early English Literature*, 1887, p. 245.

Hazlitt* quotes an edition of 1663, without specifying the number of it and without the collation. He adds the note: "This seems to have arrived at a 12th edition in 1654." The twelfth edition which I have seen is dated 1659, while 1654 is the date of the fourth edition.

40. There is a later edition in the British Museum, 7462. a. 31.:

A | *Choice Manual*, | OR, | Rare and Select | SECRETS | IN | PHYSICK | AND | CHIRURGERY: | Collected, and practifed by | the Right Honourable the Coun- | tefs of KENT, late deceafed. | Whereto are added feveral | Experiments of the Vertue of | Gafcon powder, and Lapis con- | tra Yarvam by a Profef- | for of Phyfick. | As alfo moft exquifite waies | of Preferving, Conferving, | Candying, &c. | The Eighteenth Edition. |

London, Printed for *Henry Mort-* | *lock* at the Phœnix in St. *Paul's* | Church-yard. 1682. |

24°. A to R in twelves, S six; or, pp. [18] 233 [1 blank] [5, 1 blank]; [19, 1 blank] 140. A1 is a portrait of the Countess. The title is surrounded by a double line.

L11 is the title of the second tract:

A | *True Gentlewomans* | DELIGHT. | Wherein is contain'd all | manner of COOKERY. |

Together with { *Preferving*, *Conferving*, *Drying*, and *Candying*.

Very neceffary for all Ladies | and Gentlewomen. | Publifhed by *W. G.* (sic) Gent. |

LONDON, | Printed for *Henry Mortlock*, | at the Phœnix in St. *Paul's* | Church yard, 1682. |

The title is enclosed in a double line, like the first title.

41. It may be of interest to give a summary of the editions of this book which are on record:

First (?),	R. Norton,	1653
Second,	G. D.,	1653
Second,	G. D.,	1653
Fourth,		1654
?		1656
Eleventh,		1659
Twelfth,		1659

* *Hand-Book to the Popular, . . . Literature of Great Britain*, 1867, p. 121.

Thirteenth, .	Gartrude Dawson, .	1661
Fourteenth, .	Gartrude Dawson, .	1663
Sixteenth, . .	. . . . . .	1670
Eighteenth, .	Henry Mortlock, .	1682
Nineteenth,* .	. . . . . .	1687
Twenty-first,† .	. . . . . .	1708

If there be an edition later than this, I have not had the luck to see it, or any mention of it. This may have been the last, but the career of the book during the half century of its currency is noteworthy. It has been remarked more than once by me that new editions of receipt books appeared every four or five years. In this case, however, three editions appeared in one year, the fourth in 1654 and the twelfth in 1659, or eight editions in five years; the sixteenth in 1670, or four in eleven years; the nineteenth in 1687, or three in seventeen years, and the twenty-first in 1708, or two in twenty-one years. Not many books of the kind at the present time could display such a record.

42. There are other receipt books by and for the ladies, of which various examples will present themselves. That which follows is in the British Museum (E. 1528. (2.)):

> THE | LADIES COMPANION, | Or, | A Table furnifhed with fundry | forts of *Pies* and *Tarts*, gracefull at | a Feaft, with many excellent re- | ceipts for Preferving, Confer- | ving, and Candying of all | manner of Fruits, with the | making of *Marchpain*, | *Marmalet*, and | *Quindenis*. | By Perfons of quality whofe names | are mentioned. |
>
> [Device: a thistle.]
>
> *LONDON*, | Printed by *W. Bentley*, and are to be fold by | *W. Shears*, at the fign of the *Bible*, in | S. *Pauls* Church yard. 1654.
>
> 12mo. A to D in twelves; or, pp. [4 ought to be 6] 82 [7, 1 blank].

The British Museum copy (E. 1528. (2.)) has a fly-leaf marked A and a title leaf, but it wants a third leaf, either a half-title or a frontispiece. On the title-page is the date Nov. 18, and the year is corrected 1653. The book therefore has been post-dated by two months.

---

* *Transactions of the Glasgow Archæological Society*, 1890, N.S., I., p. 318 (Reprint, 1888, IV., p. 20).

† *Ibid.* 1903, N.S., IV., p. 105 (Reprint, 1900, Supp. IV., § 26).

This is strictly a cookery book, but its title resembles so many others, and the receipts about candying and preserving link it on to so many others, that it cannot very well be excluded. Indeed books of practical receipts and those on cookery coalesce, and it is only by adhering to an arbitrary rule that one can avoid including cookery books in the lists of the others. I have, however, striven to do this, because bibliographies of cookery are already extant, and I have no wish to trench upon the subject.

43. If one may judge from portraits, satires, protestations, allusions, as well as from actual guides to the subject, the personal decoration of women by paint, powder and patches engaged much attention in the seventeenth and eighteenth centuries. The receipt books, of which some have been already quoted and others are to follow, in truth concern us most, but, as a vindication of the practice itself may be a desideratum, it is supplied by the following treatise. Though it does not deal practically with the secrets of the toilet, it is a sort of running commentary on the books which do, and it is withal of so curious a nature, that it deserves rescuing from oblivion. Moreover, though anonymous, it seems to have been well known and has been assigned to such distinguished authorship that even on that account it claims recognition, for those who have been credited with it are John Gauden, Bishop of Exeter and Worcester, Jeremy Taylor and Obadiah Walker.

Anthony à Wood* puts it under Gauden, though his editor, Dr. Bliss, in a note to Jeremy Taylor, says it seems rather to be by Obadiah Walker. Watt,† also, has it under both Gauden and Taylor. Lowndes‡ places it under Taylor, and adds: "attributed likewise to Dr. Gauden," and Halkett and Laing§ assign it to Gauden, though they state that it is often ascribed

* Wood, *Athenae Oxonienses*, 1721, II., col. 314 (Gauden); ed. Bliss, 1817, III., col. 617, for Gauden, and III., col. 790, for Jeremy Taylor.

† Watt, *Bibliotheca Britannica*, 1824, Authors, I., 403 *s*, for Gauden; II., 896 *l*, for Taylor, where he quotes the 1662 edition and adds: "attributed by some to Dr. Gauden."

‡ Lowndes, *The Bibliographer's Manual*, 1864, V., p. 2586 (1662, 8vo, Mitford, pt. ii. 7s.—1692, 18mo).

§ Halkett and Laing, *A Dictionary of the Anonymous and Pseudonymous Literature of Great Britain*, 1882, I., col. 649.

to Taylor and occasionally to Walker. Obviously the authorship lies between Gauden and Taylor, but there is not much certainty beyond that.

The book is mentioned by the Rev. Alex. Gordon * under Jeremy Taylor, and the reasons for assigning it to him are thus stated in the article there:

> 17. 'A Discourse of Auxiliary Beauty,' 1656, 8vo (anon.); 2nd edit. 1662, 8vo (by J. T., D.D.; ascribed to Taylor by Kennett; includes a defence of face-painting; the phrase on title, 'artificial handsomeness,' is also in 'Ductor Dubit[antium].' ii. 3, 6).

44. There seem to have been only three separate editions of the work. According to Bliss it was included also in a collection of tracts.

> A | DISCOURSE | OF | Auxiliary Beauty. | OR | Artificiall Hanſomeneſſe. | In point of| CONSCIENCE. | *Between* | Two *LADIES.* | [Device.]
>
> Printed for *R. Royſton*, at the Angel in *Ivie-* | *Lane*, 1656. |
>
> Small 8°. A four, B to N in eights, O four; or, pp. [8] 200. The device is engraved and consists of the sun with the motto, QVI SEQVITUR ME NON AMBULAT IN TENEBRIS. All enclosed in a scroll border.

This is the first edition.

The second edition has a modified title, but was also published by Royston:

> A | DISCOURSE | OF | Artificial Beauty, | In point of | CONSCIENCE, | BETWEEN | Two LADIES. |
>
> With ſome Satyrical Cenſures on the | Vulgar Errors of these Times. |
>
> *LONDON*, | Printed for *R. Royſton* at the Angel in *Ivy-lane.* | MDCLXII. |
>
> Small 8°. 2 leaves, A four, B to R in eights, S four; or, pp. [12, including the frontispiece] 262 [2 blank].

Then the next edition:

> A | DISCOURSE | OF | *Artificial Beauty*, | In Point of | CONSCIENCE, | BETWEEN|Two LADIES. |
>
> *LONDON*, | Printed by *J. L.* for *Luke Meredith*, | at the *Star* in St. *Paul's* Church-| Yard, MDCXCII. |
>
> 12°. A to L in twelves; or, pp. [24, including the frontispiece] 238 [2, of book advertisements].

As usual, the first edition is the handsomest; the last is in smaller type and in commoner style.

47. When "Heneretta Maria Late Queen of England" "pleased to descend to private Recreations" we are told, on the authority of the title-page of

* *Dictionary of National Biography*, 1898, Vol. LV., p. 428.

The Queens Closet opened, that she "had in Esteem" certain of the "Incomparable Secrets in Physick, Chyrurgery, Preserving and Candying, &c.," which are contained in that little volume. The edition of this work formerly mentioned* was dated 1683, and was printed for Obadiah Blagrave, at the sign of the Black Bear in St. Paul's Churchyard.

The remainder, or a portion, was reissued with a new title-page the following year:

THE
QUEENS CLOSET
OPENED.
Comprehending ſeveral Hundreds of Experienced Receipts, and Incomparable Secrets in PHYSICK, CHYRURGERY, PRESERVING, CANDYING, COOKERY, &c.
WHICH
Were preſented to the QUEEN by the moſt Eminent Doctors in Phyſick, Chyrurgions, Oculiſts and divers Perſons of Honour, whoſe Names are all . . . ed to their Receipts, many whereof were had in . . ſteem, when She pleaſed to Deſcend to Private Recreations.
CONTAINING
I. The QUEENS PHYSICAL CABBINET, or Excellent Receipts in *Phyſick, Chyrurgery, &c.*
II. The QUEENS DELIGHT, ; Or the Art of Preſerving, Conſerving, Candying; As alſo, A Right Knowledge of making Perfumes and Diſtilling the moſt Excellent Waters.
III. The COMPLEAT COOK; Or, Directions for Dreſſing all ſorts of Fleſh, Fowl and Fiſh, Ordering of Sauces, and making of Paſtry, according to the *Engliſh*, *French*, *Spaniſh* and *Italian* Mode.

The Laſt EDITION Corrected and Enlarged with many New and Late Additions.

LONDON,
Printed for *Benjamin Crayle* at the *Lamb* in *Fleet-ſtreet*, next *White-Fryers Gate*. 1684.

* *Transactions of the Glasgow Archæological Society*, N.S., 1890, I., p. 321 (Reprint, 1888, Part IV., p. 23).

As the leaf is frayed along the inner margin I am not sure of the first syllable of *added* (?), and if "Efteem" should have a capital.

48. Recently other editions have come under my notice, which may be described so as to complete the list as far as possible. Some are in my own possession, while others are in the British Museum.

The first to be noticed are copies of what is presumably the first edition.

THE | QUEENS CLOSET | OPENED. | Incomparable Secrets in | *Phyfick*, *Chirurgery*, *Pre-* | *ferving*, *Candying*, and *Cookery*; | As they were prefented to the | *QVEEN* | By the moft Experienced Perfons of | our Times, many whereof were honou- | red with her owne practice, when fhe | pleafed to defcend to thefe more private | Recreations. | *Never before publifhed.* | Tranfcribed from the true Copies of her | MAJESTIES own Receipt-Books, | by *W. M.* one of her late fervants. | *Vivit poft funera virtus.* |

Printed for *Nathaniel Brook*, at the *Angel* | in *Cornhill*, 1655.

12°. A 6 leaves; no sig. B ; C to O in twelves, P 4 leaves, Q 12 leaves; A to E in twelves, F 6 leaves; or, pp. [12] 296 [24]; 123 [7, 2].

A1 is the portrait of Queen Henrietta Maria signed by Faithorne, and "fold by Nat: Brooke att yᵉ Angell in Cornhill." The later portraits are inferior productions.

A2 is the title leaf; A 3-6 W. M.'s Epistle to the reader. C to K12: The Pearl of Practise, or, pp. 1-191 [192 blank].

L1 [193-194]: title of the second tract:

A | QUEENS | DELIGHT, | OR | The ART | OF | *Preferving*, *Conferving*, | and *Candying*; | AS ALSO | A right knowledge of | making *Perfumes*, | and *Diftilling* the moft | *Excellent Waters.* | Never before Publifhed. |

Printed for *Nathaniel Brook*, at the | *Angell* in *Cornhill*, 1655.

L2 to P4 the text, pp. 195 (misprinted 295) to 296; Q1–11, the Table; Q12 book advertisements.

The third tract has a title-page, and new signatures and pagination:

THE | COMPLEAT | COOK. | Expertly prefcribing the | moft ready wayes, |

Whether, { *Italian*, *Spanifh*, or *French*. }

For drefsing of *Flefh*, and *Fifh*, | Ordering of *Sauces*, or making | OF | PASTRY. | *LONDON*, | Printed for *Nath. Brook* at the | *Angel* in *Corn-hill* | 1655. |

A1 the title-leaf; A2 to F2 *recto* the text, pp. 3-123; F2 *verso* to F5 the Table, pp. [7], F6 blank.

49. The copy which follows is to be carefully compared with this one, for it is quite different.

THE | QUEENS CLOSET | OPENED. | Incomparable Secrets in | *Phyfick, Chirurgery, Pre- | ferving, Candying,* and *Cookery*; | As they were prefented to the | *QVEEN* | By the moft Experienced Per- | fons of our Times, many whereof | were honoured with her own pra- | ctice, when fhe pleafed to defcend to | thefe more private Recreations. | *Never before publifhed.* | Tranfcribed from the true Copies | of her MAJESTIES own Receipt | Books, by *W. M.* one of her late fervants. | *Vivit poft funera Virtus.* |

Printed for *Nathaniel Brook* at the *Angel* in | *Cornhill*, 1655. |

12°. A5, no sig. B, C to O in twelves, P four, Q twelve; or, pp. [10] 296, the Tables [21, 1 blank] [2 of book advertisements].

Faithorne's portrait of Queen Henrietta Maria is wanting in the British Museum copy (E. 1519.). It is mentioned by Henry Bromley* and by Hazlitt.

A contains the title and W. M.'s Epistle.

C to K 12 contains "The Pearl of Practice," pp. 1-191 [192 blank].

L 1 [193-194] is the title to the second part:

A | Queens Delight, | OR | The ART | OF | *Preferving, Conferving,* | and *Candying*; | AS ALSO | A right knowledge of | making *Perfumes*, and | *Diftilling* the moft | *Excellent Waters.* | Never before Publifhed. |

Printed for *Nathaniel Brooke* at the | *Angell* in *Cornhill*, 1654. |

Ends P4; or, pp. 195-296. The table follows on Q as above.

The third tract has the following title-page:

THE | COMPLEAT | COOK. | Expertly prefcribing the | moft ready wayes, |

Whether, { *Italian*, *Spanifh*, or *French*. }

For dreffing of *Flefh*, and *Fifh*, | Ordering of *Sauces*, or making | OF | PASTRY. |

LONDON: | Printed for *Nath. Brook* at the | *Angel* in *Corn-hill*, 1655. |

12°. A to E in twelves, F5; or, pp. 123 [Table, 7].

This is the third part of the preceding work, but from its date, as compared with that given by Hazlitt, it is a second edition or issue with change of date. There is a copy of 'The Compleat Cook,' 1655, in the British Museum, E. 1531.

In this 1655 edition, the editor W. M. makes an elaborate explanation and apology for having published the Secrets without the Queen's permission, the cause being that he gave a transcript to 'a person of Honour,' who lent it or lost it, and then he was surprized to find two other copies, and on the advice of his friends he dispatched his "Original copy to the

* *A Catalogue of Engraved British Portraits*, London, 1793, Period IV., Class I., p. 65.

Presse to prevent those false ones." That seems to be the usual course of secrets.

This edition of 1655 is quoted by Hazlitt,* only he says *The Compleat Cook* is dated 1654; and also by Lowndes.† Both mention Faithorne's portrait.

50. Hazlitt's title differs in detail from that just given:

> 14. The Queen's Closet opened, or the Pearl of Paradise; Incomparable Secrets in Physick, Chirurgery, Candying, and cookery. London, Printed for N. Brook, 1655. 8vo. With a portrait of Q. Henrietta-Maria by W. Faithorne. This includes (with separate titles) *The Queen's Delight*, dated 1654, and *The Compleat Cook*, dated 1654.

"The Pearl of Paradise" is a peculiar phrase which does not occur in any copy that I have seen, though it has a curious resemblance to "The Pearl of Practise," which is the sub-title of the first part. If it be correct and the dates be correct, one would surmise that if not the first edition it certainly precedes that just described.

Hazlitt adds an edition of 1656:

> (*b*) The Queen's Closet Opened, &c. With Additions. Transcribed from the true Copies of her Majesties own receipt-Books, by W. M., one of her free servants. Lond. 1656. 8vo.

As he gives no particulars, he presumably has seen no copy of it. One desiderates some authority for considering it distinct from those preceding.

51. This presumably first edition of this book also is somewhat of a puzzle, for there seem to be three issues of it, like that of the Countess of Kent's.

The edition in the British Museum has the general title dated 1655, "The Queens Delight" 1654, and "The Compleat Cook" 1655.

In my own copy all three tracts are dated 1655, and the title-pages differ in details from those in the Museum copy.

---

* *Hand-Book to the Popular, . . . Literature of Great Britain*, London, 1867, p. 121. Hazlitt quotes the 1655 edition again in his *Second Series of Bibliographical Collections and Notes*, London, 1882, p. 374. His collation is A-P6 in twelves, P5-6 blank, Q12. With a portrait by Faithorne, which is A 1, obviously.

† *The Bibliographer's Manual*, ed. Bohn, London, 1865, IV., p. 2023. He quotes Nassau, pt. ii., 119, 4s.

Again, the copy described by Hazlitt differs from both the preceding by the date of "The Compleat Cook" being 1654, as well as by differences in the general title, provided always that there has been no misprint in the transcription. According to dates this should be the earliest issue of the three, and one might be justified in inferring that the first edition of the book would have the date 1654 on all three title-pages. I have not seen any notice of such an edition. It is possible that the 1655 titles were printed after the rest of the book was finished, or that they were cancels.

In order to distinguish the two 1655 editions of "The Queens Closet Opened," it will be necessary to extract a few parallel passages.

All three title-pages are different in arrangement or other details.

B.M. E. 1519.	My own Copy.
The running title in C and D (*i.e.* pp. 1-48) is: *Accurate Medicinal* ‖ *and Physical Receipts.* After that it is: *The Pearl of Practise. Choise* ‖ *Physical & Chirurgical Receipts.*	The running title all through is: *The Pearl of Practise. Choise* ‖ *Physical & Chirurgical Receipts.*
p. 10, last lines: ufe. Take two fpoonfuls in a morning, and afmuch in the afternoon, fafting two hours after it.	Take two fpoonfuls in a morning, and afmuch in the afternoon, fafting two hours after it.
p. 34, last lines: ufe. Take a good draught of it in the morning, and at four a clock in the afternoon.	keep it in a glafs for your ufe. Take a good draught of it in the morning, and at four a clock in the afternoon.
p. 82, last lines: good Sugar to your liking, and drink a good draught thereof morning and evening towards the quantity of a pint.	good Sugar to your liking, and drink a good draught thereof morning and evening towards the quantity of a pint.
Sig. L, p. 196: The running title *The Pearl of Practife. Choife* &c. is carried on through this sheet to p. 216, although the tract is *A Queens Delight.* It is not till M begins, p. 217, that the correct headline appears.	The headline is *A Queens Delight* throughout, correctly.

B.M. E. 1519.	My own Copy.
pp. 196-216 are consecutively numbered.	The numbering of sheet L is as follows: 295, 196, 197, 298, 299, 200, 201, 302, 303, 204, 205, 306, 307, 208, 209, 310, 311, 212, 213, 314, 315, 216.
p. 222, l. 6: ſtrain-er. Be ſure you put Sugar enough to them at the firſt, and it will keep a year good, if it be ſet up well.	Be ſure you put Sugar enough to them at the firſt, and it will keep a yeare good, if it be ſet up well.
p. 269, line 5: it to dry in the Sun; and when it is dry, wash it over with this ſpirit of Wine. This must be done in hotteſt time of Summer.	in the Sun; and when it is dry, waſh it over with this ſpirit of Wine: This muſt be done in hotteſt time of Summer.

THE COMPLEAT COOK.

p. 21, last lines: times. you muſt brew this in *Aprill*, and let it ſtand till *June*, then draw them clea-rer, then waſh the hogſheads cleane and	hot out of the Oven, and put into each hogſhead a loafe, you muſt uſe this foure times, you muſt brew this in *Aprill*, and let
p. 100, line 14: three or foure Orin-ges, when you put all theſe therein, ſtir the meat and cover it againe, and ſerve it to the Table.	juice of three or four Oranges, when you put all theſe threin (*sic*), ſtir the meat and co-ver it again, and ſerve it to the Table.

There are like variations throughout the editions, showing that though they have the same date they are quite different from one another.

52. There is an edition of The Queen's Closet opened, London, Printed for Peter Dring, 1661, with a Queen's Delight, Printed by R. Wood for Nath. Brook, 1660, 12°, 2 parts in one volume. It wants the Compleat Cook. I have not seen this edition, which is contained in Pickering & Chatto's *Book-lover's Leaflet*, July 1907, No. 154. 1232.

53. The next edition which I have seen is in the British Museum, 1037, e. 4.:

> THE | QUEENS | CLOSET | OPENED: | Incomparable Secrets in | Phyſick, Chyrurgery, Preſer- | ving and Candying, &c. | Which were preſented unto the | QUEEN | By the moſt Experienced Perſons of the | Times, many whereof were had in | eſteem, when ſhe pleaſed to deſcend | to private Recreations. |
>
> Corrected and Reviewed, with many | Additions: together with three exact | Tables. | *Vivit poſt funera Virtus.* |

LONDON, | Printed for *Chrift. Eccleſton*, and are to | be fold at his Shop in St. *Dunſtans* | Church-Yard, 1662. |

12°. 6 leaves without mark; C to P in twelves, Q six; or, pp. [12] 300 [The Tables 23, 1 blank]. The first leaf contains a fine impression of the portrait of the Queen by Faithorne.

'The Pearle of Practice, Choice Phyfical and Chyrurgical Receipts,' runs from C to K12, pp. 1-192.

'A Queens Delight,' with a title-page, from L to P6, pp. 193-300, The Tables P7 to Q6. It was 'Printed by *R. Wood*, for *Nath. Brooke*, at | the Angel in *Cornhill*, 1662.' |

The Compleat Cook, A to F in twelves; or, pp. 123 [the Table 7; List of books printed for Nath. Brook, 14]. It has a separate title, and was printed for Nath. Brooke.

54. Another edition was published in 1668. The title is a copy of that of 1662, somewhat differently arranged.

I transcribe it because it differs from that given by Hazlitt.

THE | QUEENS | CLOSET | OPENED. | Incomparable Secrets in Phy- | fick, Chirurgery, Prefer- | ving and Candying, &c. | Which were prefented unto the | QUEEN: | By the moft Experienced Perfons of the | Times, many whereof were had in efteem, | when She pleafed to defcend to private Recreations. | Corrected and Reviewed, with many New and large | Additions: together with three exact Tables. | *Vivit poſt Funera Virtus.* |

LONDON, | Printed by *J. W.* for *Nath. Brooke*, at the | Angel in *Greſham*-College, near the *Ex-* | *change* in *Biſhops-Gate-Street*. 1668. |

12°. A6 (but A1, the portrait? wanting), C to K in twelves, L4; [A, B wanting] C to E8 in twelves; A to F6 in twelves; or, pp. [12, wanting the portrait?]. The Pearle of Practice, C to K12, L4, pp. 191 [1 blank, the table 8]. A Queens Delight wants A-B, pp. 1-46, has C to E8, pp. 47-106, the table [4].

In the Museum copy the e at the end of 'large' is indicated by a mere point, so that it looks as if it had been rubbed out, or been hardly printed.

The title page of 'A Queens Delight" is wanting, but that of "The Compleat Cook" is present, as follows:

The Compleat | COOK: | Expertly Prefcribing | The moft ready Wayes,

Whether { *Italian*, *Spaniſh* or *French*, }

For dreffing of *Fleſh* and | *Fiſh*, ordering of *Sauces*, | or making of | PASTRY. | *LONDON*, | Printed by *J. Winter*, for *Nath. Brooke*, at | the Angel in *Greſham*-College. 1668. |

A to F6, pp. 123, the table [7, book advertisements 2].

The copy in the British Museum (07944. de. 15.) is therefore imperfect and is soiled and in poor condition.

Hazlitt says: A 1 blank. He seems to have had a complete copy of the 1668 edition, but his account presents differences:

> The Queens Closet Opened . . . Corrected and Amended, with many New and larg (*sic in Hazlitt*) Additions: together with three exact Tables . . . London, Printed by J. W. for Nath. Brooke, . . . 1668. 12°. A, 6 leaves, A 1 blank: C-K in twelves: L, 4 (with the Table): no B: [a new title:] A Queens Delight, Or, The Art of Preserving, Conserving, and Candying. As also, A right Knowledge of making Perfumes, and Distilling the most Excellent Waters. Never before Published. Printed by J. Winter, for Nat. Brook, . . . 1668. A-E 8 in twelves, E 7-8 with the Table: [a third title:] The Compleat Cook: Expressly Prescribing the most ready Wayes, Whether {Italian, Spanish, or French} For Dressing of Flesh and Fish, ordering of Sauces or making of Pastry. London, Printed by J. Winter, for Nath. Brooke, . . . 1868 (*sic in Hazlitt*) A-F 6 in twelves.
>
> Doubtless the three pieces were published to sell either together or separately at option.*

55. After the usual interval came another edition, which requires some notice. The title runs thus:

> THE | QUEENS | CLOSET | OPENED. | Incomparable Secrets in Phy- | ſick, Chirurgery, Preſer- | ving and Candying, &c. | Which were preſented unto the | QUEEN: | By the moſt Experienced Perſons of | the Times, many whereof were had in | eſteem, when She pleaſed to deſcend to pri- | vate Recreations. | Corrected and Reviewed, with many New | and large Additions: together with three | exact Tables. | *Vivit poſt Funera Virtus.* |
>
> *LONDON*, | Printed by *E. Tyler*, and *R. Holt*, for *Nath.* | *Brooke*, at the Angel in *Corn-Hill*, near the | Royal *Exchange*. 1671. |
>
> 12°. A 6, C to K in twelves, L 4; or pp. [12] 191, 1 blank [8]; A to D in twelves, E 8, or pp. [2] 106 [4]; A to E in twelves, F 6, or pp. 123 [7] [2].

A 1 contains a portrait of the Queen Henrietta Maria. There is no name, but below it to one side are the words: *Sold by Nat. Brooke.*

---

* *Third and Final Series of Bibliographical Collections and Notes*, 1887, p. 298. I do not feel certain about the tracts being printed to sell separately. In every copy of every edition I have seen they have been together, and I have never met with a separate copy of any one of the three. It is true that the copy of "The Compleat Cook," 1655, in the British Museum (E. 1531. (1.)) is separated from "The Queens Closet Opened," 1655 (E. 1519.), but that is an accident of binding. The books are catalogued together.

A 2 is the title-page. A 3-4 is the preface to the Reader, signed W. M. A 5-6 is the list of the anthorities for the receipts. Sig. B is omitted. C to L4 contains *The Pearle of Practice.*

A 1 is the title-page to the second tract:

A | QUEENS | Delight; | OR, | The Art of Preferving, | Conferving and | Candying. | As alfo | A right Knowledge of making | Perfumes, and Diftilling the | moft Excellent Waters. | Never before Publifhed. |

LONDON, | Printed by *E. Tyler* and *R. Holt*, for *Nath.* | *Brooke*, at the Angel in *Corn-Hill*, near the | Royal *Exchange*. 1671.

This tract ends on E8.

A 1 is the title of the third tract:

The Compleat | COOK: | Expertly Prefcribing | the moft ready Wayes,

Whether { *Italian* / *Spanifh* or / *French*, }

For dressing of *Flefh* and | *Fifh*, ordering of *Sauces*, | or making of | PASTRY. |

LONDON, | Printed by *E. Tyler*, and *R. Holt*, for *Nath.* | *Brooke*, at the Angel in *Corn-Hill*, near the | Royal *Exchange*. 1671. |

This tract ends on F 5. F 6 contains book advertisements.

56. Of the edition of 1671, there was another issue of the same date with a slightly different title-page and a different imprint:

THE | QUEENS | CLOSET | OPENED. | Incomparable Secrets in Phy- | fick, Chirurgery, Prefer- | ving and Candying, *&c.* | Which were prefented unto the | QUEEN: | By the moft Experienced Perfons | of the Times, many whereof were had | in efteem, when She pleafed to defcend to | private Recreations. | Corrected and Reviewed, with many New | and large Additions: together with three | exact Tables. |

Vivit poft Funera Virtus. |

LONDON, | Printed for *N. Brooke*, and are to be | fold by *Charles Harper* at the *Flower-* | *de-luce* over againft St. *Dunftans* | Church in *Fleet-ftreet*. 1671. |

12°. In three parts: A6, C to L4 in twelves, or pp. [12] 191 [1 blank, 8]; A to E8 in twelves, or pp. [2] 106 [4]; A to F5 in twelves, or pp. 123 [7] [2].

The portrait is the same as in the other issue of this date. The title-pages and imprints of "A Queens Delight" and "The Compleat Cook" are the same as in the other issue, from which it would appear that the above title-page is either a cancel, or a duplicate printed for Charles Harper. The two issues are identical down to the end of sheet E of "The Compleat Cook." The concluding sheet, F, however, has been set up anew, and differs from the corresponding sheet in the first issue throughout.

There is a copy of this issue in the British Museum (1037. e. 5.) quoted by Arber.* One of the 1671 issues is in Lowndes,† quoted from White Knights, 3427, 5s. I have copies of both the 1671 issues.

57. THE | Queens Clofet | OPENED. | Incomparable Secrets in | Phyfick, Chyrurgery, Pre- | ferving and Candying, *&c.* | which were Prefented unto the | QUEEN: | By the moft Experienced Perfons of | the Times, many whereof were had | in Efteem, when She pleafed to de- | fcend to private Recreations. | Corrected and Revived, with many | New and Large Additions: toge- | ther with three exact Tables. | *Vivit poft Funera Virtus.* |

LONDON, Printed for *Nath. Brooke,* and are | to be Sold by *Tho. Guy* at the Corner Shop | of *Little Lumber-ftreet* & *Cornhill.* 1674. |

12°. A to S in twelves, T 10 [2 leaves wanting, blank? or advertisements?]; or, preliminary A 1-6, pp. [12].

Pearl of Practice: A7-I9, pp. 190 [the Table 8].

A Queens Delight: I 10-O5, pp. Title [2] 106 [the Table 4].

The Compleat Cook: O6-T 10, pp. 123 [the Table, 7].

In this edition the signatures run through, but the pagination is distributed among the three tracts.

A1 is the portrait, as in 1671. The second and third tracts have title-pages and the imprint:

LONDON, Printed by *F. Leach,* for *Nat. Brooks* (*sic*), and are to be Sold by *Tho. Guy,* at the Corner Shop in *Little Lumber-ftreet,* and *Corn-hill.* 1675.

The portraits in the editions of 1671 and 1674 are the same, and have below to one side: "*Sold by Nat. Brooke.*" That in the 1679 edition is better, and wants the imprint. They are poor copies of Faithorne's in the 1662 edition.

58. The next one was printed at London for Obadiah Blagrave at the Sign of the Black Bear in St. Pauls Churchyard in 1679.‡ It also is said to be "Corrected and Revived, with many new and large Additions: together with three exact Tables," but it does not specify what edition it is. It is a 12mo, and contains A to T in twelves. The pagination,

---

* *The Term Catalogues,* London, 1903, I., p. 92.

† *The Bibliographer's Manual,* ed. Bohn, London, 1865, IV., p. 2023.

‡ The 1679 edition, 18mo, portrait, is quoted by Lowndes, *The Bibliographer's Manual,* ed. Bohn, London, 1865, IV., p. 2023.

however, is distributed among the three tracts of which the book consists, as in the edition of 1674: [12]; Pearl of Practice, 190, [8]; A Queens Delight, with separate title [2], 106, [4]; The Compleat Cook, with separate title [1-2], 3-113 (*sic* for 123), [7] 4 blank. The present copy contains the portrait of Queen Henrietta, which is a sharper and better impression than that in the edition of 1683. The editions of 1679 and 1683 correspond line for line and page for page, but they are not identical; that of 1683 is an exact reprint of 1679, and is not the remainder of it with a new title-page.*

59. The other copy is called distinctly the tenth edition, and it was printed in 12mo, for E. Blagrave, probably a relation and successor of the elder man.

The title is arranged as follows:

THE | Queens Closet | OPENED. | BEING | Incomparable Secrets in | *Phyfick, Chyrurgery, Preferving,* | and *Candying*, &c. | Which were Prefented to the | QUEEN | By the moft experienc'd Perfons of the times, | many whereof were had in Efteem when | fhe pleafed to defcend to *private Recreations.* |
The Tenth *Edition, Corrected, with many new and* | *large Additions; together with Three exact Tables.* |
Vivit poft Funera Virtus. |
LONDON, | Printed for *E. Blagrave*, and are to be fold by | the Bookfellers of *London* and *Weftminfter*, 1696. |

Like the previous one, it is in three parts: the "Pearl of Practice," "A Queens Delight," and "The Compleat Cook," though, unlike it, the parts are not paged separately, but continuously. It contains signatures A6, B to I in twelves, K6, L to S in twelves, T6; or, pp. [10] Part I. 1-163, [164-168]; title-leaf of Part II. dated 1696, [H12, or 169-170] 171-264 [265-268]; title-leaf of Part III. dated 1695, [N8, or 269-270] 271-401 [402-409; 410 blank]. The portrait, included in sig. A, is the same as in the 1679 edition, but is a fainter impression. Below it is the inscription: HENERETTA MARIA | *Late Queen of England.* | The N in Heneretta is inverted.†

Assuming the edition of 1655 as the first of this book, there are more than enough of editions to justify that of 1696 being called the tenth.

---

*Arber quotes a 1679 edition, and says that it was printed for O. Blagrave at the Black Bear in St. Paul's Churchyard, and R. Harford at the Angel in Cornhill. *The Term Catalogues*, London, 1903, I., p. 363. There is a copy in the British Museum (1037. e. 7.) with the portrait. Hazlitt, *Second Series of Bibliographical Collections and Notes*, 1882, p. 374, quotes the 1683 edition.

†The 1696 edition is quoted by Arber. *The Term Catalogues*, 1905, II., p. 572.

Besides there is quite time for ten issues to appear in these forty years, for, as there has been already occasion to remark, one every four or five years seems to have been the average rate at which editions of these books of receipts and secrets were produced in the seventeenth century, when they were wanted in greatest numbers.

60. This tenth edition was reissued a couple of years later:

THE | Queens Closet | OPENED. | BEING | Incomparable Secrets in *Phyfick, Chy- | rurgery, Preferving, Candying,* and | *Cookery,* &c. | Which were Prefented to the | QUEEN | By the moft experienc'd Perfons of the times, | many whereof were had in Efteem when fhe | pleafed to defcend to *private Recreations.* |

The Tenth *Edition, Corrected, with many new and | large Additions; together with Three exact Tables.* |

Vivit poft Funera Virtus. |

LONDON, | Printed for *J. Philips* at the *Kings-Arms* in St. | *Paul's* Church-yard, *H. Rhodes* at the *Star,* the | corner of *Bridelane* in *Fleetftreet, J. Taylor* | at the *Ship* in St. *Paul's* Church-yard, and *K.* | *Bently* in *Covent-Garden.* MDCXCVIII. |

12mo. A6, B to S in twelves, T6; or, pp. [10] 1-163 [164-170], 171-264 [265-270] 271-401 [402-409, 410 blank].

A1 is a worn portrait of the Queen, identical with that in the 1696 edition. H 12, p. 169 is the title to A Queens Delight: . . . London, Printed in the year 1696. N 8, p. 269 is the title to the Compleat Cook: . . . London: Printed in the Year 1695. This, therefore, is simply a remainder reissue, with a new title-page.

61. The next edition is in the British Museum, [1037. e. 8. (1-2.)], and is in a style entirely new.

THE | Queen's Closet | OPENED. | BEING | Incomparable Secrets in PHYSICK, | SURGERY, PRESERVING, | CANDYING, and | COOKERY. | After the neweft Mode now practi- | fed in *England,* | By the moft experienc'd Perfons of the Times. |

The Eleventh Edition, *Corrected, with many* New | *and* Large ADDITIONS *throughout the whole.* |

Vivit poft Funera Virtus. |

LONDON: | Printed for *J. Phillips,* at the *King's-Arms* in | St. *Paul's* Church-yard; *H. Rhodes* at the *Star,* | the corner of *Bride-lane* in *Fleet-ftreet*; and | *J. Taylor,* at the *Ship* in *Pater-nofter-Row,* | MDCCX. |

12mo. 3 leaves. B to L in twelves, M3; A to F in twelves, G3.

The 3 leaves contain the first title, List of the Prescribers of the remedies, and a short Address to the Reader by W. M. The Pearl of Practice, B to G8, pp. 1-127, the Table [128-136].

Then G9, p. 137, is the title of A Queen's Delight:

A | QUEEN'S DELIGHT: | OR, A | SYSTEM | OF THE | *Confectionary* ART. | SHEWING | How to preferve all Sorts of FRUIT, | both Liquid and Dry, according to the | beft Methods now in ufe; as alfo how | to make |

Clear-Cakes,	*Bifcotins,*
Jellies,	*Bifkets,*
Compotes,	*March-panes,*
Conferves,	*Macaroons,*
Candy'd Confits,	*Wafers,*
Marmalets,	*Liquors of feveral forts,*
Quiddanies,	*Diftilled Waters,*
Paftes,	*Effences,*
Paftils,	*Perfumes,*
Sugar-works,	*Syrups,* &c.

LONDON: | Printed for *J. Phillips, H. Rhodes,* and | *J. Taylor,* 1710. |

which goes on to M3. Text G10-L12, pp. 139-240. Table M1-3, pp. [6].

Then follows the third Tract:

The Compleat | COOK: | Prefcribing | The moft Ready WAYS | FOR | Dressing *Flefh,* and *Fifh,* | Ordering | *Sauces, Pickles, Jellies,* &c. | And Making | PASTRY | After the *Neweft Manner.* |

LONDON: | Printed for *J. Phillips, H. Rhodes,* and | *J. Taylor,* 1710. |

Sigs. A to G3 in twelves; pp. [2], the text 138, the table [9, book advertisements 1].

The 1710 edition is quoted by Lowndes.*

62. The remainder of this edition was reissued in 1713 with a new and inferior title-page; but the "List of the Prescribers" and W. M.'s Preface "To the Ingenious and Courteous Reader" are placed at the end. Otherwise the two editions are identical. The 1713 title-page may be given, so as to help in identifying these different editions.

THE | QUEEN'S CLOSET | OPENED, | In Three PARTS. |

Part I. Containing above five Hundred | choice *Phyfical* and *Chirurgical* Receipts. |

Part II. Shewing how to *Preferve* all Sorts | of *Fruits,* both *Liquid* and *Dry*: Also | how to make all Sorts of *Sweet-Meats.* |

Part III. The *Compleat Court* and *Country* | *Cook*; prefcribing the moft ready Ways | for Dreffing *Flefh* and *Fifh,* Ordering | *Sawces, Pickles, Jellies,* &c., and making | *Paftrey,* after the neweft Manner now | practif'd in *England,* by the moft Ex- | perienc'd Perfons of the Times. |

The Eleventh Edition, Enlarged with many | *New Additions* throughout the whole. |

LONDON: Printed and Sold by *W. Taylor,* at | the *Ship* in *Pater Nofter-Row,* 1713.

* *The Bibliographer's Manual,* ed. Bohn, 1865, IV., p. 2023.

12mo. 2 leaves, B to L in twelves, M4; or, pp. [4] 1-127 [128-138] 139-240 [8]; A-F in twelves, G3; G4 wanting; or, pp. [2] 138 [10] [2 blank?]. There is no allusion to Queen Henrietta Maria, and this copy wants her portrait if the edition ever contained one, of which I am very doubtful.

The first part is called The Pearl of Practice, and runs from B to G8, pp. 1-[136]. The second part begins on G9, p. [137], with the following title:

A | QUEEN'S DELIGHT: | OR, A | SYSTEM | OF THE | *Confectionary* ART. SHEWING | How to preferve all Sorts of FRUIT, | both Liquid and Dry, according to the | beft Methods now in ufe; as alfo how | to make |

Clear-Cakes,	*Bifcotins,*
Jellies,	*Bifkets,*
Compotes,	*March-panes,*
Conferves,	*Macaroons,*
Candy'd Confits,	*Wafers,*
Marmalets,	*Liquors of feveral forts,*
Quiddanies,	*Diftilled Waters,*
Paftes,	*Effences,*
Paftils,	*Perfumes,*
Sugar-works,	*Syrups,* &c.

LONDON: | Printed for *J. Phillips, H. Rhodes,* and | *J. Taylor,* 1710. |

This tract goes from G9 to M4, pp. 137-[248].

The third part has separate signatures and pagination and different type. The title-page is as follows:

The Compleat | COOK: | Prefcribing | The moft Ready WAYS | FOR | Dreffing *Flefh,* and *Fifh,* | Ordering | *Sauces, Pickles, Jellies,* &c. | And Making | PASTRY | After the *Neweft Manner.* |

LONDON: | Printed for *J. Phillips, H. Rhodes,* and | *J. Taylor.* 1710. |

A to F in twelves, G 3, or pp. [2] 138, [10].

In binding this book, "The Preface" "To the Ingenious and Courteous Reader" has been put at the end of the second part, just before The Table, instead of immediately after the Title of the first part.

63. As in the case of the Countess of Kent, a list of the editions of the Queen's book may be given, but it is to be regretted that they were not numbered consecutively.

1654-55	N. Brook.
1655	N. Brook.
1656	
1660-61	P. Dring & N. Brook.
1662	C. Eccleston.
1668	N. Brook.
1671	N. Brook.

1671	N. Brook.
1674-75	N. Brook.
1679	O. Blagrave.
1683	
1684?	B. Crayle.
1696, called the tenth.	E. Blagrave.
1698, 1696, 1695, called the tenth.	J. Phillips.
1710, called the eleventh.	J. Phillips.
1713, called the eleventh; reissue of the 1710 edition.	W. Taylor.

It is interesting to compare the two books. They ran neck and neck for fifty years, one from 1653 to 1708, the other from 1655 to 1713. But the former had a much larger circulation; for of it there are twenty-one numbered editions as against sixteen of the other. These books, it must be remembered, are only two out of the cataract of similar little works which poured from the press during this half century. I have enumerated a few of them, but what has come of all the others, and what of the editions of this and other books of which apparently not a single copy survives? Presumably they served their day and generation, and are now fallen on sleep. It is a strange and significant chapter in the literary history of the arts and technology.

64. A book of the secrets of Natural History, published in 1657, has a certain interest for its contents, but still more for itself. Its title is as follows:

NATURE'S | CABINET | UNLOCK'D. | Wherein is Difcovered | The natural Caufes of Me- | tals, Stones, Precious Earths, | Juyces, Humors, and Spirits, | The nature of PLANTS in ge- | neral; their Affections, Parts, | and Kinds in Particular. | Together with | A Defcription of the Individual Parts | and Species of all Animate Bodies, Simi- | lar and Diffimilar, Median and Organi- | cal, Perfect and Imperfect. With a com- | pendious Anatomy of the Body of Man, | As alfo the Manner of his Formation in | the Womb. |

All things are Artificial, for Nature is the | Art of God. |

By *Tho. Brown* D. of Phyfick. |

London, Printed for *Edw. Farnham* in *Popes-* | *head* alley near Cornhil. 1657. |

12°. Title leaf, C to F in twelves, G and H12, I to R in twelves, but R12 wanting; or, pp. [2] 331 [2, 1 blank] [1 leaf, blank? wanting].

65. The question has been raised as to the authorship of this book. The author calls himself Dr. Thomas Brown, but whether that was actually his name, or whether it was assumed in order to make the book pass for the composition of Sir Thomas Browne, author of the "Enquiry into Vulgar

Errors," is open to discussion. It will be noticed that he omits the final e in his surname, and it is possible that he was the real author, but was otherwise unknown. Such, however, was not the belief of contemporaries, and at a later date, of Anthony à Wood, who seems to have regarded the publication as an attempt to foist the book upon the Norwich physician.

In the *Garden of Cyrus*, by Sir Thomas Browne, published in 1658, a year later than the present book, it was thought worth while for the publisher —I presume, Henry Brome, or Broome—to repudiate this production on behalf of Sir Thomas Browne. He does so in the following by no means clear terms:

> The Stationer to the Reader.
>
> I cannot omit to advertise, that a Book was published not long since, Entituled, *Natures Cabinet Vnlockt*, bearing the Name of this Authour: If any man have been benefited thereby this Authour is not so ambitious as to challenge the honour thereof, as having no hand in that Work. To distinguish of true and spurious Pieces was the Originall Criticisme; and some were so handsomely counterfeited, that the Entitled Authours needed not to disclaime them. But since it is so, that either he must write himself, or Others will write for him, I know no better Prevention then to act his own part with lesse intermission of his Pen.

Anthony à Wood * makes the following observations:

> "The reader may be pleased now to know that there hath been published under Dr. Thomas Browne's name a book bearing this title:
>
> "*Nature's Cabinet Unlocked, wherein is discovered the natural Causes of Metals, Stones, Pretious Earths, &c.*, printed 1657, in tw. A dull worthless thing, stole for the most part out of the *Physics* of Magirus by a very ignorant person, a plagiary so ignorant and unskilful in his Rider, that not distinguishing between *Lævis* and *Levis* in the said Magirus, hath told us of the liver, that one part of it is *gibbous* and the other *light*; and yet he had the confidence to call this scribble *Nature's Cabinet, &c.* an arrogant and fanciful title, of which our author's (Browne) true humility would no more have suffered him to be the father, than his great learning could have permitted him to have been the author of the said book. For it is† certain that as he was a a philosopher very inward with nature, so was he one that never boasted his acquaintance with her."
>
> "See a discourse by way of introduction to *Baconiana; or certain genuine Remains of Franc. Visc. S. Albans*, Lond. 1679, 8vo, pp. 76, 77. Written by Tho. Tenison, D.D."

---

* *Athenae Oxonienses*, 1721, II., col. 714; ed. Bliss, 1820, IV., col. 57. It is to be observed, however, that this criticism is not Wood's own, but is Tenison's, and is copied by Wood, with some verbal alterations, from the discourse he mentions.

Even on the narrowest grounds Sir Thomas Browne—or Hen. Broome, his publisher, for him—is not to be blamed for declining the authorship of this book, for to be credited with the opinions expressed in it, after having taken such pains to expose so many common errors, was more than could be endured. Not merely are the original views erroneous, but the comments on them, which were supposed to elucidate obscurities and give an insight into the secrets of Nature, are in many cases devoid of all meaning whatsoever. The book is, therefore, somewhat of a curiosity for its origin and its contents.

66. The secrets and receipts for the ladies, which passed under the name of Sir Kenelm Digby enjoyed great popularity. Several editions in English, French and German have been already enumerated, but the number is not exhausted, for I have here another, a copy of the seventh edition of the *Nouveaux Secrets Experimentez*, in two volumes, printed at the Hague in 1715. The sixth edition was published at the same place in 1700, in the same style. Both contain the discourse on the powder of Sympathy. A Dutch translation printed at Amsterdam in 1727 was different. It contained not only Digby's Discourse on the Powder of Sympathy, but other tracts on the same subject by Papin and Rattray. At the end there is a translation of the Medical Secrets of Digby as published by Hartman. Besides these there are some separate editions of the famous Discourse, one the excessively rare first edition in English, dated 1658, and a French reprint at Paris, 1660. Them, however, I pass by, because they belong to the subject of the Powder of Sympathy and Magnetic Cure of Wounds, on which a paper was communicated to the Society* a long time ago, but to which I hope to return.

67. In the same part of these "Notes" another work by Digby was alluded to: The Closet . . . opened, the third edition, London, 1677. I can now show what appears to be the first edition:

THE | CLOSET | of the Eminently Learned | Sir *Kenelme Digbie* K[t.] | OPENED : | Whereby is DISCOVERED | several ways for making of | *Metheglin, Sider, Cherry-*

* *Transactions of the Glasgow Archæological Society*, N.S., 1896, II., p. 21 (Reprint, 1890, Part VI., p. 23).

Wine, &c. | *TOGETHER WITH* | Excellent Directions | FOR | COOKERY: | As alfo for | *Preferving, Conferving, Candying*, &c. | Publifhed by his Son's Confent. |

London, Printed by *E. C.* for *H. Brome*, at | the Star in *Little Britain*. 1669. |

Small 8°. 2 leaves, B to X in eights, Y two leaves; or, pp. [4] 312, Index [11, 1 blank]. The title is surrounded by a narrow border.

68. Another candidate for the patronage of the ladies was Mrs. Hannah Woolley, or Wolley, *alias* Chaloner, who during the last quarter of the seventeenth century published works on the usual subjects. I may describe those which I have seen.

THE | LADIES DIRECTORY, | In Choice | Experiments & Curiofities | OF | PRESERVING & CANDYING | both Fruits and Flowers. | ALSO, | An Excellent way of making Cakes, | and other Comfits: | With Rarities of many Precious Wa- | ters (among which are feveral Confump- | tion Drinks, Approved of by the Doctors) | and Perfumes. | By *Hanna Woolley*, | Who hath had the Honour to perform | fuch things for the Entertainment of His late | MAJESTY, as well as for the Nobility. |

To prevent Counterfeits: |

Take Notice, That thefe Books are no where to be | had, but from the Authrefs, and at *Peter Drings* | at the Sun in the *Poultery*, Book-feller; and at | *Tho. Milbourns*, Printer, in *Jewen-Street* near *Al-* | *derfgate-Street*: Who alfo fells for her, a moft ex- | cellent Cordial-Powder, and very Soveraign Pils. |

London, Printed by *Tho. Milbourn* | for the Authrefs, 1661. |

Small 8°. A to H in eights; or, pp. [16] 111 [1 blank].

There is no frontispiece in this copy, which is in the British Museum (7944. a. 58.).

69. The book was issued again the following year.

THE | LADIES DIRECTORY, | In Choice | Experiments & Curiofities | OF | PRESERVING | in JELLIES, | and CANDYING both Fruits and Flowers. | Alfo, an Excellent way of making Cakes, | Comfits, and Rich Court-Perfumes. | With Rarities of many Precious | WATERS; | Among which, are Doctor *Stephens*'s Water, | Dr. *Matthias*'s Palfie-Water; and an excellent | Water againft the Plague: With feverall Con- | fumption Drinks, Approved of by the Ableft | Phyficians. | By HANNA WOLLEY, | Who hath had the Honour to perform fuch things | for the Entertainment of His late MAJESTY, | as well as for the Nobility. |

LONDON, Printed by *T. M.* for *Peter* | *Dring*, at the *Sun*, next Dore to the *Rofe* | Taverne in the *Poultry*, 1662. |

Small 8vo. A to H in eights, 1 leaf of advertisements at the end extra; or, pp. [16] 111 [1 blank, 2].

The frontispiece (extra) represents a lady engaged in distilling. The title gives a sufficiently accurate notion of the contents. There is a copy in the British Museum (7944. a. 24.).

The difference between the two issues is in the address to the Ladies. In the 1661 edition it ends A4 *verso*, and has two paragraphs, omitted in that of 1662. The first paragraph contains the notice of "a most Soveraign Antidote in Powder" and "excellent Pils." The second states: The Price of this Book is Six Shillings; I befeech you grudge it not, fince there is in it, many Pounds worth of fkill imparted to you, by, Ladies, your very Humble Servant, Hanna Woolley."

The address in the 1662 edition ends A4 *recto* and the *verso* is blank. This edition seems to be the remainder of that of 1661 with a new title-page and altered address.*

70. The next little treatise by Mrs. Wolley which I have seen is the following:

THE | COOKS GUIDE: | OR, | RARE RECKIPTS | FOR | COOKERY. | PUBLISHED | And fet forth particularly for Ladies | and Gentlewomen; being very benefi- | cial for all thofe that defire the true way of dref- | fing of all forts of Flesh, Fowles, and Fifh; the beft | Directions for all manner of Kickfhaws, and the | moft Ho-good Sawces: Whereby Noble Perfons | and others in their Hofpitalities may be gratified | in their Gufto's. *Never before Printed.* |
By HANNAH WOOLEY. |
LONDON, | Printed for *Peter Dring* at the *Sun* in the | *Poultry*, next door to the *Rofe-Tavern* 1664. |
Small 8vo. A8, a4, B to G in eights, H3; or, pp. [24] 100 [1, 1 blank].

The book is dedicated separately to Lady Anne Wroth, and Mistress Mary Wroth her daughter, and there is an address to the Ladies in general, in which she refers to "The Ladies Directory." "Ho good Sawces," by the way, and "Gusto's" are charming phrases. There is a copy in the British Museum, 1037. a. 22.

71. THE | Ladies Delight: | OR, A | RICH CLOSET | OF CHOICE | Experiments & Curiosities, | Containing the ART of | Preserving & Candying | both Fruits and Flowers: | Together with | The Exact Cook; Or, The ART of | Dreffing all Sorts of *Flefh*, *Fowl*, | and *Fifh*. |
By HANNAH WOOLLEY. |
To which is Added: |

* Both editions are quoted by W. Carew Hazlitt in an article entitled "Gleanings from Recent Book-Sales," in *The Antiquary*, 1889, XX., p. 273.

The Ladies Physical Closet: Or, Ex- | cellent *Receipts*, and Rare *Waters* for | Beautifying the *Face* and *Body*. |

London, Printed by T. Milbourn, for | N. Crouch, in *Exchange-Alley* over | againſt the Royal-Exchange in | *Cornhil*, 1672. |

12°. A2, B to N in twelves, O6, P12, Q4; or, pp. [4] 1-137 [1 blank]; [139-146] 147 (misnumbered 138)-256; [257-258] 259-300 [32].

Frontispiece, poorly executed, of a lady distilling, extra. A1, title. A2, address. B1 to G9 *r.*, text of "The Art of Preserving and Candying," *v.* blank. G10 *r.* [p. 139] title-page of the second tract:

THE | EXACT COOK: | OR, | Excellent Receipts | FOR | COOKERY, | BEING | The True way of Dreſſing all ſorts | of *Fleſh*, *Fowls*, and *Fiſh*, with the | beſt Directions for making of all | manner of *Cickſhawes*, *Pyes*, and | the moſt curious *Sauces*; whereby | Noble Perſons, and others in their | *Hoſpitalities*, may be gratified in | their *Guſto's*. | By HANNAH WOLLEY. |

London, Printed by T. Milbourn, | for Nathaniel Crouch, | 1672. |

G11 *r.*, Dedication to Lady Anne Wroth. G12 *v.*, Address to the Ladies. H2, p. 138 (for 147), the Text begins, and ends M8 *v.* p. 256. M9 *r.* [p. 257], Title:

THE | LADIES | Phyſical Cloſet. | OR, | Choice Receipts and | Experiments in *Phyſick* | and *Chyrurgery*. |

With ſome Excellent *Wa-* | *ters* for Beautifying and A- | dorning the Face and Body. |

And ſome Rare Receipts | in *Cookery*; with Bills of *Fare* | for all the *Seaſons* in the year. |

London, Printed by *Tho. Milbourn* | for *Nath. Crouch*, 1672. |

M10 *r.*, p. 259, Text, ends O6 *v.* p. 300. P1 *r.*, Bills of Fare, to P3 *r.* P3 *v.*, The separate contents of the three parts, to Q3 *v.* Q4, Note from H. W.

The first tract is a reprint of "The Ladies Directory" of 1661, and the second of "The Cooks Guide" of 1664; of the third tract I have not seen an earlier separate edition.

72. It was from this edition that the following German translation was made (British Museum, 7945. a. 28.):

Frauen-Zimmers | Zeit-Vertreib, | Oder | Reiches Gemach von auſserleſe- | nen Experimenten und Curioſitäten | betreffend die rechte | Præſervir- und Candier- | Kunſt, | Wie man beydes Früchte und | Blumen in Zucker einmachen, und | lange Zeit erhalten könne, | Sampt einem | Vollkommenen Koch, | Zeigend die rechte Art und Weiſe, wie | man allerhand Fleiſch, Fiſche und Geflügel | künſtlich zu bereiten könne. | Dazu auch angehencket iſt eine kleine | Frauen-Zimmers Apothecke, | oder herzliche Recepten und rare Waſſer, ſo zum | Schmincken dienlich. | Anfänglich in Engliſcher Sprache befchrieben | von | ANNA WOLLEY | Jetzo aber dem Teutfchen Frauen-Zimmer zu Nutzē in ihre | Sprache verfetzet. | Hamburg, | auff Gottfried Schultzens Koften, im Jahr 1674. |

12mo. [A] B to O in twelves, P6. No pagination.

[A1] is the engraved title, which is a modified copy of that in the English edition of 1662, reversed. [A2] Printed title, as above. [A3] Anna Wolley's address to the ladies. [A4] The text begins, and G iiij *r.* ends. G iiij *v.*, Inhalt, to G viij *v.* G viiij *r.*, Title of the second tract:

Der vollkommene | Koch, | oder | Fürtreffliche Anweifungen | zur | Koch Kunft, | Worinnen gelehret wird die | rechte Art und Weife allerhand ar- | ten Fleifch, Vögel und Fifche zuzu- | richten, imgleichen allerley Schau- | Effen, Pafteten | und wohlfchme- | ckende Brühen Künftlich | zu bereiten, | Allen Adelichen und andern Stan- | des Perfohnen folche bey ihren Gä- | ftereyen und Mahlzeiten zu | gebrauchen. | Zufondern gefallen an den Tag gegeben | Von | Hanna Wolley. |

G viiij *v.*, H. Wolley's address. G x *r.*, Text begins, M xij *r.* ends. M xij *r.*, Register begins, N iiij *r.* ends. *v.* blank. N v *r.*, Title of the third tract:

Frauen Apotecke | oder | Aufserlefene Recepten und | Experimenten zur Leib und | Wund-Artzney ge- | hörig. | Nebenft einen herlichen | Schminck-Waffer | und ungemeiner | Koch-Kunft, | wie auch einer Lifta der Spei- | fe und Tifch-gängen nach | den vier Zeiten des | Jahres. |

N vj *r.*, Text begins, P j *r.* ends. P j *r.*, Verzeichnüfs der Speifen. P iij *v.*, Register. P vj *r.*, Note to the Ladies by A. W.

73. Mrs. Wolley's next work is not to be confused with "The Queens Closet opened," already described.

THE | QUEEN-LIKE CLOSET; | OR, | RICH CABINET: | Stored with all manner of | RARE RECEIPTS | FOR | Preferving, Candying & Cookery. | Very Pleafant and Beneficial to all | Ingenious Perfons of the | FEMALE SEX. | By HANNAH WOLLEY. |

LONDON, | Printed for *R. Lowndes* at the | *White Lion* in *Duck-Lane*, | near *West-Smithfield*, | 1670. |

Small 12mo. Signatures: A6, B to S in twelves, T6; or, pp. [12] 383 [1 blank] [32] [1, 3? blank].

Signature A contains the License by Robert L'Estrange, dated Nov. 16, 1669 (wanting in my copy), the engraved title, in six compartments, representing the lady engaged in cooking, baking, distilling, preserving, etc., the printed title as above, the authoress' dedication to Mrs. Buzby, two to all ladies and gentlewomen, one in prose, the other in verse. The text is in two divisions. The first part (B to I6 *recto*, *verso* blank, pp. 1-179 [180]) is occupied chiefly with preserving and candying fruits, baking of cakes, making of cordials, possets, waters and other confections. The second is concerned with cookery.

It begins on I7 *recto* (p. [181]) with a title-page (given below), and the text goes on from I8 *recto* to Q7 *recto* (pp. 183-349). Then follow bills of fare, Q8 to R5 *recto* (pp. [351]-369), directions to servants, R5 *verso* to R12 *recto* (pp. 370-383, R12 *verso* is blank), the Contents of the two parts, S to T3 *verso*, a postscript in commendation of the second part, and advertisements of certain medicines, T4 and 5.

The title-page of the second part runs as follows:

I7. p. 181.

The SECOND PART of the | QUEEN-LIKE CLOSET; | Having an Addition of what hath | already been treated of, and di- | recting a very true and excellent way | for all manner of | COOKERY, | BOTH | FISH, FLESH & PASTRY. | Shewing the true Seafoning of all things for | COMPLEAT TABLES. | Alfo all Kinds of SAUCES and PICKLES | in a very brevious Way. | Here is to be noted, that in divers of thefe | Receipts there are Directions for two or three | feveral Things in one, not confounding the | Brains with multitudes of Words to little or no | purpofe, or vain Expreffions of things which are | altogether unknown to the Learned as well as the | Ignorant; this is really imparted for the good of | all the FEMALE SEX. | By *Hannah Wolley*, alias *Chaloner*. |

London, Printed for *R. L.*, 1670. |

This is the rare first edition, and there is a copy of it in the British Museum, 1037. d. 42.*

74. There is there (7957. a. 42.), also, a copy of the second edition,† to which is added a Supplement: this edition is dated 1672-74, and a description follows:

The Queen-like Clofet, | OR | RICH CABINET: | Stored with all manner of | RARE RECEIPTS | FOR | *Preferving*, *Candying and Cookery*. | Very Pleafant and Beneficial to all | Ingenious Perfons of the | FEMALE SEX. | By HANNAH WOLLEY. | The Second EDITION. |

LONDON | Printed for *Richard Lowndes* at the | *White Lion* in *Duck-Lane*, | near *Weft-Smithfield*, | 1672. |

12mo. A6, B to Q in twelves, R6; or, pp. [12] 344 [26, 2 book advertisements].

A1, (containing the licence?) is wanting. A2, Engraved title-page in six compartments. A3, Printed title-page as above. A4, Dedication to Mrs. Buzby. A5-6 *v.*, Addresses to the Ladies. B1, Text, to Q4 *verso*. H7, Title-page to the second part. Q5 *r*, Contents, to R4 *verso*. R5, Postscript. R6, Advertisements.

---

* See also Arber, *The Term Catalogues*, 1903, I., p. 27.

† The second edition appears under Mich., 1671, in *The Term Catalogues*, ed. Arber 1903, I., p. 91.

Then follows:

A | SUPPLEMENT | TO THE | *QVEEN-LIKE* | CLOSET; | OR | A LITTLE | OF | EVERY THING. | PRESENTED | To all Ingenious Ladies, | and Gentlewomen. | By *HANNA WOOLLEY.* |

LONDON, | Printed by *T. R.* for *Richard Lownds,* | and are to be Sold at the Sign of the | *White Lion* in *Duck-Lane,* 1674. |

12mo. A8, B to I in twelves, K4; or, pp. [16] 200.

Sig. A is irregularly arranged—three leaves having been inserted between A2 and A3, thus: A1. Title. A2. Verses to the Ladies. An advertisement, one leaf. To the Reader, two leaves, concluding on A3 *recto*; *verso* blank. A4-5. The table. B to K4 *r*, the Text. K4 *v*, lines to the Ladies, by H. W.

75. I have copies of the third and fourth editions, both of which are also in the British Museum.*

The Queen-like Closet, | OR | RICH CABINET: | Stored with all manner of | RARE RECEIPTS | FOR | *Preserving, Candying and Cookery.* | Very Pleasant and Beneficial to all | Ingenious Persons of the FEMALE SEX. | To which is added, | A SUPPLEMENT, | *PRESENTED* | To all Ingenious LADIES, | and GENTLEWOMEN. | By *Hannah Wolley.* | The Third EDITION. |

LONDON, | Printed for *Richard Lowndes* at the *White* | *Lion* in *Duck-Lane,* near *West-* | *Smithfield,* 1675. |

12°. A6, B to Q in twelves, R6; or, pp. [12] 344 [Contents 24, Postscript 2, Advertisements 2]. A1 contains the Licence. A2 is the frontispiece in six compartments. A3-6 contain the dedications. H7 contains the title of the second part. In it her name is spelled Hannah Wolly.

Supplement, dated 1674, A8, irregular; B to I in twelves, K4, or pp. [16] 200. This is identical with that just described.† Her name here is Hanna Woolley.

76. **The Queen-like Closet,** | OR | RICH CABINET: | Stored with all manner of | RARE RECEIPTS | FOR | *Preserving, Candying, and Cookery.* | Very Pleasant and Beneficial to all | Ingenious Persons of the *Female Sex.* | *To which is added,* | A SUPPLEMENT, | *PRESENTED* | To all Ingenious LADIES, | and GENTLEWOMEN. | By *Hannah Wolley.* | The Fourth EDITION. |

LONDON, | Printed for *R. Chiswel* at the *Rose* and | *Crown* in St. *Paul's Church-yard,* and | *T. Sawbridge* at the *Three Flowers-de-* | *luce* in *Little Britain,* 1681. |

12mo. A6, B to Q in twelves, R6; or, pp. [12] 343 (misprint for 344) [28]; H7 is the title-page of the second part. Supplement: A8, irregular; B to I in twelves, K4 [or, pp. 16] 200. The Supplement is dated 1680, and her name is given as Hannah Woolley.

---

* Arber, *The Term Catalogues,* 1903, I., p. 423, Mich., Nov. 1680, 4th edition, and again, p. 446, Easter, May 1681, 4th edition.

† "A Supplement to the Queen-like Closet." Trin., July 6, 1674, in Arber, *The Term Catalogues,* 1903, I., p. 179.

77. In the fifth edition the title is the same as in its immediate predecessor, and it was published by the same firms in 1684.* It is a 12mo, and has the usual frontispiece and licence. The "Queen-like Closet" contains six leaves, and B to M in twelves, N6; or, pp. [12] 254 [misnumbered 264] [22]. G1 contains the title of the second part. Unfortunately the Supplement is wanting in the copy I possess, but I am able to supply it from that in the British Museum (1037. e. 17.):

> A | SUPPLEMENT | TO THE | *QVEEN-LIKE* | CLOSET, | OR | A LITTLE | OF | EVERY THING. | PRESENTED | To all Ingenious Ladies, and | Gentlewomen. | By HANNAH WOOLLEY. |
>
> *LONDON*: | Printed for *R. Chifwel* at the *Rofe* and | *Crown* in St. *Paul's Church-Yard*, and | *T. Sawbridge* at the *Three Flower-de-* | *Luces* in *Little-Britain*, 1684. |
>
> 12°. A, Bb to Gg in twelves; or, pp. [24] 144.

Comparison with the preceding edition of 1681 shows that this one is much more closely printed.

78. Another work by Mrs. Woolley has the following title:

> THE | [Gentlewomans Companion;] | OR, A | GUIDE | TO THE | [Female Sex:] | CONTAINING | Directions of Behaviour, in all Places, Companies, | Relations, and Conditions, from their Child- | hood down to Old Age: | *VIZ.* As, |
>
> Children to Parents. Scholars to Governours. Single to Servants. Virgins to Suitors. Married to Husbands. } { Hufwifes to the Houfe. Miftreffes to Servants. Mothers to Children. Widows to the World. Prudent to all.
>
> With [LETTERS] & [DISCOURSES] | upon all Occafions. |
>
> Whereunto is added, *A Guide for Cook-maids, Dairy-* | *maids, Chamber-maids*, and all others that go to Service. | *The whole being an exact Rule for the Female Sex in General.* | By [*HANNAH WOOLLEY.*] |
>
> [*LONDON*,] | Printed by [*A. Maxwell*] for [*Edward Thomas*,] at the | [*Adam*] and [*Eve*] in [*Little-Brittain*,] 1675. |
>
> 8°. A8, *b*4, B to S in eights; or, pp. [23, 1 blank] 262 [Advertisement 2], List of Books [7, 1 blank].

---

* This fifth edition of 1684 is quoted by W. C. Hazlitt, *Handbook to the Popular, . . . Literature of Great Britain*, 1867, p. 121.

In *The Term Catalogues*, ed. Arber, 1905, II., p. 119, the fifth edition appears in Hil. Feb. 1685.

A 1 is an engraved portrait by W. Faithorne, of a lady, with a coat of arms. This portrait is intended for Mrs. Wolley, but belongs to Mrs. Gilly. The same portrait on wood, much inferior, with certain alterations, and omission of the shield, is prefixed to *The Accomplished Lady's Delight*, and, as a matter of fact, the woodcut title of the latter is to some extent a copy of that of *The Queen-like Closet* of Mrs. Wolley. The title is red and black. The red is placed here within black brackets. In my copy, A 1, the portrait, is wanting, and S4 to 8 appear to have been cancelled.

This work would hardly have fallen within the scope of these notes had there not been some sections devoted to secrets of medicine and chirurgery, receipts for preserving and candying, as well as a number of cookery and carving receipts and hints to various classes of servants. These correlate the work with several others described in the present part.

79. A subsequent edition * is in the British Museum (8416. a. 61.):

THE | [Gentlewomans Companion;] | OR, A | [GUIDE] | TO THE | [Female Sex:] | CONTAINING | [Directions of Behaviour, in all Places] | Companies, Relations, and Conditions, | from their Childhood down to Old Age: |

[*VIZ.* As,] |

Children to Parents.	Hufwifes to the Houfe.
Scholars to Governours.	Miftreffes to Servants.
Single to Servants.	Mothers to Children.
Virgins to Suitors.	Widows to the World.
Married to Hufbands.	Prudent to all.

[With LETTERS and DISCOURSES] | upon all Occafions. | [Whereunto is added, *A Guide for Cook-maids,*] | *Dairy-maids, Chamber-maids*, and all others that | go to Service.

The whole being an exact Rule for the Female Sex in General. | The THIRD EDITION. | [By *HANNAH WOOLLEY.*] |

[*LONDON,*] Printed by [*T. J.*] for [*Edward Thomas*] | at the [*Adam*] and [*Eve*] in [*Little Brittain,*] 1682. |

12mo. A to R in twelves, S six; or, pp. [24] 371 [1]. The frontispiece, which is mounted, is the same engraved portrait by Faithorne as that in the previous edition. Title red and black; the red is within black brackets.

80. An amusing section is that in which the authoress describes her qualifications and abilities for giving advice to others. They certainly lose nothing in her telling of them. According to her own account, she was skilled in all needlework, shell work, moss work, bugle work, feather work,

* This edition is mentioned by W. C. Hazlitt, *Collections and Notes*, 1876, p. 467.

cookery, making of ointments, waters, cordials, washing, perfumes, writing and arithmetic. At the age of fifteen she kept a school. She knew Italian, could sing, dance, and play on several sorts of musical instruments. She became the companion of ladies, looking after all their affairs, conducting their correspondence, reading books to them, plays, poems, and all the newest French works. She practised carving, was ready to help in physick and chirurgery, learned how to speak and conduct herself with the greatest ease and grace. "In short time," she adds, "I became skilful, and stayed enough to order an house, and all offices belonging to it; and gained so great an esteem among the Nobility and Gentry of two Counties, that I was necessitated to yield to the importunity of one I dearly lov'd, that I might free myself from the tedious caresses of a many more." *

What can the Grand New Woman offer us as a set off to such attractions? She cuts a very sorry figure indeed.

One can see after reading—with increasing wonder—such a catalogue of attainments, that Hannah was perfectly genuine in her statement: "It is no ambitious design of gaining a Name in print (a thing as rare for a Woman to endeavour, as obtain) that put me on this bold undertaking; but the meer pity I have entertain'd for such Ladies, Gentlewomen, and others, as have not received the benefits of a tythe of the ensuing Accomplishments." What would be the depth of Mrs. Wolley's commiseration now?

I wish I could have given some extracts from her advices to young gentlewomen, but they would be out of place in their present surroundings. One is less impressed by the soundness of the advice than by the rudimentary state of manners in London at the time Mrs. Wolley flourished.

She had a pretty wit too, and a turn for epigram, not without a spice of sarcasm; but her remarks are flavoured also with the peculiar diction of the time. They could, therefore, hardly be addressed to the young person of the present day, even if brought up on the realistic novel and the problem play, and familiar with the society and morals of mean streets. The language of the time, judging by specimens of dialogues Hannah

* "Importunity," "tedious caresses," poor Mr. Chaloner.

See p. 59.

gives, was a mixture of fulsome compliment, inflated hyperbole, and coarse brutality.

I have not been able to find any reference to Mrs. Wolley's revelations of the state of breeding in England in the seventeenth century in Mr. Buckle's *History of Civilization*. But perhaps it did not fall within his scope to show that there was ever any want of it in England. Mrs. Wolley might have convinced him of the contrary.

81. So far as I can judge from these books, Mrs. Wolley did not allow her portrait to appear in them. She may have had good reason for this. Apparently her portrait was never taken, for no less notable a collector than Horace Walpole was unsuccessful in getting it. In a letter addressed to "Mr. Cole, at Waterbeche, near Cambridge," from Strawberry Hill, August 12, 1769, he says:*

> ". . . The *Countess of Kent* is curious too. I have lately got a very dirty one, so that I shall return yours again. *Mrs. Wooley* I could not get high nor low."

From this juxtaposition, both authoresses of receipt books, I have little doubt our Hannah is meant. But why did she dislike having her portrait taken? Was she afraid that the artist would have done her less or more than justice, or did the "dearly loved one" object? It is singular, I may remark in passing, that the portraits of the Countess of Kent, which I have seen, are like Mr. Walpole's, "very dirty," with one exception.

82. Granger† mentions one in octavo by Faithorne, but adds that he had seen the same head with the name of Mrs. Gilly affixed to it. In the interleaved copy of Granger's book, formerly belonging to W. Musgrave and annotated by him, now in the British Museum, Musgrave's note states that "the arms engraved under the print belong to the family of 'Gilly' and are *not* those of Wolley. This furnishes reason to believe that the print was originally designed for Mrs. Sarah Gilly, who died at Totenham high Cross, 14 Augt. 1659, and was buried at Islington." This print is here given.

---

*John Nichols, *Literary Anecdotes*, London, 1812, IV., p. 706.

† *A Biographical History of England*, 1769, II. ii., pp. 384-385.

Granger and Musgrave seemed to think that Mrs. Wolley was a predecessor of the immortal Mrs. Harris. But, then, who wrote the books, and why was the true name concealed?

Bromley * says:

> Sarah GILLY. Ob. 1659, æt. cir. 48—prefixed to her "Receipts," 1662, 8vo. The name of HANNAH WOOLEY appears upon the latter impreffions; the former is very fcarce.
>
> Painter or Defigner (Lely). Engraver or Printfeller, W. Faithorne.

A "Lamentation" or "Threnodia" on Sarah Gilly was written by Henry Woolnough: "Fideles Aquæ: or some Pious Tears dropped upon the Hearse of the Incomparable Gentlewoman Mrs. Sarah Gilly . . . London, 1661, 8vo." She died at Tottenham High Cross, Augt. 14, 1659, but no biographical details are given. A great deal is said about her beauty and mental and moral qualities, but there is nothing that could be construed into the authorship of receipt books or of having pursued practical occupations therewith connected. One wishes that Mr. Woolnough had stated some facts about Mrs. Sarah's life and works and prefixed her portrait to his fantastic rhapsodies and silly acrostics. Bromley ascribes the 1662 *Receipts* to her. Whether by Mrs. Wolley or not, they are not by Mrs. Gilly. How the portrait of Mrs. Gilly got inserted in Mrs. Wolley's books I am unable to explain. It is curious, showing how persistent ignorance is, that this portrait is produced more or less accurately both in "The Accomplish'd Ladies Delight" (§§ 95-101) and in John Shirley's "The Accomplished Ladies Rich Closet of Rarities," to be described afterwards.

83. One more book of popular medicine in the British Museum (1038. c. 18. (4.)) is attractive from its very title, which throws light on the ways of the time:

> THE | Countrymans Friend, | AND | No circumventing Mountebanck | But a rare method of Chyrur- | gery and phyfick, teaching the coun- | try people excellent cures, the | like was never laid open in any | age before. |
>
> Befides here are four arts, three | of them concerning horfes, and the | fourth an art to keep a Field of corn | from any manner of Fowles, that

* Henry Bromley, *A Catalogue of Engraved British Portraits*, London, 1793, p. 112. Period IV., Class IX., Second Subdivision. Bromley has entirely overlooked the dates of Mrs. Wolley's works.

devour | grain, this art is only by anointing | a few Crow feathers, for nei- | ther Pidgeon, Sparrow, | Rook nor Crow will en- | dure the Field where they ftick. | By ABRAHAM MILES. |
London printed for *E. Andrews* at the *White* | *Lyon* neer Pye-corner, 1662. |
Small 8°. Sigs. A, B in eights, C in four; or, pp. [40].

The author complains in his preface of the want of charity to the poor, the unwillingness of Physitians and Chyrurgeons to "help the fick or the lame" and "many people there are who are loft for want of help, that have not a great deal of mony to give for a fmal cure"; "But they may be fick and recover as they can, for any help they fhal have, without great ftore of mony to pay for the cure: But here is an eafie way for thee to work by to help thy felf in time of neceffity of any defperate impediment, and at a very eafie rate to help thee or thy friend, and thou mayft gain both love and profit." The diseases of those days were very much the same as those now, and the author treats them with a confidence which must have had a reassuring effect upon his patients. All his descriptions and prescriptions have the magic phrase "Probatum est," at the end of them. In these receipts and the naïve remarks of the author, one gets incidentally a glimpse of aspects of country life which are enlightening as to its reputed wholesomeness and innocence.

84. A book by Robert Lovell about animals and minerals and their medicinal properties was described on a former occasion.* At the time I mentioned the companion treatise on plants, but had not seen it. I have since got a copy, which is similar in style to the other, and may be included in these notes as completing the author's review of the Materia Medica. It contains a great amount of matter, more interesting now to the folk-loreist than to the therapeutist, but of excellent use as showing the distribution of indigenous plants in those days, those imported from abroad, and the healing virtues wherewith they were credited. The book is furnished with full indices, and the introduction classifies under the different ailments, the plants which were used to effect a cure. Like

* *Transactions of the Glasgow Archæological Society*, N.S., 1890, I., p. 319 (Reprint, 1888, Part IV., p. 21).

most books of its class, it is an "aggregation" or summary of the opinions of the different authorities, ancient, medieval and modern.

There is an imperfect copy of the first edition in the British Museum (E. 1858.) which may be included:

ΠΑΜΒΟΤΑΝΟΛΟΓΙΑ. | *SIVE* | ENCHIRIDION BOTANICUM. | *OR* | A COMPLEAT | HERBALL | Containing the Summe of what hath | hitherto been publifhed either by Anci- | ent or Moderne Authors both Galenicall | and Chymicall, touching Trees, Shrubs, | Plants, Fruits, Flowers, &c. In an Al- | phabeticall order: wherein all that are | not in the Phyfick Garden in Oxford are | noted with afterifks. | Shewing their Place, Time, Names, | Kindes, Temperature, Vertues, Ufe, | Dofe, Danger and Antidotes. | Together with |

An { Introduction to Herbarifme, &c. |
Appendix of Exoticks. |
Universall Index of plants: fhewing |
what grow wild in England.

By ROBERT LOVELL *St. C. C. Ox.* | Φιλοθεολογιατρόνομος. |
OXFORD, | Printed by WILLIAM HALL, | for RIC. DAVIS. *An.* 1659. |

12°. *, **, ***, in twelves, **** in six; A to O in twelves; or, pp. [84] 1-336. This ends at the word *Onion*, and all after that is wanting. The title-page is surrounded by a border of small fleurs-de-lis. Print: $4\frac{1}{8}\frac{3}{8}$", not including the catchwords, etc., × $2\frac{9}{16}$".

85. The book was reissued as follows:

ΠΑΜΒΟΤΑΝΟΛΟΓΙΑ. | *SIVE,* | ENCHIRIDION BOTANICUM. | *OR,* | A COMPLEAT | HERBALL, | Containing the Summe of Ancient and | Moderne Authors, both *Galenical* and *Chy-* | *mical,* touching Trees, Shrubs, Plants, Fruits, | Flowers, *&c.* In an Alphabetical order: | wherein all that are not in the Phyfick Gar- | den in *Oxford,* are noted with Afterisks. | Shewing their Place, Time, Names, | Kinds, Temperature, Vertues, Ufe, | Dofe, Danger and Antidotes, | Together with |

An { Introduction to Herbarifme, &c.
Appendix of Exoticks.
Univerfal Index of Plants: fhewing
what grow wild in *England.*

The *Second Edition,* with many *Additions* men- | tioned at the end of the *Preface.* |
By ROBERT LOVELL, | Φιλοθεολογιατρόνομος. |
OXFORD: | Printed by *W. H.* for *Ric. Davis.* 1665. |

12mo. Signatures: A, b, c in twelves, d six, A to Z, Aa to Ee in twelves, Ff four; or, pp. 84; 672; [8]. Ff1 *recto* is paged 675. The print is 5" × $2\frac{1}{4}\frac{3}{8}$."

At first sight it seemed possible that this was an issue of the remainder with a new title-page and preliminary matter, but the different size of the printed page shows that this is a genuine second edition, which is to Lovell's credit. The increased dimensions of the page in this second edition have caused the section on the "Onion" to fall on the last page of signature N, p. 312, instead of the same page of signature O, p. 336, so that a whole sheet has been saved. But the book is not nearly so well printed, in fact it is coarse and inferior in all respects. There are, however, very considerable additions to the introduction and text throughout, and numerous references from synonyms to the heading under which the description is given. There is no doubt, however, that the first edition is the more presentable book of the two.

86. Of William Salmon's *Polygraphice* a notice was given formerly of the second edition, 1673.* I have since seen the copy in the British Museum (7857. aa. 57.). It contains an engraved title and fourteen engravings. The good fortune also has befallen me of finding a copy of the first edition in the University Library, Cambridge, from which the following account has been made. It seems to be a rare book, and the one copy I have seen for sale, I failed to secure.

POLYGRAPHICE; | OR | The Art of Drawing, Engrav- | ing, Etching, Limning, Paint- | ing, Wafhing, Varnifhing, | Colouring and Dying. | IN THREE BOOKS. |
I. Shews the Drawing of Men, and o- | ther Animal Creatures, Landskips, Countries, | and Figures of Various Forms. |
II. The way of Engraving, Etching, and Limn- | ing, with all their Requifits and Ornaments. |
III. The way of Painting, Wafhing, Varnifh- | ing, Colouring, and Dying, according to the | Method of the best Authors now Extant. | Exemplified in the Painting of the Antients, | Wafhing of Maps, Globes, or Pictures; Dy- | ing of Cloth, Silks, Bones, Wood, Glafs, | Stones and Metals: together with the way | of Varnifhing thereof according to any Pur- | pofe or Intent. *The Like never yet Extant.* | By *W. S.* a Lover of Art. |
LONDON, | Printed by *E. T.* and *R. H.* for *Richard Jones* | at the *Golden Lion* in *Little-Britain.* 1672. |

Small 8vo. A4, B to T in eights, V4; or, pp. [8] 293 [3]. There is besides an engraved title in four compartments. The printed title is enclosed within two lines. Sig. A1 contains the printed title and on the reverse the Imprimatur, dated September 11th, 1671; A2 the Preface to the Reader ending A4 *verso*; B to V3 *recto* (pp. 1-293) the text; V3 *verso*, Books sold by Richard Jones; V4 Salmon's dedication: To the Honourable and truly Noble Peter Stanley of Alderly. [See ADDENDA, §86.]

---

* *Transactions of the Glasgow Archæological Society*, N.S., 1890, I., p. 227 (Reprint, 1885, III., 42).

87. In the British Museum (1043. b. 1.) there is an imperfect copy of the third edition with a fragment of the title-page. The imprint is:

London, Printed by [*Andr. Clark*,] for [*John Crumpe*,] at the | Sign of the three Bibles, in St. *Pauls* Church-Yard. 1675. |

The title is in red and black. 8° pp. [8] 407 [1 blank] [Contents, Index and Errata, 16]. There are a portrait by W. Sherwin, from life, the engraved title, dated 1675, and fifteen engravings, all extra. [See ADDENDA, § 87.]

88. As the 1678 edition is a line for line and page for page reprint of that of 1675, it might be anticipated that the illustrations would also be the same. As a matter of fact they each contain 15, but of these the 1675 edition contains three, a group of faces, a philosopher, and a landscape, not in the edition of 1678, while it contains three, a nude man wielding a hammer, a group of children, and two nude female figures, not in the 1675 edition. Therefore the full number is either 18, in which case the copies examined want three each, or 15, in which case three in the 1675 edition have been replaced by other three in the 1678 edition. This I suspect is the more likely explanation. The remaining twelve are the same in both. [See ADDENDA, § 88.] Alterations and additions were introduced in successive editions, some of which may be enumerated.

89. [POLYGRAPHICE :] | OR | The Arts of Drawing, Engraving, | Etching, Limning, Painting, Wafh- | ing, Varnifhing, Gilding, Colouring, | Dying, Beautifying and Perfuming. | [IN FOUR BOOKS.] |

Exemplified, in the *Drawing of Men*, *Women*, *Landskips*, *Countreys*, and Figures of various forms; The way of *Engraving*, *Etching*, and *Limning*, with all their Requifites and Ornaments; The *Depicting of the moft eminent Pieces of Antiquities*; The *Paintings of the Antients*; *Wafhing* of Maps, Globes or Pictures; The *Dying* of Cloth, Silk, Horns, Bones, Wood, Glafs, Stones and Metals; The *Varnifhing*, *Colouring* and *Gilding* thereof, according to any purpofe or intent: The *Painting*, *Colouring* and *Beautifying* of the Face, Skin and Hair; The *whole Doctrine of Perfumes* (never publifhed till now,) together with the *Original*, *Advancement* and *Perfection* of the Art of Painting.

To which is added | [A Difcourfe of **Perfpective** and **Chiromancy**.]

The Fourth Edition, with many large Additions: Adorned with | Sculptures: The like never yet extant. |

[By *WILLIAM SALMON* Φιλαλήθης,] | Profeſſor of Phyſick. | *Non quot, ſed quales.* |

London, Printed by [*Robert White*,] for [*John Crumpe*:] at the Sign of the | Three Bibles in St. *Paul's* Church-Yard, 1678. |

8°. A to Z, Aa to Dd, in eights; or, pp. [8] 407 [1 blank]; Contents, Table, Advertisements [16]. Title red and black [the red portions are within black brackets]; Engraved title-page dated 1675; 15 Engravings by W. Sherwin, W. Vaughan and by F. H. van Houe. The copy I have seen wants the portrait, one engraved plate at least, and C3-6.

This edition is quoted by Hazlitt.*

90. POLYGRAPHICE: | OR | The Arts of Drawing, Engraving, Etching, Limning, | Painting, Waſhing, Varniſhing, Gilding, Colou- | ring, Dying, Beautifying and Perfuming. |

IN SEVEN BOOKS. |

Exemplified, in the *Drawing of Men, Women, Landskips, Countreys*, and Figures of various Forms; The way of *Engraving, Etching* and *Limning*, with all their Requiſites and Ornaments; The *Depicting of the moſt eminent Pieces of Antiquities*; The *Paintings of the Antients*: *Waſhing* of Maps, Globes or Pictures; The *Dying* of Cloth, Silk, Horns, Bones, Wood, Glaſs, Stones and Metals; The *Verniſhing, Colouring* and *Gilding* thereof, according to any purpoſe or intent; The *Painting, Colouring* and *Beautifying* of the Face, Skin and Hair; The *whole Doctrine of Perfumes* (never publiſhed till now,) together with the *Original, Advancement* and *Perfection* of the Art of Painting: And a Diſcourſe of Perſpective, Chiromancy and Alchymy.

To which alſo is added,

I. The one hundred and twelve Chymical Arcanums of *Petrus Johannes Faber*, a moſt learned and eminent Phyſician, Translated out of *Latin* into *Engliſh*.

II. An abſtract of *Choice Chymical Preparations*, fitted for Vulgar Uſe, for curing moſt Diſeaſes incident to Humane Bodies.

The fifth Edition: Enlarged with above a thouſand | conſiderable Additions. Adorned with XXV. Copper Scul- | ptures; the like never yet extant. |

By *WILLIAM SALMON, Profeſſor of Phyſick*, | living at the *Blew Balcony by* Fleet-Ditch, *near* Holborn-Bridge, | *LONDON* |

Non quot, ſed quales.

London, Printed for *Thomas Paſſinger* at the *Three Bibles* on *London-* | *Bridge*; and *Thomas Sawbridge* at the *Three Flower de Luces* in | *Little-Brittain.* MDCLXXXV. |

---

* *Bibliographical Collections and Notes*, London, 1903, p. 338.

8°. A; *a*, b, c; B to Z, Aa to Zz, Aaa to Ccc, in eights; or, pp. [64] 767 [1 blank]. Portrait of Salmon by W. Sherwin, rather worn; engraved title, by Sherwin, also worn; and 23 engraved plates by Sherwin, Wm. Vaughan, F. H. van Hove, Tho. Crofs, Junr., all extra. (B.M. 1043. d. 9.) *

91. *POLYGRAPHICE*: | OR, | *The Arts of* Drawing, Engraving, | Etching, Limning, Painting, Ver- |nifhing, Japaning,|Gilding, *&c.* | In Two Volumns. | CONTAINING, |

I. *The Arts of* Drawing Men, Women, Landskips, *&c.*
II. *Of* Engraving, Etching, *and* Limning.
III. *Of* Painting, Wafhing, Coloring, Gilding.
IV. *Of the* Original, Advancement *and* Perfection *of* Painting, *with the* Various Paintings *of the* Ancients.
V. *Of the Arts of* Beautifying *and* Perfuming.
VI. *Of the Arts of* Dying *and* Staining.
VII. *Of* Alchymie, *and the* Grand Elixir *of* Philofophers.
VIII. *Of the* 112 Chymical Arcana *of* Peter Faber.
IX. *Of* Chiromantical Signatures.
X. *Of* Staining *and* Painting Glafs, Enamel *and* Gems.
XI. *Of* Vernifhing, Japaning, *and* Gilding.

The Eighth Edition.

Enlarged, with above Five Hundred *confiderable* Additions | *thro' the whole Work; and the Addition of almoft five whole* | *Books, not in any of the former Impreffions: Adorned with* | XXV Copper Sculptures, *the like never yet Extant.* |

By *WILLIAM SALMON*, M.D.

Non Quot, fed Quales.

London, Printed for *A.* and *J. Churchill*, at the *Black Swan* | in *Paternofter-Row.* And *J. Nicholfon*, at the *King*'s-*Arms* | in *Little-Britain.* MDCCI. |

8°. A, a, B to P in eights; Aa to Ll in eights; or, pp. [32] 1-224; 301-475 [1 blank]. Fine portrait of Salmon by M. v. Gucht prefixed, and XX. engraved plates at the end. The title is surrounded by two lines. [Plate XXI. is wanting?]

The second volume has the following title:

POLYGRAPHICE: | OR, | *The Arts of* Drawing, Engraving, | Etching, Limning, Painting, Ver- | nifhing, Japanning, Gilding, *&c.* | The Second Volumn. CONTAINING, | The Latter Five Books. | *VIZ.* |

VII. *Of* Alchymie, *and the* Grand Elixir *of* Phi- | lofophers.
VIII. *Of the* 112 Chymical Arcana *of* Peter | Faber.
IX. *Of* Chiromantical Signatures.
X. *Of* Staining *and* Painting Glafs, Enamel, | *and* Gems.
XI. *Of* Vernifhing, Japanning, *and* Gilding.

The Seventh *and* Ninth Books, *are almost all* | *of them, and the* Two laft Wholly New. |

By *WILLIAM SALMON*, M.D.

---

* Quoted also by Hazlitt, *Bibliographical Collections and Notes*, London, 1903, p. 338.

LONDON, | Printed for *A.* and *J. Churchill*, at the *Black Swan* | in *Paternoster Row.* And *J. Nicholson*, at the | *King's-Arms* in *Little-Britain.* MDCCI. |
8°. Title-leaf; Mm to Zz, Aaa to Rrr; or, pp. 477-939 [1 p. of the Advertisement of Dr. Salmon's Pills]. Engraved title and plates XXII. and XXIII.

This copy belonged to W. Musgrave.

The eighth edition was just referred to in my first paper on the present subject.*

92. William Salmon, who ultimately became M.D., at least he assumed that title, was born in 1644, and drifted into the empirical practice of medicine, combining with it, as was the custom in the 17th century, pharmacy, astrology and alchemy. He wrote and published a great deal, and he had the skill to put books in successful circulation, notwithstanding the jeers and satires of opponents among his contemporaries, who may not have been so clever in securing popularity. His books were not without merit, and the present work, at all events, is quite reasonable and contains much that is still useful in an empirical way. He seems to have lived down the odium which was at first his portion, for in the preface to the sale catalogue of his library after his death in 1713, he is spoken of in complimentary terms.†

Bromley‡ gives the following list of his portraits:

æt. 36, 1681,	"Synopsis Medicinæ," 8°.	Burnford.
... ...	... ...	v. Hove.
... ...	... ...	vr. Gucht.
æt. 26, 1670,	"Polygraphice," 8°.	W. Sherwin ad vivum.
æt. 23, 1667,	"Herbal," fol.	R. (*sic*) White ad vivum.
no date	"Practise of Physick," 8°.	Id. 1700.
æt. 42,	"Polygraphice," 1685, 8°.	

---

* *Transactions of the Glasgow Archæological Society*, 1883, II., p. 196 (Reprint, 1896, Part I., p. 19). It is mentioned also by Hazlitt, *Bibliographical Collections and Notes*, London, 1903, p. 338.

† *Bibliotheca Chemica*, 1906, II., p. 319. The portrait in the *Practise of Physick*, 1700, is repeated in the 3rd edition, 1716, "*R. White delin. & sculp.*" Bromley puts his death in 1712, æt. 68; *The Dictionary of National Biography*, 1897, L. p. 210, in 1713.

‡ *A Catalogue of Engraved British Portraits*, London, 1793, Period VII., Class VII., p. 237.

93. Salmon also deserves to have the editions of his book of receipts enumerated. They are:

First,	1672.
Second,	1673.
Third,	1675.
Fourth,	1678.
Fifth,	1685.
Eighth,	1701.

The sixth and seventh editions will presumably be dated about 1690 and 1695 respectively. If there were later editions I do not know them, but, whether or not, for the thirty years of its existence, it flourished as healthily as any of the books already mentioned. If it had seen its jubilee, doubtless it would have been in its thirteenth or fourteenth edition.

Of these editions Lowndes* quotes only the third (the Gordonstoun copy, 2127, 5s.), the fifth, and the eighth, 2 vols. 10s. 6d.

Watt † enumerates the same editions: 1675, 1685 and 1701. He seems to call this last the tenth edition, although it is not quite clear that he does not mean the edition of 1685. It is wrong in either case.

94. A rather unusual set of secrets is contained in the following work, the title-page of which gives an ample survey of its contents:

> *Mathematical Recreations*: | OR, | A Collection of many | PROBLEMS | Extracted out of the | Ancient and Modern Philofophers: | AS, | SECRETS and EXPERIMENTS | IN | Arithmetick, Geometry, Cofmography, Horologio- | graphy, Aftronomy, Navigation, Mufick, Opticks, | Architecture, Statick, Mechanicks, Chymiftry, | Water-Works, Fire-Works, *&c.* | Not vulgarly manifest till now. | Written firft in *Greek* and *Latin*, lately compil'd in | *French* by *HENRY VAN ETTEN*, and now | in *Englifh*, with the *Examinations* and *Augmenta-* | *tions* of divers *Modern MATHEMATICIANS.* | Whereunto is added, | The DESCRIPTION and USE | OF The Double HORIZONTAL DYAL, | AND | The General HOROLOGICAL RING: | Invented and written by *William Oughtred.* |

* *The Bibliographer's Manual*, ed. Bohn, 1865, IV., p. 2180. In Sherwin's portrait prefixed to *Polygraphice*, 1675, he is said to be 27. The same portrait and date are in the 1685 edition. In the 1701 edition, v.d. Gucht's portrait is not dated. The portrait by R. White, dated 1700, and somewhat worn, is prefixed to the third edition of Salmon's *Praxis Medica*, London, 1716.

† *Bibliotheca Britannica*, 1844, Authors II., 829 *h-i*.

LONDON: | Printed for *William Leake*, and *John Leake*, at the Crown in | *Fleet-ſtreet*, between the Two Temple-Gates. 1674. |

8vo. 2 leaves, *a*, *b*, in eights; B to U in eights; or, pp. [36] 280; [2] 19 [1 blank]. (T5 blank? is wanting.) Engravings in the text. The "Description of the Horizontal Dial" has a separate title-leaf, T6. There is an engraved title, which has been removed from the present copy.

95. No one who has followed my lists to this point but will admit that whatever may be the case with female education at the present day, the instruction in household affairs for the convenience of the ladies was not neglected in the seventeenth century. Closet after closet was thrown wide open to them, secret after secret was unveiled for their inspection and use; their eagerness for information was met by a never-failing readiness on the part of those who knew to tender them good advice and to supply the desiderated illumination. It cannot be said that the light came in a dazzling form, but it was not stinted, and it burned steadily. Here are some more lights and delights:

The Accompliſh'd | LADY'S DELIGHT | In Preſerving, Phyſick, Beautifying, | and Cookery. | *CONTAINING*, |

I. The ART of PRESERVING, | and CANDYING Fruits and Flow- | ers, and the making of all ſorts of Con- | ſerves, Syrups, and Jellies. |

II. The PHYSICAL CABINET, | Or, Excellent Receipts in Phyſick and | Chirurgery; Together with ſome Rare | Beautifying waters, to adorn and add | Lovelineſs to the Face and Body: And | alſo ſome new and Excellent Secrets | and Experiments in the ART of *AN-* | *GLING*. |

III. The COMPLEAT COOKS GUIDE | Or, directions for dreſſing all ſorts of Fleſh, | Fowl, and Fiſh, both in the *Engliſh* and *French* | Mode, with all Sauces and Sallets; and the mak- | ing Pyes, Paſties, Tarts, and Cuſtards, with the | Forms and Shapes of many of them. |

The Second Edition Enlarged.

LONDON, | Printed for *B. Harris*, and are to be Sold at his | Shop, at the *Stationers* Arms in *Swithins* Rents | near the Royal Exchange, 1677. |

12mo. A to Q in twelves; or, pp. [4] 380.

Besides a general engraved title, there are a frontispiece to the "Secrets of Angling," an engraved plate of the fish and two engraved plates of the forms of pies, not included in the signatures or paging. There is a printed title-page to the "Phisical Cabinet," p. 117; one to the "Secrets of Angling," p. 191; and one to the "Cook's Guide," p. 223; pp. 345-349 contain Terms of Carving; pp. 350-356, Bills of Fare (which would astonish a present day cook and diner out); pp. 357-380, the alphabetical tables to the three parts. This copy does not contain any portrait. The title-pages of the tracts here follow.

The title of the second tract runs thus (p. 117):

> THE | PHISICAL | Cabinet: | CONTAINING | Excellent Receipts | IN | Phyfick and Chyrurgery, | For Curing moft Difeafes Incident to | the Bodies of Men and Women. | TOGETHER | VVith fome Rare Beautifying | Waters, Oyls, Oyntments and Pow- | ders, to Adorn and add Lovelinefs | to the FACE and BODY. | AS ALSO | Some New and Excellent Secrets | and Experiments in the Art of | *ANGLING.* |
> *London* Printed in the Year 1677. |

p. 191:

> New and Excellent | EXPERIMENTS | AND | SECRETS | In the ART of | Angling: | BEING | Directions for the whole | ART. |
> *LONDON,* | Printed in the Year 1677. |

p. 223:

> THE | Compleat Cook's | GUIDE. | OR | Directions for the Dreffing of | all forts of Flefh, Fowl, and Fifh, | both in the *Englifh* and *French* Mode; | with the preparing of all manner of | Sawces and Sallets proper thereunto. | TOGETHER | With the making of all Sorts of | Pyes, Pafties, Tarts, and Cuftards; | with the Forms and Shapes of many | of them. | WITH | Bills of Fare, both for Ordinary, and | Extraordinary. |
> *London,* Printed in the Year 1677. |

The Epistle Dedicatory to the Ladies and Gentlewomen is signed "T. P.," under which initials it is placed in the British Museum Catalogue. It is little more than a repetition of the title-page. It appears that at this time angling was a lady's recreation. This is a more attractive book than any of the following. The type is larger, all the plates are engraved, and the extra title-pages give it a liberal and generous appearance. There is a copy in the British Museum (7907. aa. 40.).

96. I have no information about the third edition, but I have ascertained that there are two quite distinct issues entitled the fourth edition. One is here, the other is in the British Museum, and there may now be indicated a few of the differences by which they can be distinguished from one another.

See p. 71.

The Accomplifh'd | LADIES DELIGHT | IN | **Preferving, Phyfick, Beautify- | ing, and Cookery.** | CONTAINING |

I. The ART of PRESERVING, | and CANDYING Fruits and Flowers, and | the making of all forts of Conferves, Syrups, and | Jellies. |

II. The PHYSICAL CABINET, | Or Excellent Receipts in Phyfick and Chirurgery, | Together with fome Rare Beautifying Waters, to | adorn and add lovelinefs to the Face and Body: | And alfo fome New and Excellent Secrets and Ex- | periments in the ART of *ANGLING*. |

III. The Compleat COOKS GUIDE, | Or, directions for dreffing all forts of Flefh, Fowl | and Fifh, both in the *Englifh* and *French* Mode, | with all Sauces and Sallets; and the making Pyes, | Pafties, Tarts, and Cuftards, with the Forms and | Shapes of many of them. |

The Fourth Edition Enlarged. |

LONDON, | Printed for *Benjamin Harris*, at the *Stationers* Arms | and *Anchor*, in the *Piazza*, at the Royal-*Exchange* | in *Cornhil*, 1684. |

12mo. A to I in twelves, K six; engraved title extra [2]; or, pp. [4] 208 [12]. There are no titles to the different parts, and the pictures of angling, fish and forms of pies, etc., are inferior woodcuts, included in the signatures and pagination. This copy (B.M. 1037. e. 22) contains no portrait.

The preceding description applies also to my copy, except that it has a portrait of the "accomplish'd Lady," presumably in a state of delight, which is here reproduced. Its resemblance to that of Mrs. Gilly or Wolley, is obvious. Nevertheless, the two issues are not identical, as the following comparison of passages, taken at random, will show:

<table>
<tr><th>B.M. Copy.</th><th>My own Copy.</th></tr>
<tr><td>p. 1. No Sig.</td><td>Sig. A 3.</td></tr>
<tr><td>Receipt 2. l. 2. bery</td><td>very</td></tr>
<tr><td>l. 5. fugar</td><td>Sugar</td></tr>
<tr><td>Receipt 3. l. 1. rofes</td><td>Rofes</td></tr>
<tr><td>last lines. put it up in a Gally | pot and fet it in the fun for a fortnight. 4: To</td><td>put it up in a Gal- | ly pot and fet it in the Sun for a Fortnight. 4. To</td></tr>
<tr><td>p. 31. Receipt 127. To make Banbury Cakes.</td><td>To make Banbury Cakes.</td></tr>
<tr><td>p. 54. Receipt 225. l. 9. barrel</td><td>Barrel</td></tr>
<tr><td>226. l. 2. Ounce</td><td>ounce</td></tr>
<tr><td>227. Title. Chips</td><td>chips</td></tr>
<tr><td>228. l. 4. pour it upon a Pye-plate, |</td><td>pour it upon a Pye- | plate</td></tr>
<tr><td>229. l. 2. loaf-Sugar</td><td>Loaf-Sugar</td></tr>
<tr><td>l. 7. Ufe</td><td>ufe</td></tr>
</table>

p. 127. Receipt 1. *To make a Lamb-Pye.*	*To make a Lamb Pye.*
First, cut your Lamb into Pieces, and then fea-fon it with Nutmegs, Cloves, and Mace, and fome Salt with Currans, Raifins of the Sun, and fweet butter; and if you will eat it hot, when it is baked put in fome Yolks of Eggs, with Wine-Vinegar, and Sugar beaten together, but if you will eat it cold, put in no Eggs, but only Vinegar and Sugar.	First, cut your Lamb into Pieces, and then Seafon it with Nutmegs: Cloves, and Mace, and fome Salt with Currans, Raifins of the Sun, and Sweet butter; and if you will eat it hot, when it is baked, put in fome Yolks of Eggs, with Wine-Vinegar and Sugar beaten together, but if you will eat it cold, put it (*sic*) no Eggs, but only Vi-negar and Sugar.
p. 161. Receipt 127. Title occupies a separate line.	Occupies part of a line only.

And so on, showing that they differ throughout.

Similar parallelisms have been noted in the case of Albertus Magnus, the fifth edition of *Arts Treasury*, the fifth edition of White's *Rich Cabinet*, etc.

Westwood and Satchell * quote this book, and say that it was attributed to Mrs. Hannah Wolley. The editions they mention are of 1675, 1677, 1683, 1684, 1685, 1686, 1719, 1720, etc. They give a reference to another book, *The Compleat Servant-maid, or the Cooks Guide*, undated, which they announce as a re-issue of this work, with a new title. If so, it must not be confused with the work, with a similar title, mentioned in § 101 below. They are obviously different. Arber quotes the 1675 edition, and, so far as one can gather from the Index, ascribes it to Mrs. Wolley.†

97. The next edition (B.M. 1037. e. 24.) has its title identical with that of the Museum copy of the fourth edition, except that it is dated 1685, and is stated to be "The fifth edition enlarged." It, too, contains A to I in twelves, K six; or, pp. [4] 208 [12]; and it has not only the engraved title, and the woodcuts of fishing, of fish and of the forms of pies, but it has the portrait as well, though rather worn. Prefixed to this copy are two tables of the dumb alphabet and "A True Lover's Knot," but they do not form part of the book. Comparisons of the quotations above given, with the corresponding passages in this so-called fifth edition, show that it agrees exactly with the copy in the British Museum, and not with that which I have. I conclude, therefore, that the fifth edition consists

* *Bibliotheca Piscatoria*, London 1883, p. 128.

† Arber, *The Term Catalogues*, 1903, I., p. 198.

neither more nor less than of copies of the 1684 edition, with the alteration from fourth to fifth, and from 1684 to 1685. On the other hand, I should not be surprised if my copy of the fourth edition was a similar re-issue of surplus copies of the third edition, which I have not as yet had an opportunity of inspecting.

98. The seventh edition has made its appearance, and, all things considered, it would have been much better for it if it had never emerged from its concealment. It is such a disreputable-looking production. An account of it in detail may be given, but the ensemble of the book is beyond description. It will be observed besides that this edition is printed *by* B. Harris and not *for* him, and that his premises are different from those in the previous editions. The main point, however, is the inferior execution of the book. The portrait is different from that in the earlier editions, while the engraved title-page is an inferior copy. The portrait and title-page were copied in the tenth edition of 1719, but the portrait differs in expression and various details, and there are some differences in the title-page also. Both, too, are printed much more heavily and blacker. One might almost believe that the rather faded and worn plates of the seventh edition had been gone over again for the tenth. The title-page is as follows:

99. The ACCOMPLISH'D | Ladies Delight, |

In { *PRESERVING, PHYSICK, BEAVTIFYING,* and *COOKERY.*

CONTAINING

I. The Art of *Preferving* and *Candying* Fruits | and Flowers; and the making of all forts of | Conferves, Syrups, and Jellies. |

II. The *Physical Cabinet:* Or, Excellent Re- | ceipts in Phyfick and Chirurgery; together | with fome Beautifying Waters, to Adorn | and add Lovelinefs to the Face and Body: | And alfo fome New and Excellent Receipts | relating to the *Female Sex:* And for the | general good of Families, is added the True Receipt for making that Famous *Cordial Drink* |

Daffy's Elixir Salutis. |

III. The Compleat *Cook*'s Guide: Or, Di- | rections for Dreffing all forts of Flefh, Fowl and Fifh, both | in the Englifh and French Mode; with all forts of Sauces and | Sallets: And the making Pyes, Pafties, Tarts, and Cuftards, | with the Forms and Shapes of many of them. |

The Seventh EDITION Enlarged. |

LONDON, | Printed by *B. Harris,* in *Maiden-head*-Court | in Great *Eaſt-Cheap* 1696. |

12mo. 2 leaves, A, 6 leaves; B, C, 12 leaves; D, E, 6 leaves; F, 12 leaves, 2 leaves; G-I, 12 leaves; K4; or, pp. [4, 2] 1-106 [4], 107-178, [8]. The title is surrounded by double lines.

The two preliminary leaves contain the portrait and engraved title. The portrait (of which a copy is given) is after that in the earlier editions, but is different in features and expression. Below it is the inscription: "*Sold by William Hunt at* | *the Ball in St. Pauls Alley* | *at the West End of St. Pauls.*" | The title-page contains, in compartments, representations of the lady engaged in Preserving, Physick, Beautifying and Cookery. In the centre is the short title. Both are poor prints. A1 is the title as above, and on the reverse is T. P.'s letter to the Ladies. The text begins on A2, p. 1, and runs on to I12 *verso*, p. 178. K contains the tables of contents of the different parts. The two leaves between F and G contain the drawings of the shapes of pies, etc. Some miscalculation seems to have been made by the printer, for D3 (misprinted A3) to E6 *recto*, pp. 63-81 inclusive, are printed in double columns and in much smaller type than the rest of the book. It has in consequence a very patched and shabby appearance.

This is a poor chap-book, in the original covers of brown sheep, unlined.

100. The tenth edition is in the British Museum (1037. e. 23.), and it may be included to complete the tale of these books and to display their vitality:

The Accompliſh'd | **Lady's Delight,** | IN | **Preserving, Phyſick, Beautifying,** | **Cookery, and Gardening.** | CONTAINING,

I. The Art of *Preſerving,* and *Candying.* | Fruits and Flowers, and making all ſorts of | Conſerves, Syrups, Jellies, and Pickles. |

II. The *Phyſical Cabiuet* (sic): Or, excellent | Receipts in *Phyſick* and *Chirurgery.* Alſo | ſome New Receipts relating to the Fair Sex, | whereby they may be richly furniſh'd with all | manner of *Beautifying Waters,* to add Love- | lineſs to the Face and Body.

III. The *Compleat Cook's Guide*: Or, Di- | rections for Dreſſing all ſorts of Fleſh, Fowl, | and Fiſh, after the Neweſt Faſhion now in | Uſe at the *Britiſh* Court; with the making | Sauces, of Pyes, Paſties, Tarts, Cuſtards, *&c.*

IV. The *Female Angler,* inſtructing Ladies | and others, in the various Methods of taking | all manner of Fiſh, in the Fiſh-Pond or River. |

V. The *Lady's Diverſion in her Garden*: | Or, the compleat Floweriſt, with the Nature | and Uſe of all ſorts of Plants and Flowers. |

The Tenth Edition Inlarged.

LONDON, Printed, for *Daniel Pratt,* at | the *Bible* and *Crown* againſt *Hungerford* | *Market* in the *Strand.* 1719. |

12° 2 leaves; A to G12 in twelves, H6; or, pp. [4] [6] 177, Table [7].

See p. 74.

The two preliminary leaves contain the portrait—below which is: "printed for Dan: pratt | att the Bible in the Strand" |—and the engraved title, both of which have been printed very black. The portrait resembles that in my seventh edition.

The diagrams of the pies, etc., are given, but not the drawings of the fish. This is an inferior chap reprint.

101. In the British Museum (1036. c. 19.) is another collection with a similar title, but it is a mere selection from the present work. It has no date, but it is much later than any of the preceding editions. Its title is as follows:

> THE | Accomplifhed LADY's | Delight in Cookery; | OR, THE | Complete Servant's-Maid's (*sic*) | GUIDE. | [Vignettes].
> WOLVERHAMPTON: Printed by J. SMART. |
> 12°. No signature. pp. 24.

The upper vignette represents the lady rolling out paste. The under vignette a banquet with four men at table and a fifth waiting on them. Both are rude woodcuts. It is a chap-book. It contains receipts to dress flesh (8), fish (5), rabbits, hares or leverets, etc. (6), tame or wild fowl etc. (16), baked meats, puddings, etc. (11), a table of terms used in carving, and cautions about marketing.

102. With this work may be compared *The Ladies Companion*, *The Ladies Cabinet Opened*, *The Queens Closet Opened*, *The Ladies Directory*, *The Queen-like Closet*, *The Accomplished Female Instructor*, *The Accomplished Ladies Rich Closet of Rarities*, and so on.

ADDENDA.

10th December, 1909.

86. The copy of the first edition of Salmon's *Polygraphice* in the University Library, Cambridge, is in the original sheepskin covers, unlined, and shows the book as it came from the publisher. It is in the usual small 8vo size which was current at the time, and in which so many of these receipt books appeared. The subsequent editions were all larger octavo volumes. The engraved title in four compartments represents four men in the costume of the day engaged in etching, engraving, painting, and washing a map. This engraved title is wanting in the British Museum copy (1269. a. 25.).

Comparison of the two copies shows that the dedication which is printed on V4, and is the last leaf of the book in the Cambridge copy, has been removed and placed after the preface and before the text in the British Museum copy, with the effect of disturbing the register. In the Cambridge copy it looks as if a leaf after the preface had been torn out, but it seems to be only the edge of the engraved title which has been turned over for sewing purposes.

87. A perfect copy of the 1675 edition, in Cambridge University Library, has enabled me to complete the description of this third edition of Salmon's *Polygraphice*. The full title runs thus:

> [POLYGRAPHICE:] | OR | The Arts of Drawing, Engraving, | Etching, Limning, Painting, Wafhing, | Varnifhing, Gilding, Colouring, | Dying, Beautifying and Perfuming. | [IN FOUR BOOKS.]
>
> Exemplified, in the *Drawing of Men, Women, Landſkips, Countries*, and Figures of various forms; The way of *Engraving, Etching* and *Limning*, with all their Requifites and Ornaments; The *Depicting of the moſt eminent Pieces of Antiquities*; The *Paintings of the Antients*; *Waſhing* of Maps, Globes, or Pictures; The *Dying* of Cloth, Silk, Horns, Bones, Wood, Glafs Stones, and

Metals; The *Varnifhing*, *Colouring* and *Gilding* thereof, according to any purpofe or intent; The *Painting*, *Colouring* and *Beautifying* of the Face, Skin and Hair; The *whole Doctrine of Perfumes* (never publifhed till now,) together with the *Original*, *Advancement* and *Perfection* of the Art of Painting.

To which is added,

[A Difcourfe of **Perfpectibe** and **Chiromancy**.]

The Third Edition, with many large Additions: Adorned with Sculptures: The like never yet extant.

[By *WILLIAM SALMON* Φιλαλήθης.]

Profeffor of Phyfick.

Non quot, fed quales.

London, Printed by [*Andr. Clark*,] for [*John Crumpe*,] at the Sign of the three Bibles, in St. *Pauls* Church-Yard. 1675.

8vo. A to Z, Aa to Dd in eights; or, pp. [8] 407 [1 blank]; The Contents, the Table, and Errata [16]. Title in black and red; the red portions are here enclosed in black brackets.

Engraved title and Portrait of Salmon, aged 27 in 1675, by W. Sherwin from life. Eighteen engravings, namely:

The Muscles.
Diagrams for drawing faces and heads.*
Heads of women and children.*
Drawings of eyes, nose, mouth and chin, and ears.*
Faces, Female and male.*
Hands, arms, legs.*
Hands and feet.*
Nude Children.*
Nude man wielding a hammer.†
Two nude females.
Nude warrior with helmet, shield and sword.†
Draped male figure, or philosopher.*
Medallion portraits of Charles II. and Catherine.
Draped female figure.*
Beasts and birds.
A Landscape: a ruin with trees, a lake, and a castle.
The method of enlarging or reducing a picture by squares.
Lines of the hand for Chiromancy.‡

These engravings are by W. Sherwin (*), W. Vaughan (†), F. H. Van Houe (‡). The others have no name affixed to them.

88. I have thought it worth while to give a list of these drawings, because in this paragraph it is pointed out that the British Museum copy of the

1675 edition contains fifteen pictures only, and the reprint of 1678 also fifteen. They are not identical, however, in the two editions, and accordingly the alternative inference drawn is that either the complete set numbers eighteen, or else that three plates in the 1675 edition have been replaced by a different three in that of 1678, and I thought this the more probable explanation of the difference. But as the Cambridge copy actually contains eighteen, that inference seems to be erroneous, and the presumption is that both copies just mentioned need, each of them, the three pictures in the other (all of which are in the Cambridge 1675 copy) to make them complete.

ERRATA.

The following errata, which I have detected from time to time in the previous Supplements, may be enumerated for correction. There may be others which have escaped notice:

SUPPLEMENT II. TRANSACTIONS, Vol. III., pp. 175-213.

p. 193, note 11, l. 3. Insert 'or' before 'Hardervici.'
p. 194, last line. Delete the quotation marks before 'Gotha.'
p. 198, l. 10 from the bottom. For 'decimoocta | no' read 'decimoocta | uo.'
p. 199, line 11 from the bottom. Insert a full stop after 'Thott.'
p. 203, note *, l. 3. Insert 'editions' after 'above.'
p. 212, l. 13 from the bottom. For 'it' read 'this edition.'

SUPPLEMENT III. TRANSACTIONS, Vol. III., pp. 360-425.

p. 380 (Reprint, p. 23), l. 8. After 'apparatus' add 'are.'
p. 383 (Reprint, p. 26), l. 19. After *dignité* insert *de*.
p. 388 (Reprint, p. 31), l. 5 from the bottom. For 'compass' read 'compasses.'

SUPPLEMENT IV. TRANSACTIONS, Vol. IV., pp. 95-120.

p. 111 (Reprint, p. 19), l. 16. For 'they' read 'the receipts.'
p. 119 (Reprint, p. 27), l. 8. For 'sixteenth' read 'seventeenth.'

SUPPLEMENT V. TRANSACTIONS, Vol. V., pp. 125-185.

p. 151 (Reprint, p. 27), l. 18, etc. The question is raised about an apparent conflict of dates, which need never have existed if I had only considered that at the time indicated the year began on the 25th of March. The licence, dated 21st March 1638, was therefore granted before the expiry of the year, while Dr. Read dated his preface on 25th March 1639, the beginning of the new year. All the doubts and difficulties and suppositions put forward in the paragraph therefore disappear, for there is nothing to explain.

INDEX

TO

BIBLIOGRAPHICAL NOTES ON HISTORIES OF INVENTIONS AND BOOKS OF SECRETS.

SUPPLEMENTS I.-VI.
READ TO THE ARCHÆOLOGICAL SOCIETY OF GLASGOW,
1894-1908.

BY
JOHN FERGUSON, LL.D., F.S.A. LOND. AND SCOT.,
REGIUS PROFESSOR OF CHEMISTRY IN THE UNIVERSITY OF GLASGOW

GLASGOW:
Printed at the University Press by
ROBERT MACLEHOSE & CO. LTD.
1910.

NOTE.

SIMILAR to the alphabetical Index that was compiled for the original six parts of these 'Notes,' I give now one for the six supplements which include the acquisitions of some years past. They have involved a number of questions biographical, bibliographical and critical. Many of the persons named are unknown, and their lives are without record. The bibliographical points relate to irregularity of dates and editions, to identical editions of different dates, to identical dates of different editions, to identical books with different title-pages, to identical editions by different publishers.

As before the index numbers refer to the paging of the reprint, but they can be applied without much difficulty to the respective volumes of the Society's 'Transactions' by the following table:

Supplement I.	New Series, 1896,	Vol. II.		pp. 364-404.
II.	1899,	III.		175-213.
III.	1899,	III.		360-425.
IV.	1903,	IV.		95-120.
V.	1908,	V.	part III.	125-185.
VI.	1910,	VI.	part I.	1- 79.

GLASGOW, *October* 17, 1910. J. F.

INDEX.

BIBLIOGRAPHICAL NOTES

ON

HISTORIES OF INVENTIONS

AND

BOOKS OF SECRETS

SEVENTH SUPPLEMENT

BY

JOHN FERGUSON, LL.D., F.S.A. LOND. AND SCOT.

REGIUS PROFESSOR OF CHEMISTRY IN THE UNIVERSITY OF GLASGOW

GLASGOW

Printed at the University Press by

ROBERT MACLEHOSE & CO. LTD.

1915

Read to the Archæological Society of Glasgow, 20th April, 1911.
100 *Copies Reprinted, May* 1915.

No. I.

BIBLIOGRAPHICAL NOTES ON HISTORIES OF INVENTIONS AND BOOKS OF SECRETS.

(*SEVENTH SUPPLEMENT.*)

BY

JOHN FERGUSON, LL.D., F.S.A.

(*Read at a Meeting of the Society, held on Thursday, 20th April,* 1911.)

1. The English books to be produced to-night, in continuation of those already shown, bring these notes so far on their way towards the end of the year 1700, a date at which I have been trying to arrive for some time past, but which seems to recede as it is approached. That it has taken so long even to come within sight of it, proves how many of the earlier books dealing with practical matters have been hitherto undescribed, and by claiming notice have delayed the advance of the research. The sixteenth and seventeenth centuries were prolific in books on almost every technical subject, so that notwithstanding what has been already recorded, there must be numbers of these books which I have not yet encountered. If so, it will be no hardship for them to wait for recognition till their acquaintance has been made, and, meanwhile, those of a later period which are known may have consideration.

2. Before proceeding, however, to the books subsequent to 1650, the year with which the previous supplement closed, a notice must be inserted of a few works, of which examples of other editions have been already enumerated. These copies came recently into my hands from various sources, and their advent not only helps to fill up the lists, but enlarges our knowledge of the science and medicine of the time, inasmuch as some of them, at least, are unknown or unnoticed.

3. Of a volume of medical receipts by a certain T. C., two editions have been described.* I have since got other two, both of them rarities. One is an early edition, possibly the second, and, as it is not in the British Museum, the title of it may be given at length.

¶ An Hoſpitall for the
diſeaſed.
Wherein are to bee founde moſte excellent and approued
Medicines, aſwell Emplaſters of ſpeciall vertue,
as alſo notable Potions or Drinkes, and o-
ther comfortable Receptes, bothe for
the Reſtitution and the Pre-
ſeruation of bodily
healthe.
*Very neceſſary for this tyme of common Plague
and immortalitie, and for other tymes
when occaſion ſhall require.*
With a newe addition.
Gathered by T. C.
ECCLESIAS 38. 4.
The Lorde hath created medicine of the earth,
and he that is wiſe will not abhorre it.

¶ *Imprinted at London for Edward*
White, at the little Northdore of Paules
Churche, at the ſigne of the Gun,
and are there to be ſolde.
1579.

Small 4to; two leaves, B to K in fours, L two leaves; or, pp. [4] 76. The text is in black letter, with two ornamental capitals. The running title, and titles of the sections are in italic.

---

* *Transactions of the Glasgow Archæological Society*, N.S., 1908, V. 144 (Reprint, Supplement, V., p. 20.)

This edition of 1579 is mentioned by Hazlitt.* He had Herbert's copy for collation, which wanted the last leaf.

4. The other copy has this title:

AN
HOSPITALL FOR
THE DISEASED:
VVherin are to be found moſt excellent
approved Medicines, as well Emplaiſters of
ſpeciall vertue, as alſo notable Potions or Drinks
and other comfortable Receites, for the reſti-
tution and preſervation of bodily health.
Very neceſſary for the time of common Plague and
mortality: and for other times when
occaſyon (*sic*) ſhall ſerve.
Newly augmented and enlarged.
Gathered by *T. C.*
Eccleſiaſticus. 38. 4.
The Lord hath created Medicines of the earth, and he that is
wiſe will not abhorre it
[Ornament]
LONDON:
Printed by *I. B.* for *Robert Bird,* and are to be ſold at
his houſe in S^t. *Lawrenſe Lane,* 1638:

Small 4to. A to G in fours. No leaf or page numbers. Black letter, except the title-page, address to the reader, headlines, and titles of the sections, in roman. Does I. B. stand for Iane Bell?

Like the previous one, this edition is not represented in the British Museum, nor has it been noted by Hazlitt, so it, too, must be uncommon. There is a great difference between the two, for while the earlier one is nicely got up, with liberal spacing and broad margins, the other is printed on inferior paper, crudely and crowdedly, for hawking about the country, so that it is an example of how shabby book-production had become in the seventeenth century.

5. So far as I have compared the different issues, the latest, of 1638, is a reprint of the earlier, without alteration, though affirming that it is enlarged. When this book was referred to before, it prompted the

* *Second Series of Bibliographical Collections and Notes*, London, 1882, p. 76.

remark that medicine had been making but little progress, when this very book published in 1595 could be reprinted unchanged in 1630. But the stagnation is still more conspicuous by this later reissue in 1638, when the book actually came out in 1579, or probably four or five years earlier. A treatise on medicine, sixty years old, would not be of much account at the present day, but three hundred years ago medical practice seems to have been more esteemed the longer it had existed unchanged, and had thus apparently been confirmed by unbroken experience.

6. To complete what has been already said about Leonard Mascall,* there may be added here a notice of his treatises on the art of planting and grafting trees, and on cattle and horses. An edition of 1580 of the former work was quoted from Maunsell and several from Hazlitt, but, at the time, I had never seen the book. Since then, I have had access to a number of editions, some of which are not mentioned in any list accessible to me, and are not in the British Museum. They may, therefore, be described and compared with the others that are known.

How far Mascall's directions would satisfy the forester and orchardist of the present day cannot be discussed here, but it is not improbable that some useful hints might be gleaned even now from his pages.

7. The first edition to be noticed was printed by Denham, and the copy of it here is in exceptional state for a book of the kind and period.

A Booke of the Arte and maner,
howe to plant and graffe all ſortes of
trees, howe to ſet ſtones, and ſowe Pe-
pines to make wylde trees to graffe on, as al-
ſo remedies and mediicnes (*sic*). VVith diuers o-
ther newe practiſe, by one of the Abbey of
Saint Vincent in Fraunce, practiſed with his
owne handes, deuided into ſeauen
chapters, as hereafter more
plainely ſhall appeare.
With an addition
in the ende

* *Transactions of the Glasgow Archaeological Society*, N.S., 1908, V., part iii., pp. 133-138, 178 (Reprint, Supplement V., pp. 9-14, 54).

of this booke, of certaine Dutch practifes,
ſet forth and Engliſhed, by Leo-
nard Maſcall.
[Vignette]

In laudem incifionis diſtichon,

Heſperidum Campi quicquid Romanaque tellus,
Fructificat nobis, incifione datur.

Imprinted at London by Henrie
Denham, for Iohn Wight.

Small 4to. A to P in fours, Q two leaves; or, pp. [24] 90 [10]. Black letter, with some roman. The vignette is a man graffing. It is repeated on Liij *verso*. Ciiij contains a drawing of the implements used in graffing, and there are other woodcuts on Kiij *v.*, Niij *v.*, Niiij *r.* and *v.*, Oiij *r.*

There is no date on the title-page, but at the end of the "Exhortation to the Planter and Graffer," there is printed the date 1572, which is probably that of the book. If so, this is doubtless the first edition; anyhow, I have not seen or heard of an earlier one. It is quoted by Hazlitt.* There is also a copy among Sir Joseph Banks's books in the British Museum (453. a. 2.), and, in it, the date, 1572, stands on the title-page in old manuscript.

8. The other editions of Mascall's book which I have seen are reprints of this 1572 edition, with variations in the arrangement of the title-page and in the number of leaves in the book.

The following notes, without giving all the title-pages in full, will suffice to differentiate the editions from one another. Most of them are in the British Museum.

¶ A Book of the Arte and maner
how to Plant and Graffe all ſortes of
trees, . . . [Vignette]

Imprinted at London for
Iohn VVight, 1575

Small 4to. A to P in fours; or, pp. [15, 1 blank, 6] 88 [10]. Ciij is the plate of the graffing implements. Black letter, with some roman. The "Exhortation" is dated 1575. Compare § 10, below. British Museum, 233. d. 3. It is mentioned by Hazlitt.†

* *Handbook to the Popular, . . . Literature of Great Britain*, London, 1867, p. 382. *Second Series of Bibliographical Collections and Notes*, London, 1882, p. 387.

† *Second Series of . . . Collections and Notes*, London, 1882, p. 387.

9. At this point comes in the edition of 1580, included by Maunsell, and referred to above (§ 6).

There is a copy in the British Museum Catalogue (966. i. 22. (1.)) dated there: "(London, 1580?)," which is apparently the same as mine of 1589 (§ 11, below), but as the Museum copy wants four leaves at the beginning and two at the end, it is difficult to say. What, however, is of chief significance is that there is no date at the end of the "Exhortation," whereas mine is dated 1589. I have not detected any other difference, and, in the absence of the title-page, one can say nothing positive. Perhaps the 1580 edition had the title-page dated and not the "Exhortation," whereas it is the other way about in the reprint of 1589.

10. The next one also is in the British Museum (7076. aa. 49. (1.)).

¶ A BOOKE | of the Arte and maner how to Plant
and Graffe all ſortes of Trees, . . . [Vignette]

¶ Imprinted at London for Ihon Wight.
Anno M.D.LXXXII.

Small 4to. A to P in fours; or, pp. [24] 84 [10, 2 blank]. Black letter, with some roman. Ciiij contains the graffing implements. The "Exhortation" is dated 1575, as in the edition in § 8, above. It is quoted by Hazlitt.*

11. The next edition is of some interest, for it is one of those not in the Museum and not mentioned by Hazlitt. Curiously enough it was the first of this group of books which fell into my hands and led me to the examination of them, with the present results. As the book is nowhere recorded, its title may be given.

¶ A booke of the Art and | maner howe to plante and graffe | all ſortes of trees, how to ſet ſtones, and ſow Pe- | pynes to make wylde trees to graffe on, as alſo remedies | *and medicines, with diuerſe other newe practiſes, by one* | of the Abbey of S. Vincent in *Fraunce*, prac- | tiſed with his owne handes, deuyded in- | to ſeuen Chapters, as hereafter more | playnly ſhall appeare. With an | addition in the ende of | this booke, of certayne | Dutche practi- | ſes, ſette foorth and Engliſhed, by | *Leonarde Maſcal.* | [Vignette.] | In laudem inciſionis diſtichon, | Hesperidum Campi quicquid

* *Collections and Notes*, London, 1876, p. 282.

Romanaq; tellus, | Fructificat nobis, incifione datur. | ¶ *Imprinted at London by Henry | Bynneman for Iohn Wight.*

Small 4to. A to P in fours, Q two leaves; or, pp. [24] 90 [10]. Black letter, but the epistle, running title, headings of chapters, are in roman; other headings are in italic. Ciiij *v.* contains drawings of the tools for graffing. The title-page is not dated, though 1589 stands on it in old manuscript, but at the end of the "Exhortation to the Planter and Graffer," there is printed the year 1589, which, it may be assumed, is the date of the edition. The woodcuts correspond with those of the edition of 1572, of which this one is practically a reproduction. It also agrees with the edition dated (1580 ?) in the British Museum, referred to in § 9, above.

12. There is still another edition of which I have a copy. It is referred to shortly by Hazlitt,* but is not in the British Museum. It is as follows:

A BOOKE OF THE
Arte and maner how to Plant and
Graffe all ſorts of Trees, how to ſette
Stones & ſow Pepins, to make wild trees
to graffe on, as alſo remedies & medicines. With
diuers other new practiſes, by one of the Abbey
of S. Vincent in Fraunce, practiſed with his owne hands:
deuided into vij. Chapters, as hereafter more plainly ſhall
appere, with an addition in the end of this booke, of
certaine Dutch practiſes, ſet forth and Engli-
ſhed, by *Leonard Maſcall.*
[Vignette]

In laudem inciſionis diſtichon.
Heſperidum Campi quicquid Romanaque tellus,
Fructificat nobis, inciſione datur.

Imprinted at London by T. Eſte, for
Thomas Wight. 1590.

Small 4to. A to P in fours, of which Piiij is blank; or, pp. [24] 84 [10, 2 blank]. Black letter, but the title, headlines, chapter headings, preliminary matter and the table are in roman. The vignette represents a man graffing, Ciiij the graffing tools, and the text contains the usual illustrations. The colophon is a repetition of the imprint. The "Exhortation" is not dated.

13. The next two editions are in the British Museum, that of 1592 in 966. i. 22. (3.), and that of 1596 in 1147. c. 25.

---

* *Second Series of . . . Collections and Notes,* London, 1882, p. 387.

A BOOKE OF THE | Arte and maner how to Plant and | *Graffe all ſorts of Trees,* . . . [Vignette]

Imprinted at London by T. Eaſt, for
Thomas VVight. 1592.

Small 4to. A to O in fours, P three leaves (Piiij wanting) or, pp. [24] 84 [10]. Ciiij contains the graffing tools. Black letter, with the preliminary matter in roman. The "Exhortation" is not dated. After the table, Piij *v.*, is the colophon:

¶ *Imprinted at London by T. Eſte, for*
Thomas Wight. 1592.

14.

☛ A BOOKE OF THE | Arte and manner how to plant and | Graffe all ſorts of Trees, . . . [Vignette]

☛ Imprinted at London.
1596.

Small 4to. A to O in fours, P three leaves (Piiii wanting) or pp. [24] 84 [10]. Ciiii contains the graffing tools. The text is in black letter, but the preliminary matter and the table at the end are in roman.

At the top of Cii *v.*, within a border, is the date 1596. Below is a device: a bird, apparently a pheasant, lying on a fillet. It is projected on a diapered dotted background inside a scroll frame with a wreath, angels, a cherub's head and other devices.

Mascall knew himself to be gifted with poetical powers, as the following stanza proves, which forms the "Authours Conclusion of this Table" (*i.e.* the table of Contents):—

To God be praises on hie
in all our worldly planting,
And let us thanke the Romaines also,
For the Art of Graffing.

The "also" here is good.
This 1596 edition is quoted by Hazlitt.*

15. There was another issue by Este, which must be rare, as it does not occur in the catalogue of the British Museum or in Hazlitt's lists. The title-page is the same as the others, but differently arranged.

---

* *Second Series of . . . Collections and Notes*, London, 1882, p. 387. Piiii is blank, according to him.

A BOOKE OF THE
Arte and manner how to Plant and
Graffe all ſorts of Trees, how to ſet
Stones and ſow Pepins, to make wilde Trees
to graffe on, as alſo remedies and medicines.
With diuers other new practiſes, by one of the Ab-
bey of S. Vincent in France, practiſed with his own hands,
deuided into vij. Chapters, as heereafter more plaine-
ly ſhall appeare, with an addition in the ende
of this booke, of certaine Dutch practi-
ſes, ſet foorth and Engliſhed by
Leonard Maſcall.
[Vignette]

In laudem inciſionis diſtichon,
Heſperidum Campi quicquid Romanaque tellus.
Fructificat nobis, inciſione datur.

¶ Imprinted at London by T. Eſte,
for Thomas Wight. 1599.

Small 4to. A to P in fours, Piiij blank; or, pp. [24] 84 [10, 2 blank]. Black letter; but the title, preliminary matter, running titles, chapter headings and table are in roman and italic. The vignette is the usual one of a man graffing. Ciiij contains the graffing tools and there are the other illustrations in the text. The "Exhortation" is not dated.

The text runs from Di *r.* to Oij *v.*, followed by the "Table (by Alphabete)" Oiij *r.* to Piij *v.* Piiij is blank. The colophon is curious:

¶ *Imprinted at London by T. Eſte, for*
Thomas Wight. 1592.

I do not know whether this is a misprint for 1599, or whether it indicates that this is a copy of the 1592 edition with a new dated title-page. I suspect the latter is the true explanation. It is noteworthy that, if the dates are reliable, this form of the book passed through nine editions at least in twenty-seven years, which implies a brisker demand than usual.

16. After a long break of fifty years, the book reappeared with a new title in the degenerate style of the middle of the seventeenth century. Here again may be noticed an instance of the resuscitation of a book to serve a new generation. It would be interesting to ascertain

why this took place from about the close of the sixteenth century on to the middle of the seventeenth.

The copies in the Museum may be quoted.

17.

THE | Country-Mans | NEW ART OF | PLANTING | AND | GRAFFING : | Directing the beſt way to make | any ground good for a Rich Orchard : | With the manner how to Plant and Graffe | all ſorts of Trees, to Set and Sow Curnels ; As | alſo the Remedies and Medicines concerning | the same ; With divers other New Experiments : | *Practiſsed by* LEONARD MASCALL | Publiſhed by Authority | [Vignette]

London, Printed by *Jane Bell*, and are to be ſold at the Eaſt | end of *Chriſt Church*. 1652 (? or 3 ?)

Small 4°. A & B 6 leaves, C to L in fours ; or, pp. [12] 70 [2 blank]. The graffing tools form the frontispiece. The vignette is a very rude reproduction of the earlier one. Roman letter. The Epistle and other preliminary matter are omitted, as well as the Alphabetical Index at the end.

The date is uncertain because it has been shaved off in the binding, but it is given as 1652 in the Catalogue (966. i. 22. (2.)).

18. Another edition followed at the customary interval :

THE | Counrty (*sic*) - Mans | NEW ART OF | PLANTING | AND | GRAFFING : | Directing the beſt way to make | any ground good for a Rich Orchard ; | . . . [Vignette]

London Printed, by *Jane Bell*, and are to be ſold at the Eaſt | end of *Chriſt-Church*, 1656.

Small 4to. A four, B two, C to K in fours, L three leaves ; or, pp. [12] 70. The graffing tools form the frontispiece (A1). Roman lettei. The book is poorly printed on thin hard paper, the woodcuts are very rude, and the book is nothing more than a chap book. All the preliminary matter is omitted, except the Contents, and the index at the end is also wanting, if it has not been omitted designedly.

This copy has the signature of W. Musgrave (B.M. 967. k. 13.).

19. Besides these two editions there is another of earlier date, not in the British Museum. It is quoted by Hazlitt, and I have a copy which, so far as can be judged, agrees with his account and differs from the other two.

THE
Country-Mans
NEW ART OF
PLANTING
AND
GRAFFING:
Directing the beſt way to make
any ground good for a Rich Orchard:
With the manner how to Plant and Graffe
all ſorts of Trees, to Set and Sow Curnels; As
alſo the Remedies and Medicines concerning
the ſame; With divers other New Experiments:
Practiſed by LEONARD MASCALL.
Publiſhed by Authority.
[Vignette.]

London, Printed by *I. Bell*, and are to be ſold by *Iohn* . . .

Small 4to. [A1 and 2] B1 and 2, A3 and 4, six leaves, C to L3, L4 blank? wanting; or pp. [12] 70. Roman letter with some italic. [A1] is the frontispiece, a picture of the graffing tools. [A2] the title-page has a vignette of a man graffing. It is a very rude reproduction of the original. The other woodcuts in the text are equally rough. Part of the preliminary matter and the alphabetical table at the end are omitted, but the Instructions to the Planter and Graffer and the Table of Contents are retained.

The date is uncertain, because in this copy also the last line has been cut off by the binder, but according to Hazlitt * it runs: and are to be ſold by Iohn | Wright, . . . 1651. His collation is: A, 2 leaves, B-L in fours, with a frontispiece of implements.

There are thus, at least, a dozen editions of this book, a run of appreciation which it would hardly have enjoyed if it had not proved itself useful.

20. Reference has been already made † to Mascall's treatise on cattle and horses, and copies have been described of 1662 and 1610-09, the latter not in the British Museum. To these can now be added others of which one, dated 1587, seems to be the first of this treatise. It is

* *Bibliographical Collections and Notes. Third and Final Series. Second Supplement.* London, 1892, p. 55.

† *Transactions of the Glasgow Archæological Society*, N.S., 1908, V., 136, 179 (Reprint, Supp. V., 1908, pp. 12, 55).

in the British Museum (C. 27. e. 3.), and a detailed description follows. It is rather a bother that the early editions contain three title-pages, one for each division of the subject, which seems a bibliographical superfluity.

21. The firſt booke of
Cattell:
Wherein is ſhewed the gouernment of
Oxen, Kine, Calues, and how to vſe Bulles and
other cattell to the yoake, and fell. With diuers ap-
proued remedies, to helpe moſt diſeaſes among cat-
tell: moſt neceſſarie for all, eſpecially for huſ-
bandmen, hauing the gouernment of
any ſuch cattell.
Gathered and ſet forth by Leonard Maſcall.
Thou Heardman, keeper of thy beaſtes,
when any beaſt is ſickly:
Search herein, and thou ſhalt find,
of proued remedies quickly.
[Fleur-de-lys]
LONDON
Printed by Iohn Wolfe,
1587.

Small 4to. A to N in fours, O to T in eights, V four leaves, Aa to Cc in eights, Dd ten leaves (with two leaves, one marked *Dd6, and one not marked inserted for the Table), Ee and Ff in eights, Gg four leaves; or pp. [8] 1-92 [93-98] 99-195 [196 (blank)-203] 204 (misprinted 190)-285 (*sic*) [2 of table, 1 blank]. The pagination in the last four sheets is irregular. The actual number of pages as it stands seems to be [8] 308. Black letter, except the titles, dedication, verses, headlines, chapter titles, rubrics and tables, which are in roman. Some ornamental capitals.

O *recto*, p. [97]:

THE
Second booke
intreating of the gouernment of
Horſes, with the approued remedies
againſt moſt diſeaſes.
Verie profitable for all men, ha-
uing a charge and gouerment (*sic*) thereof,
and chieflie for huſbandmen: with
diuers other remedies practi-
ſed in this lande.
Gathered by L. M.

Although the Learned haue reuealde
the helpes for horſe great ſtore,
Yet practiſers therein againe:
haue found for them much more.

LONDON
Printed by Iohn Wolfe.
1587

Aa1 *recto*, p. [201]:

THE
Third booke
intreating of the ordering of ſheep
and goates, hogs and dogs: with ſuch
remedies to helpe moſt diſeaſes as
may chance vnto
them.
Taken forth of learned Authors, with
diuers other approued practiſes, verie
neceſſarie for all men, eſpecially thoſe
which haue any charge and go-
uernement thereof.
Gathered by L. M.
LONDON
Printed by Iohn Wolfe
1587.

The first leaf is blank except for the signature A. The first title-page is plain. The other two have an ornamental border, perfect in both. The 1591 edition is a page for page reprint of this 1587 edition.

22. The editions of 1587 and 1591 are both in the British Museum, and are regarded as of special rarity and value, but a fine unused copy of the later edition has come to hand Its title-pages differ in a few details of arrangement from those of the earlier one, but it is unnecessary to give them in full The imprint, however, which occurs in all three, may be reproduced.

23.

The firſt Booke of | Cattell: | VVherein is ſhewed the gouerne- | ment of Oxen, Kine, Calues, and howe to | vſe Bulles and other cattell to the yoake, and fell. | . . .

LONDON
Printed by Iohn Wolfe, and are to be ſolde
by John Harriſon the elder, at the ſigne of the White
Grayhound in Paules Church yard.
1591

Small 4to. A, B to T in eights, V four leaves; or pp. [8] 1-92 [93-98] 99-195 [196 blank-203] 204-301 [302] for [304], because R7 [pp. 259-260] has been dropped in the numbering. The first leaf is blank except for the signature A. The text is in black letter, but the titles, tables, preface, headlines, chapter titles, marginal notes and verses in the text are in roman. There are some ornamental capitals.

On G5 *recto*, p. [97] is the title of the second book:

THE | Second booke | intreating of the gouernment of | *Horſes, with the approued remedies* | againſt moſt diſeaſes. | . . . | LONDON . . . 1591.

On O *recto*, p. [201] is the title of the third book:

THE | Third Booke | intreating of the ordering of ſheep | *and goates, hogs and dogs: with ſuch* | remedies to helpe moſt diſeaſes as | may chaunce vnto | them. | . . . | LONDON . . . 1591.

The first title-page is plain, but the other two have each a broad border with scrolls, wreaths, and two female figures; at the top the Royal arms, and, at the bottom, a phœnix. In the third title-page the scroll behind the Lion supporter at the top is broken. The British Museum copy (C. 27. c. 38.) has what seems to be the author's autograph on the *verso* of Ai, opposite the title-page.

24. In the British Museum (235. g. 30.) there is an imperfect copy of the Treatise on Cattle of date 1600, but by comparing it with the reprint of 1605 its defects can be ascertained. It is not possible, of course, to put forward a guaranteed register, but the following one will probably be found correct:

Small 4°. A to T in eights, V four leaves; or pp. [8] 1-92 [93-98] 99-195 [196-203] 204-301 [302].

Black letter, except title-pages, running titles, chapter headings, rubrics, tables, verses, and the wanting dedication (?) in roman. There are some ornamental capitals. The device on the second and third title-pages is a flaming sun, a bird and a rabbit, with the words: DVM SPERO FERO, all within a scroll border On G4 *recto*, p. [95], at the end of the table of the first book, is a small woodcut, which seems of earlier date than the volume. It represents a person carrying a cup in which is a small black dragon, head erect, mouth open, wings outspread, and a curled tail. In the background there is a rose (?) and another flower not identified. In the 1605 reprint this cut is replaced by a characterless scroll ornament. Is there any allusion here to an antimony cup?

The Museum copy wants A1, 2, and 3, N3 and 6, and V2, 3, and 4.

There are besides small portions of the text defective where it has been necessary to repair the leaves. The collation may be given.

A1 wanting, was either blank, or more probably contained the signature, as in the edition of 1605.
A2 wanting, the title of the Booke of Cattell.
A3 wanting, dedication to Sir Edward Montague.
A4 Mascall *To the Reader*, and *To the Husbandman*.
A5, p. 1. The text begins and goes on to V1 *verso*, p. 296.
V2-4 wanting. From the 1605 edition it may be inferred that these leaves contained the following matter:

V2 *recto*, p. 297. On mad dogs and remedies for their bites, to V3 *recto*, p. 299.

V3 *verso*, p. 300. On remedies and medicines for mangy dogs, ends V4 *recto*, p. 301. After this begins the table referring to hogs and moules, and the remedies for mad and mangy dogs. This ends on the following page, 302, which is not numbered.

The second title-page is on G5 *recto*, p. [97].

THE
SECOND BOOKE IN-
TREATING OF THE GO-
VERNMENT OF HORSES, WITH
THE APPROVED REMEDIES
AGAINST MOST DISEASES. | . . .
Gathered by Leonard Mascall. | . . .
AT LONDON
Printed by Iohn Harison, and are to
be sould in Pater noster Rowe, at
the signe of the white Graye hound.

On Oi, p. [201], is the third title-page:

THE
THIRD BOOKE,
Intreating of the ordering of Sheepe,
Goates, Hogges, and Dogs: . . .
AT LONDON
Printed by Iohn Harison, and are to
be sould in Pater noster Rowe, at
the signe of the white Grayhounde.
1600.

This is the title which supplies the date of the edition. The device is the same as in the second book, but here it is upside down.

25. The 1605 edition, which is in the British Museum (7076. aa. 49. (2.)), is a page for page, if not strictly a line for line reprint of that of 1600. The collations, as will be seen, are identical, and the woodcut of the ringed boar's head appears in both on p. 276. But, in the interval, Harison had modified his device, and had had it cut more clearly. The following is a short account of the edition:

> THE | FIRST BOOKE OF | CATTELL. . . . *Gathered and ſet forth by Leonard Maſcall*. . . . [Vignette.]
>
> At London, | Printed by Iohn Hariſon, and are to be | ſould in Pater Noſter Rowe, at the | ſigne of the white Grayhounde. | 1605. |
>
> Small 4to. A to T in eights, V four leaves; or pp. [8] 1-92 (misprinted 94) [93-98] 99-195 [196 (blank)-203] 204-301 [1]. A1 contains the signature only.
>
> G5 *recto*, p. [97], is the title of the second book "intreating of the Government of Horses," and it is said to be "printed for Iohn Hariſon."
>
> O1 *recto*, p. [201], is the title of the third book "intreating of the ordering of Sheepe, Goates, Hogges, and Dogs," and this is "printed by Iohn Hariſon."
>
> Black letter, except the title-pages, dedication, verses, headlines, chapter titles, rubrics, and tables in roman. Some ornamental capitals.
>
> The vignette is the sun over a cloud, under it a rabbit with an ear of wheat, and in the corner a saint (?) carrying a clasped book. The whole within a scroll border, with the words: DVM SPERO FERO and, below, the initials I. H.

26. Next comes the edition of 1610, a copy of which has been already described.* It is practically a page for page reprint of that of 1591, and it was printed for the same John Harison, or Harrison, at the White Grayhound, but by Nicholas Okes.

Harison seems to have been a publisher of books of this kind, and his name appears again on the analogous treatises of Gervaise Markham as late as 1649. It is more likely, however, that the John Harison of Markham's time was a descendant, or else that the name was retained merely for business purposes. That, however, has nothing to do with the present subject.

27. The subsequent editions of the seventeenth century which I have seen are not attractive, and there is the usual falling off in quality

---

* See § 20 above, and the reference there given.

from the first edition, which is almost always better than its successors. In these later issues the ornamental borders disappear, the paper is not so firm, the type not so clear, the margins not so broad, the title-page altered so as to present a summary of the contents, and, in some, additions are made to the text. These changes will be observed in the following abstracts:

28. THE
Gouernment of Cattell.
Diuided into three Bookes.
The firſt, entreating of Oxen, Kine, and Calues: and how to vſe Bulls, and other Cattell, to the yoake or fell.
The second, diſcourſing of the Gouernment of Horſes, with approued Medicines againſt moſt Diſeaſes.
The third, diſcouering the ordering of Sheepe, Goates, Hogges, and Dogges, with true Remedies to helpe the Infirmities that befall any of them.
Alſo perfect inſtructions for taking of Moales, and likewise for the monthly husbanding of Grounds, as hath been already approued, and by long Experience entertayned amongſt all ſorts, eſpecially Huſbandmen, who haue made vſe thereof, to their great *profit and contentment,*
Gathered by LEONARD MASCAL.
LONDON:
Printed by *T. S.* for *Roger Iackson*, and are to be ſould at his Shop, neare Fleet-ſtreet Conduit. 1620.

Small 4to. A to V in eights; or pp. [8] 1-91 [92-96] 97-190 [191-195] 196-307 [308-309] and [3 blank].

In the B.M. copy A1 and A8 are wanting, and V7 *verso* and 8 are blank.

Black letter, except the title-pages, dedications, verses, headlines, chapter titles, rubrics and tables in roman. Some ornamental capitals.

G4 *recto*, p. [95], is the title of the second book. It is dated 1619, and has a vignette: an anchor, on the cross-bar of which is a lion passant, langued and gorged, with a 5-pointed star on its left shoulder, within a scroll border, having the words on it: Desir n'a Repos. 1586.

N6 *recto*, p. [195], contains the title of the third book, dated 1620, with the same vignette.

There is a copy in the British Museum, 7295. d. 31.

29. An edition of the year 1627, which I have not seen, is mentioned by Richard Weston: * Mascall, (——.)—Government of Cattle, in three parts, black letter, in 4to.

30. The next edition is in the British Museum (988. e. 4.) and is dated 1633. It is rather unfortunate that the editions were not numbered, so that one could say whether this is next in order or not. There is time for an edition between 1627 and 1633, but possibly there may not have been one, and the interval between the dates may indicate that the demand was slackening and that the book was being superseded by more modern works, such as Markham's, on the same subject.

> The Gouernment of | CATTELL. | *Diuided into three Books.* | . . . [Vignette.] LONDON, | Printed by *Thomas Harper* for *Iohn Harifon*, and are to be fold at his | fhop in Paternofter Row, at the figne of the Vnicorne. 1633. |
>
> Small 4to. A to V in eights. V8 blank ?, wanting, or pp. [8] 1-91 [92-96] 97-190 [191-195] 196-307 [308-309] [310 blank and 311-12 awanting].
>
> Black letter, except the title-pages, dedication, verses, headlines, chapter titles, rubrics, tables, which are in roman. Some ornamental capitals.
>
> G4 *recto*, p. [95], is the title of the second book, with a scroll vignette.
>
> N6 *recto*, p. [195], is the title of the third book, with the same scroll vignette.

31. Weston † refers to an edition of 1653, which I have not seen: Mascall, (Leonard.)—On Cattle, black letter, in 4to.

32. In the copy of the 1662 edition in the British Museum (451. a. 30.) there is a large folding plate of a horse, with lines drawn from his various organs to the border, where the name of the disease of the particular part is mentioned. These are numbered, and are thus referred to the margin, where directions are given for the treatment of the disease. This table has the imprint:

> *LONDON*, Printed for *William Gilbertson* at the Bible in *Giltfpur-ftreet* without *Newgate*, 1662.‡

The plate is wanting in the copy already described.

---

* "A Catalogue of English Authors, who have written on Husbandry, . . ." p. 20, in his *Tracts on Practical Agriculture and Gardening*, London, 1773, 8vo.

† *u.s.* p. 27.

‡ Gilbertson's name has come up before, as the publisher of some of John White's books. For references to the 1662 edition, see § 20 above.

33. Further alterations were made on the title of the edition which appeared a number of years later, in 1680.

The Countreyman's Jewel:
Or, the Government of
CATTEL.
Divided into three Books.

The firſt, Diſcourſing of the Go- | vernment of *Horſes*, with approved Me- | dicines againſt moſt Diſeaſes.

The ſecond, Treating of *Oxen, Kine*, and | *Calves*; and how to uſe *Bulls*, and other Cat- | tel, to the yoke or fell.

The third, Diſcourſing the Ordering of | *Sheep, Goats, Hogs* and *Dogs*; with true Reme- | dies to help the Infirmities that befall any of | them.

Alſo, Perfect Inſtructions for taking of *Moles*, | and likewiſe for the monthly Husbanding of | Grounds; and hath been already approved, | and by long experience entertained amongſt all | ſorts; eſpecially Husbandmen, who have made | uſe thereof, to their great profit and content- | ment. Alſo Directions for Gardening.

Gathered at firſt by *Leonard Maſcal*, but | much inlarged by *Rich. Ruſcam*, Gent.

London, Printed for *William Thackery* at the *An-| gel* in *Duck-Lane*. 1680.

8vo. A to Z, Aa to Dd in eights, Ee four leaves (but Ee4 is wanting in the B.M. copy); or, pp. [16] 390 [32]. Roman letter.

K2 *recto*, p. 131, is the title-page of the second book.

R4 *recto*, p. 247, is the title-page of the third book.

There is an engraved title in compartments extra, not the same as that in the 1662 edition. At the top are a horse and a bull, on the left-hand side a sheep and a boar, on the opposite side a goat and a dog, at the bottom a woodman with an axe splitting logs, and a gardener with a rake in a garden.

Prefixed to the first book is a copperplate of a horse with the diseases of different parts indicated by lines to the margin, where the names are printed. This is a copy of the 1662 plate, but the treatment is omitted.

Prefixed to the second book is a similar plate of a bull.

In the third book there is a woodcut of a sheep, p. 249; one of a goat, p. 311; one of a boar, p. 322; and one of a dog, p. 370.

This is a very commonplace sort of book. It is in the British Museum, 450. d. 38. This is the latest version of the treatise I have seen.

34. Before leaving the subject it will be well to correct one or two inaccuracies in the account of Mascall already given,* which was taken in part from Thompson Cooper's article.† It is stated in it that Mascall belonged to an old family settled at Plumstead in Sussex. Though the only Plumstead of importance I knew of was in Kent, I allowed the statement to pass, for there might be a place of that name in Sussex in spite of its not appearing in a map or gazetteer.‡

35. This statement is taken, I find, from Thomas Fuller, who, under Sussex, says: §

> "Leonard Mascall of Plumsted in this County being much delighted in gardening, man's Original vocation, was the first who brought over into *England* from beyond the seas *Carps* and *Pippins*, the one well cook'd delicious, the other cordial and restorative. For the proof hereof, we have his own[a] word and witness, and did it, it seems, about the Fifth year of the reign of King *Henry* the eighth, *Anno* Dom. 1514. The time of his death is to me unknown."
>
> [a] In his book of fishing, fowling and planting.

Cooper, who puts Mascall's burial on May 10, 1589, takes no notice of the great age he must have been, or suggests how his death then can be reconciled with his introduction of carp in 1514, when he must have been more than a mere youth, and with the publication of books in 1588, 1590, and 1596.

36. Lower, however, gives a different account.‖ The name Mascall, according to him, is a corruption of the Norman Mareschal or Marescal. One of the families of that name obtained an early footing in Sussex

* *Transactions of the Glasgow Archæological Society*, N.S., 1908, V., p. 135 (Reprint, Supplement V., §§ 17-21, and "Addenda").

† *Dictionary of National Biography*, London, 1893, XXXVI., pp. 404-05.

‡ There is a Plumstead in Norfolk.

§ *The Worthies of England*, London, 1662, folio, II., p. 113; or edited by John Nichols, no place, 1811, 4to, II., p. 399.

‖ Mark Antony Lower, *A Compendious History of Sussex*, Lewes, 1870, 8vo, II., p. 101; and *The Worthies of Sussex*, Printed for Subscribers, Lewes, 1865, 4to, pp. 52-54. This last contains a short biography of Leonard Mascall. Lower is not referred to by Thompson Cooper.

Trans. Arch. Soc. of Glasgow, 1915.

See § 37.

and lived at Plumpton Place, Plumpton (not Plumstead) * in that county, holding the manor under a crown lease. Leonard was one of the family, flourished in the sixteenth century, and wrote the books on rural affairs which have been already enumerated. Lower says that the date of his death was unknown, whereas Tanner gives it as 1589.

Like other writers, Lower is doubtful about Mascall's having introduced carp into England, seeing that this fish was mentioned by Dame Juliana Berners long before his time. He then adds:

> "He must have lived to a great age, unless indeed some of his productions were posthumously given to the world, for they appeared at intervals from the year 1578 down to the end of the century."

Apparently, therefore, Lower had the same difficulty about Mascall's dates as others who have written about him. The only conclusion one can come to, in the meantime, is that there is error or confusion somewhere, and that such data as are available at present are inadequate to dispel the difficulties which surround this writer.

37. Not to omit anything which may contribute to a knowledge of Mascall, a facsimile is here inserted of the engraved title-page, or frontispiece, of the 1662 edition of the book on cattle. It contains at the foot a portrait, which may or may not be one of the author. There is no evidence that it is meant for him. It is in this edition that Mascall is styled "Chief Farrier to King James," a post which dates make it hard to believe that the "Clerk of the Kitching" of Archbishop Parker could have held.

38. [June 15, 1911. The fifth supplement † contained an account of a collection of medical receipts entitled: *A Rich Storehouse, or Treasurie for the Diseased*, 1596, etc., by a certain A. T., who was ousted from the authorship by one G. W. in the most unblushing manner.

* Though Lower has quoted what Fuller says about Mascall's pursuits, he has made no comment whatever on Fuller's statement that Mascall was of Plumstead instead of Plumpton.

† *Transactions of the Glasgow Archæological Society*, N.S., 1908, V., p. 148 (Reprint, *Supplement* V., 1908, p. 24).

Among the editions then quoted was one called the seventh, printed by Richard Badger, with the date 1630 on the title-page and 1631 in the colophon. The copy described was in the Hunterian Museum, but I have now got another which corresponds with the description already supplied.]

39. Contemporaneous with this book of A. T.'s, and parallel with it, though different in detail, is the following treatise:

THE
CHYRVRGIANS
CLOSET:
OR:
An Antidotarie Chryurgicall. Furniſhed
with varietie and choyce of :·
Apophlegms, Balmes, Baths, Caps, Cataplasmes, Cauſticks, Cerots, Clyſters, Collyries, Decoctions, Diets, and Wound-Drinks, Defenſatiues, Dentifrices, Electuaries, Embrocations, Epithemes, Errhines, Foments, Fumes, Gargariſmes, Iniections, Liniments, Lotions, Oyles, Peſſaries, Pils, Playſters, Potions, Powders, Quilts, Suppoſitaries, Synapiſmes, Trochiſces, Vnguents, and Waters.
The greateſt part whereof were ſcatteredly ſet downe in ſundry bookes and papers; by the Right Worſhipfull M^r. THOMAS BONHAM, D^r. of Phyſick; and now drawne into method, and forme, *by* EDVVARD POETON of *Petworth*, (late and long ſeruant to the foreſaid Right Worſhipfull D^r.) and publiſhed for the benefit of his Country, and the helpe and eaſe, of young practitioners in the ancient, neceſſarie, and noble Science of Chyrurgerie.
[Ornament]
LONDON,
Printed by GEORGE MILLER, for EDVVARD BREVVSTER, and are to be ſold at the ſigne of the Bible at the North doore of *Pauls*. 1630.

Small 4to. A to Z, Aa to Zz, a to c in fours, d two leaves; or, pp. [8] 359 [1 blank] [Table 26, Errata 1, 1 blank].

The title gives an alphabetical and sufficiently complete survey of the contents. The book consists of a series of receipts, and, in the excellent and full table at the end, there is a list of the diseases with reference to the remedies described in the book. It is an elaborate compilation, and, as the authorities are quoted for each preparation, it forms an index to the literature as well.

The volume is dedicated to Frances, Countess of Exeter, who is praised for her devotion to the art and her skill in it, and for the care she has of " the sick and sore." This is another instance of the practice of medicine and surgery by a woman, of which so many examples have come forward in these notes.

This book is not mentioned by Hazlitt.*

40. Legerdemain or juggling, one of the oldest of arts, for I have little doubt that the Egyptian priests hocus-pocussed the laity,† has developed an extensive literature, of which there are numerous examples to cite, and which it is quite legitimate, I suppose, to include among books of secrets. For, if any practical art is immersed in mystery and replete with secrets, it is surely that of the conjuror, whose trade would be gone if his methods were known.

41. I have not ascertained when legerdemain literature started, but what I have seen and what I have to show, dates from about the beginning of the seventeenth century and continues till now. The books are of all sorts, from the now very costly chap-books of that century to the cheap remainders of the present year. The older editions were mere reprints one after the other, first in quarto and then in duodecimo. Next came catch-penny issues, so badly printed, on such miserable paper, as sometimes to be illegible. These were succeeded by the productions of the early years of last century, with coloured frontispieces, and finally arrived the modern treatises, with illustrations hardly if at all superior to those of a couple of hundred years ago. These various manuals and their editions will come up at their proper times. They

* For Bonham's life, see Thompson Cooper's notice in the *Dictionary of National Biography*, London, 1886, V., 345.

† There seems to be no doubt that they did; for it is said that of the structural arrangements in the temples necessary to produce the supernatural effects required, vestiges are still to be seen in ruins which remain.

all profess to describe how the tricks are done, but whether it is that the authors cannot make their meaning explicit, or that words fail them at the critical moment, or that their explanations are intentionally hazy, it is difficult to say, but the directions are far from being clear and convincing.

42. The earliest work on the subject I have come across was by a person calling himself Samuel Rid, who according to one view is Robert Greene, but according to another is Samuel Rowlands. I first made its acquaintance in the Gordonstoun Sale Catalogue,* and then I found it entered by Lowndes under both the subject and the author's name,† and the two editions of 1612 and 1614 mentioned.

The edition of 1612 is not in the British Museum, but the title of it is quoted in full by Hazlitt.‡ As this agrees exactly—except in the imprint—with that of the edition of 1614, which is given below from the Museum copy (1042. b. 36.), it will suffice if the imprint of the first edition be reproduced here from Hazlitt's version.

> Printed at London for T. B. and are to bee solde by Samuel Rand, neere Holborne-bridge. 1612.
>
> 4to, 24 leaves, black letter. §

Whether the 1614 book is a genuine second edition, or merely remainders dressed up with a new title-page and preliminary matter, could be ascertained only by comparison of copies of the two dates, and as that is not possible at the moment, all that can be done is to describe the 1614 edition as it stands.

---

* London, 1818, No. 1973. The Gordonstoun copy was of the 1614 edition and fetched £1 13s.

† *The Bibliographer's Manual*, London, 1864, III., p. 1333; IV., p. 2092, where the Gordonstoun copy is referred to along with others.

‡ *Handbook to the Popular, . . . Literature of Great Britain*, London, 1867, p. 508. In his *Collections and Notes*, London, 1876, p. 358, Hazlitt describes the 1614 edition.

§ It may be noticed that the author's initials S. R., those of the name at the end of the dedication, Sa. Rid, and of the name of the bookseller, Samuel Rand, are identical. Are they all fictitious ?

THE
ART OF IVGLING
OR
Legerdemaine.
WHEREIN IS DECIPHERED,
all the conueyances of Legerdemaine and
Iugling, how they are effected, and
wherein they chiefly conſiſt.
Cautions to beware of cheating at
Cardes and Dice.
The detection of the beggerly Art
of Alcumiſtry.
And,
The foppery of fooliſh couſoning Charmes.
All tending to mirth and recreation, eſpecially
for thoſe that deſire to haue the inſight and
priuate practiſe thereof.
By *S. R.*
Quod noua teſta capit, Inueterata ſapit.
LONDON,
Printed by GEORGE ELD. 1614
(*.*)

Small 4to. Two leaves, B to F in fours, G two leaves, no page or leaf numbers. Black letter; the title-page, dedication to his father [-in-law], William Bubb, by Sa. Rid, and Bubb's reply, address " To the Courteous Reader," running title, headings of chapters, marginal notes and special words phrases and verses in the text, in roman.

In the preliminary section the author gives a short account of the coming of the " Egiptians," the following they got in the country in a few years, and the measures taken by Parliament against them, though it did not stop their wandering. They practised sleight-of-hand which other people also acquired, and the tricks the author approves of, if " done for recreation and mirth," and not with any evil intent. The art is defined as nimbleness of hand, exhibited especially with balls, with money, and with cards, and many feats may be done with these by an expert and afford much pleasure.

He tenders advice to those who will practise the art:

the operator must have an attractive manner;

there must be no bungling, but the tricks must be cleanly and swiftly done;

constant practice is necessary so as to deceive the eye, the hand, and the ear ;
everything necessary for the trick must be ready and at hand ;
words of art, of which examples are quoted, must be used.

He then describes how tricks are performed with balls and money ; he exposes the use of false dice and cards in gambling, and concludes :

> "Therefore utterly forbeare to hazard any thing at dice and live in doubt and suspition of cheating wheresoever you play."

Next come illustrations of legerdemain with cards, tricks which require a confederate, and other juggling illusions, such as pulling laces out of one's mouth, eating a knife, cutting one's nose asunder, and such like. Two examples of cheating by alchemists are given, a few charms are quoted and instances of performing animals.

43. Some years must have elapsed before the subject was resumed and a new book appeared.

The earliest copy I have to show is dated 1658, but it is only of the fifth edition. The next is the ninth, and is dated twenty-four years later. If one could speculate on these numbers, one might surmise that the book was first published about 1630. I have seen no reference to an edition of that date, and in the British Museum there is nothing before the second edition, printed in 1635.

To Lowndes,* however, we are indebted for the notice of what may possibly be the first edition, or, at all events, an earlier dated issue of the second edition than that in the British Museum :

> Hocvs Pocvs, Junior. The Anatomie of Legerdemain. London, 1634. 4to. Frontispiece and cuts. Nassau, pl. i. 2123. 15*s*. White Knights, 1949. 1*l*. 1*s*. Hibbert, 3945, morocco, 14*s*. There are several Chapbook editions in 12mo.

Presumably some one of the copies here quoted is of the year 1634. Very surprising are the prices which they realized at the respective sales. I have never had the fortune to fall in with any of these copies.

The notes following have been made on the editions of this book,

* *The Bibliographer's Manual*, ed. Bohn, London, 1865, II., 1079.

specimens of which have come before me. The earliest, 1635, is that in the British Museum (7913. bb. 36.), as has just been said.

HOCVS | POCVS | IVNIOR. | *The Anatomie of* | LEGERDEMAIN. | *OR,* | The Art of Iugling ſet forth in his proper colours, fully, | plainly, and exactly, ſo that an ignorant person may thereby | learn the full perfection of the ſame, after a little practiſe. |

Vnto each Tricke is added the figure, where it is needfull | for inſtruction. |

The ſecond Edition, with many additions. |

Præſtat nihili quam nihil facere.

[Vignette.]

LONDON, | Printed by *T. H.* for *R. M.* 1635. |

Small 4to. Four leaves, B to H in fours. No leaf or page numbers. Roman and italic. Leaf 1 is the frontispiece. It represents a conjuror behind a table with a deep cover, and on it a cord and cups and balls. Leaf 2, the title. The vignette is the conjuror holding the "Bonus Genius," or "Nuntius invisibilis." Out of his mouth is a label with the words "Hiccius Doccius" on it. Leaf 3: "To the Reader." Leaf 4, the text begins. Numerous woodcuts in the text, fairly well executed. On E2 *recto* is a drawing of the "Bonus Genius."

This book is based on Rid's, but with considerable alterations and additions. The introductory matter about the gypsies, the nature and object of legerdemain, and the qualifications necessary for a successful juggler, is much compressed; the warnings about false dice and cards in gambling are omitted; and there is no account of the wonderful intelligence displayed by the learned donkey and the horse of knowledge, and how they are trained. On the other hand, the directions for tricks with balls are more detailed, and numerous woodcuts to illustrate them are inserted in the text. Most of the tricks described by Rid are reproduced, and several of them have woodcut illustrations added, but the larger proportion of the tricks are new. Some are of sleight-of-hand, others are illusions or deceptions, many of them with appropriate illustrations. There is only one trick founded on a physical basis, which is to freeze water by the fireside. The apparatus consisted of a stool, a quart-pot, snow, a little water, and a short stick. Water is poured on the stool, the pot is set in the water, snow is put in the pot, with some salt on the sly, and the mixture is churned with the stick. By and by the pot freezes to the stool. Then the author adds:

> "There's a naturall reason may be given for this, which he thats a scholler need not be told, and for a common Iugler I would not have so wise as to know, therefore, I omit it."

That was hardly fair to the "common Iugler," who would have been more successful if he had known the principle involved.

44. Hazlitt * quotes an edition of *Hocus Pocus Junior*, of the date 1638, 4to, which is in the Douce Collection in the Bodleian Library. I have not seen it, but it probably resembles the edition of 1635.

45. The next is in the British Museum (1042. c. 42.), and is dated 1654. The notable thing here is the long interval between the date of the second edition, 1635, and of this, the fourth, with only one edition to fill it. Assuming the above-mentioned, 1638, edition to be actually the third, it is difficult to account for the lapse of sixteen years between them, when the editions subsequent to the fourth followed the usual rate of publication with fair regularity. It is an exceptional case.

HOCUS | POCUS | JUNIOR. | *The Anatomy of* | LEGERDEMAIN. | *OR,* | The Art of Jugling . . .

The fourth Edition, with many Additions. |

Præſtat nihil (*sic*) quam nihil facere.

[Vignette.]

LONDON, | Printed by *G. Dawſon,* and are to be ſold by *Thomas Vere,* at | the *Angell* without Newgate. 1654. |

Small 4to. Four leaves, B to H in fours. No leaf or page numbers. Roman and italic. Leaf 1 is the frontispiece as before (§ 43), but of inferior execution. Leaf 2 is the title as before, but somewhat differently arranged The vignette is an inferior copy. Leaf 3, To the Reader. Leaf 4, the text begins. Woodcuts in the text.

46. The next edition followed punctually. It is a reprint of those preceding, with two leaves additional.

HOCUS | POCUS | JUINOR (*sic*). | *The Anatomy of* | LEGERDEMAIN. | *OR,* | The Art of Jugling . . .

The fifth Edition, with many Additions. |

Præſtat inhili quam nihil facere.

[Vignette.]

LONDON, | Printed by *G. Dawſon,* and are to be ſold by *Thomas Vere,* at | the *Angel* without *Newgate,* | and *Francis Grove* on *Snow-Hill,* 1658. |

---

* *Handbook to the Popular, . . . Literature of Great Britain,* London, 1867, p. 276.

Small 4to. Four leaves, B to H in fours, [1 leaf of Errata and 1 blank]. No leaf or page numbers. Roman and italic. Leaf 1 is the frontispiece of the conjuror; leaf 2 is the title, and the vignette is the conjuror with the "Bonus Genius"; leaf 3, To the Reader; leaf 4, the text begins. Woodcuts of apparatus in the text. The leaf of Errata and the blank leaf are wanting in the British Museum copy (1042. c. 18.).

47. Of the subsequent editions three have hitherto eluded my notice, but there is no reason to doubt that they were simple reprints. The book, however, may possibly have changed hands during the four-and-twenty years that had passed. The ninth edition is similar to the others, but the publisher is different.

Hocus Pocus | JUNIOR. | *The Anatomy of* | LEGERDEMAIN; | *OR,* | *THE ART OF* | JUGLING. | . . .

The Ninth Edition, with many Additions, |

Præſtat nihili quam nihil facere. |

[Vignette.]

Printed by *T. H.* and are to be ſold by *J. Deacon,* at the ſign of the | *Rainbow,* a little above St. *Andrews* Church, in *Holbrn* (sic). 1682. |

Small 4to. Four leaves, B to H in fours; or, pp. [6] 57 [1]. Leaf 1 is the frontispiece of the conjuror. This same cut is repeated on H4 *verso*. Leaf 2 is the title. The vignette is the "Bonus Genius" as before. Leaf 3, To the Reader. Leaf 4, the text begins. Woodcuts in the text. There is a woodcut of the Bonus Genius on E2 *recto*, p. 29.

48. Curiously enough the next edition came out in the following year. There are two or three alterations in the title-page, but otherwise the book is the same as its predecessors. There is a copy in the British Museum (7913. bbb. 32.).

Hocus Pocus | JUNIOR. | *The Anatomy of* | LEGERDEMAIN: | *OR,* | *THE ARTE OF* | JUGLING. | . . .

The Tenth Edition, with many Additions, |

Præſtat nihili quam nihil facere. |

[Vignette.]

Printed by and are to be ſold by *J. Deacon,* at the ſign of the | *Rainbow,* a little above St. Andrews Church, in Holborn. 1683. |

Small 4to. Four leaves, B to H in fours; or, pp. [6] 57 [1]. Leaf 1 is the frontispiece of the conjuror, repeated on H4 *verso*. Leaf 2, the title with the vignette of the "Bonus Genius." Leaf 3, To the Reader. Leaf 4, the text begins. The Bonus Genius is described on E2 *recto* p. 29. The copy I have wants the frontispiece.

49. In the next edition a new and extended title is introduced. It is as follows:

Hocus Pocus
JUNIOR.
The Anatomy of Legerdemain;
OR,
The Art of Jugling
Set forth in its Proper Colours, fully, plainly, and exactly; ſo that an Ignorant Perſon may thereby learn the full perfection of the ſame, after a little Practice.
Unto which is newly added, The *Trick* of Changing a CARD into a Live BIRD, or into any Picture; with other Additions.
As likewiſe, To each Trick the Figure needful for Inſtruction.
Licensed and Entered according to Order.
The Eleventh Edition, with many Additions.
Præſtat nihili quam nihil facere.
[Vignette.]
Printed by J. M. *for* J. Deacon, *and are to be Sold by* J. Gilbertson, *at the* Sun *and* Bible, *on* London Bridge: 1686.

Small 4to. Four leaves, B to H in fours; or, pp. [6] 58. Roman and italic. Leaf 1 is a poor woodcut of the conjuror; leaf 2, the title, with the usual vignette; leaf 3, To the Reader, and book advertisements; leaf 4, the text begins. Woodcuts in the text. On H4 *verso* (p. 58) the frontispiece is repeated.

The British Museum copy (1042. c. 7) is soiled and defective.

50. The twelfth edition has not as yet made its appearance, but by way of amends two thirteenth editions of different dates are extant. In order to distinguish these from each other and from the eleventh, it will be necessary to transcribe the titles, as they present variations.

HOCUS POCUS, Junior.
THE
ANATOMY
OF
LEGERDEMAIN:
OR,
The Art of Jugling,
Set forth in its proper Colours, fully, plainly and exactly; ſo that an ignorant Perſon may thereby learn the full Perfection of the ſame, after a little Practice.

Vnto which is newly added,
The Trick of Changing a Card into a Live Bird, or into any Picture; with other Additions. As likewiſe, To each Trick, the Figure needful for Inſtruction.
Licens'd and Enter'd according to Order.
The Thirteenth Edition, with many Additions.
Præſtat nihili quam nihil facere.
[Vignette.]
LONDON: Printed by *W. Onley*, and are to be Sold by *J. Deacon*, at the ſign of the Angel in *Guiltſpur ſtreet*, without *Newgate*. MDCXCVII.

Small 4to Four leaves, B to H in fours; or, pp [6] 58 Roman and italic. Leaf 1 is the frontispiece, a very crude impression; leaf 2, title, with the usual vignette; leaf 3, To the Reader, and book advertisements; leaf 4, the text begins. Woodcuts in the text. The frontispiece is repeated on H4 *verso*, p. 58. British Museum, 7913. d. 28.

51. The title of the other issue of the thirteenth edition is differently arranged:

HOCUS POCUS, Junior:
THE
ANATOMY
OF
LEGERDEMAIN:
OR,
The Art of Jugling.
Set forth in its proper Colours, fully, plainly, and exactly; ſo (*sic*) that an ignorant Perſon may thereby learn the full Perfection of the ſame, after a little Practice.
To which is added, The Trick of Changing a Card into a Live Bird, or into any Picture; with other Additions. As likewiſe, to each Trick, the Figure needful for Inſtruction.
Licenſ'd and Enter'd according to Order.
The Thirteenth Edition with Additions.
Præſtit (*sic*) nihili quam nihil facere.
[Vignette.]
LONDON: Printed by *W. O.* for *B. Deacon*, at the Sign of the *Angel* in *Gilt-ſpur-ſtreet*, without *Newgate*, 1706.

Small 4to. Four leaves, B to H in fours; or, pp. [6] 58. Roman and italic. Leaf 1, the frontispiece is a poor impression; leaf 2, the title, with the usual vignette; leaf 3, To the Reader, and book advertisements.; leaf 4, the text begins. Woodcuts in the text. On H4 *verso*, p. 58, the frontispiece is repeated. It is in the British Museum, 7913. d. 29., and has the Duke of Sussex's book plate.

It will be observed that these books are practically page for page reprints, deteriorating in style and value with each successive issue.

52. In this latest edition I have seen, the title is again altered (B.M. 1042. b. 8.):

> *HOCUS POCUS Junior.* | THE | Anatomy of Legerdemain : | OR, | The Art of Jugling, |
> Fully, plainly and exactly ſet forth in its proper Co- | lours, ſo that any Perſon may thereby learn the full Per- | fection of the ſame, by a little Practice. |
> Illuſtrated with ſeveral new Cuts or Figures, that | were not in the former Impreſſions, for the more full In- | ſtruction and Satisfaction of the Reader.
>
> Præſtat nihil (*sic*) quam nihil facere.
>
> [Vignette.]
>
> *London*: Printed for *B. Deacon*, at the Angel in *Guilt-ſpure-ſtreet.*

8vo. A in eight. No leaf or page numbers. Roman and italic. The vignette (very rude) is the conjuror with the Bonus Genius. There is the drawing of the Genius on A7 *recto*. The new drawings are the double funnel, A5 *verso*, the light of a candle focussed through a flask of water, and the reading of cards, held above one's head, by reflection in a drop of water. These illustrations appear also in Dean's *Legerdemain*.

As this is a bare abstract of the complete work, numerous woodcuts are omitted.

At the end is this

" Advertisement.

Printed for B. Deacon, at the Angel in Guilt-ſpur-ſtreet without Newgate.

Hocus Pocus Junior. The Anatomy of Legerdemain: or The Art of Jugling in full: containing thirty Tricks more than this Breviate, with Cuts and Figures to each Trick. Price 6d. I desire you not to be deceived, for the right Book is only Printed for as above-ſaid, under the Advertisement."

This book is not dated, but from what is said in the title about new cuts this must be after the edition of 1706.

One of the added receipts is this (A8 *recto*):

> *How to light a Candle at the ſide of a Glaſs of cold Water, or any other Liquors without the help of Fire.*
>
> Take a little Katathumptos, and ſtick a ſmall Bit on the Edge of the Glaſs, and apply a Candle new blown out to it, and it will light immediately, tho' the Snuff be quite extinguiſh'd, if warm.

Which is only prepared and ſold by Samuel Carter, *and* James Bland, *at the Sign of the Bell in* Vine-ſtreet, in *St.* Giles's in the Fields, *where any Perſon that is curious, may be furniſhed with al manner of Inſtruments, as well as inſtructed in the Art Dexterity of Hand. To be ſpoke with all from ſix till nine in the Morning, and from ſix to nine at Night.*

Presumably Katathumptos is phosphorus. If so, this book is probably subsequent to its introduction into England.

53. The preceding list is a singular one, representing as it does an output of at least fifteen editions from about 1630 to say 1712, with hardly any change in the text. One would have thought that, after eighty years, every one of the tricks would have been so familiar that even the least exacting of spectators could hardly have been impressed or mystified by them, however dexterously performed. Still, how did the book exist and persist, if there was no demand for it? It must have been well used too, seeing that of several of the editions not a copy apparently survives. But that is a question of the state of education and of society then, rather than of the bibliography and history of technical literature now.

54. The next work, though still mainly medical, is not entirely so, for, as the title declares, it includes the art of preserving and conserving, with other secrets. Of the author, Thomas Brugis, nothing is known but what he himself has stated in the preface to his *Vade Mecum*, but the book, on account of its contents, deserves a little attention. Though it runs very much on the same lines as many of its contemporaries, it differs from them by its containing an exposition of medical principles, in which, as a rule, the ordinary collections of medical secrets are deficient. This exposition serves besides as a guide to the state of medicine generally in the middle of the seventeenth century in England.

In the Preface to the Reader, Brugis defends the propriety of printing and publishing books of this sort in English, and not keeping useful knowledge wrapt up in a language which the ordinary man could not understand and therefore could not profit by. This vindication occurs often enough in medical treatises about this time to justify the inference that there was considerable opposition to such publication on the part of, at least, some of the qualified physicians. It was an innovation sure to provoke dissent, for it was a withdrawing of the veil from possible defects in medical theory and practice which the practitioners would fain have kept out of sight. Uncomplimentary epithets had been bestowed on those who had been guilty of such unprofessional conduct, like Salmon and Culpeper, just as a couple of centuries earlier the same treatment had been meted out to Paracelsus and Fioravanti, although they were none of them worse than their learned neighbours, and in some respects were better. Brugis was quite awake to the censure he had to expect, but he was indifferent to it in that he felt he was right and his critics were wrong.

He begins with a list of the usual authorities. The first twenty-one chapters contain an exposition of the principles of medicine, including the elements, the temperaments, ages, influence of the signs and planets, the humours and complexions, making up a sort of physiology; the members, a sort of anatomy; the faculties, actions, spirits, a sort of psychology; the non-naturals; air and its importance, food, labour and rest, sleep and waking, repletion and inanition, and the passions; diseases, their causes and symptoms, critical days, bleeding, issues.

The rest of the work is devoted to practical pharmacy, cosmetics, preserving and conserving. He prefaces this part with a table of weights and measures, a catalogue of apparatus required for the practical work, and an explanation of the technical terms employed. The preparations are similar to those in other books that have been enumerated.

Brugis' manual, however, is a rather striking performance, and the first part is particularly interesting and useful for understanding the old physic, the nomenclature of which is not always intelligible.

55. A description of it follows:

THE | MARROW | OF | PHYSICKE. | Or, A learned Difcourfe of the feverall | parts of mans Body. | BEING A | MEDICAMENTARY | Teaching the maner and way of making and | compounding all fuch Oiles, Unguents, Sirrups, Cata- | plafmes, Waters, Powders, Emplaifters, Pilles, &c. as | fhall be ufefull and neceffary in any private houfe, with | little labour, fmall coft, and in fhort time. |

And alſo an addition of divers experimented Medicines, | which may ſerve againſt any diſeaſe that ſhall | happen to the Body. |
Together with some rare Receipts for Beauties, and the | neweſt and beſt way of Preſerving and Con- | ſerving: with divers other ſecrets | never before publiſhed. |
Collected and experimented by the induſtry of T. B. Gen. | Practitioner in Phyſicke and Chyrurgery. |
LONDON, | Printed by RICHARD HEARNE. | M.DC.XL. |

Small 4to. *A*, b, B to M; Aa to Yy, *Zz*, *Aa*, *Bb*, in fours; or, pp. [15, 1 blank] 88; 175 [1 blank] [22, 2 blank]. Folding table of the Humours in Mans Body.
The Dedication to Mrs. Margaret Eure is signed: Tho. Brugis.

There is a copy of Brugis' book in the British Museum (778. c. 25.). It is quoted also by W. C. Hazlitt,* but he gives the register as A–Bb in fours, and b, four leaves, which does not tally with either my own copy or that in the British Museum. In Gray's *Index* to Hazlitt's collections, the date is given as 1611.

56. Watt † calls him Thomas Bruggis, and does not mention this book, but only his "Companion for a Chirurgien, London, 1651, 8vo," and "Chirurgical Vade-Mecum, with a Supplement by Ellis Pratt, London, 1689, 12mo." These are the first and seventh editions of the *Vade Mecum* respectively, and are in the British Museum. Other editions in the Surgeon-General's Library at Washington are these: ‡ second edition, 16mo, London, 1653; fourth edition, 16mo, London, 1665; seventh edition, 16mo, London, 1681. The seventh edition was apparently re-issued in 1689 without change of number. Both were edited by Ellis Prat or Pratt. Prefixed to the 1651 edition of the *Vade Mecum*, forming *a*1, is a good frontispiece, engraved and signed by J. Cross, containing in the centre a portrait of Brugis; in the upper left-hand corner a surgeon operating on a man's head, with a cavalier standing by, booted

* *Second Series of Bibliographical Collections and Notes*, London, 1882, p. 66. *A General Index to Hazlitt's Handbook and his Bibliographical Collections*, London, 1893, p. 101.

† *Bibliotheca Britannica*, Edinburgh, 1824, Authors, I. 162 *d*.

‡ *Index-Catalogue*, Washington, 1881, II., p. 499. Second Series, 1897, II., p. 864.

and spurred, and his left arm in a sling; in the right-hand corner, an apothecary at a table, apparently making pills, while behind are shelves with pots and bottles; in the lower left-hand corner distilling apparatus, and in the right-hand corner crucibles, tongs, pliers, bellows and other implements. In the 1689 edition all this is repeated in inferior style, and the portrait is quite altered.

In the Preface to the 1651 edition (omitted from that of 1689)—a remarkable composition by the way—he speaks of himself as serving as a surgeon "in these late unnaturall Civill-wars," and dates his preface "From my Chamber at Rickmersworth in Hertfordshire, this 20. of September, 1651." The dates of his birth and death are unknown, and Dr. Payne has not ascertained where he obtained his degree.* Besides the strictly surgical part, the book is largely taken up with the preparation and uses of unguents, liniments, emplaisters, oils (including oil of vitriol, oil of sulphur, and oil of brick-bats), waters, syrups, conserves, electuaries, opiates, pills, laxatives, simples, etc.

57. The copies of this book which I have are dated 1652 and 1689. That of 1652 is apparently a reissue of the first edition with the date altered. It contains Brugis' curious Preface, but the volume has had the misfortune to pass through the hands of some Grangerizing thief, who has made away with the frontispiece for the sake of the portrait. The book otherwise is in fair condition, with moderate margins, and it has the remains of a good binding which had ties.

The later edition, though enlarged, is in every respect inferior as a book, with crowded pages, meagre margins, and, worst of all, without the index. As a supplement, however, it includes a tract by Pratt himself, entitled: *Chirurgus Methodicus; or, the Young Chirurgion's Conductor, . . .*" London, 1689.

58. In 1652 Leonard Sowerby published a volume of medical receipts for ladies on a different system from any yet described. He himself says so—correctly enough—and states that "The work doth suffi-

* *Dictionary of National Biography*, London, 1886, VII., 140.

ciently commend itself, being the first and only peece of this kinde in our English tongue, containing almost as many Medicines as lines, and those consisting of pure, Elementary Simples; not compounded of a hodge podge of sophisticated Drugs (like the Cadmeian Progenie) one oppugning the other"—also correctly enough.

But, perhaps, it is just this redundancy of undefined simples, which one using the book might criticize, for the author leaves the reader without a clue as to what simple should be employed in preference to the others, in a particular case.

The book is not common, and, though there is a copy in the British Museum, that which I have here is the first that I have ever seen for sale.

It is arranged under two heads. The first deals with ailments, beginning at the head and going down to the feet. Under the ailment of each part, external or internal as may happen, is a list of the remedies that can be used, but, as has been already observed, no preference is expressed for one over the other, and, still less, a reason for the treatment at all; so that it is in the lady's option to try any or all of the medicines enumerated.

Next follows the treatment of certain troubles, such as fevers, aposthumes, wounds, ulcers, ruptures, dislocations, venom and poison, bites of animals.

The second head consists of an alphabetical list of the simples mentioned in the foregoing section, with the occasions in which they are to be employed.

The book, therefore, is complete in itself, and the lady dispenser could collect and prepare her simples and have them ready to use when a case of the disease for which they were prescribed came before her for treatment.

The cures, if ever they were effected, depended, I should suppose, more on luck than good guiding.

Once more it may be noticed that medical practice was exercised by the women of a household, as I have pointed out even in this paper, and as I may have to do hereafter.

59. The book is as follows:

THE | LADIES | Difpenfatory. | Containing | The *Natures, Vertues*, and | *Qualities* of all *HERBS*, and | *SIMPLES* ufefull in *Phyfick*. | Reduced | Into a Methodicall Order, for | their more ready ufe in any fick- | neffe, or other accident of the Body. |

The like never publifhed in Englifh.

WITH | *An Alphabeticall Table of all the* | *Vertues of each* Herb, *and* Simple. | *London* Printed for *R. Ibbitfon*, to be fold by | *George Calvert* at the Halfe-Moon in *Watling ftreet*. 1652.

Small 8vo. A three leaves (A1 blank? frontispiece? fly-title, or licence? wanting), B to Z, Aa to DD in eights, Ee three leaves; or, pp. [6] 317 [1]; the Contents [20], the Table and Errata [95, 1].

A narrow border of fleurs-de-lis runs round the title. The preface is signed by the Author: Leonard Sowerby.

This book is not mentioned by Hazlitt. The copy in the British Museum (E. 1258.) agrees with that now described.

60. In the Surgeon-General's Library * under the name of Leonard Sowerby [bracketted] there appears a work entitled: *The ladies dispensatory: or, every woman her own physician*, . . . " 2. ed. xvi. 324 pp., 4 l. 12°. London, J. Hodges, 1740.

So far as one can infer from the tenor of the title, this is quite different from the present work, for it professes to deal with all the diseases, infirmities and disorders natural or contracted, incidental to women in all situations of life, whereas the other treats of herbs and their uses. It is unlikely that the Leonard Sowerby, to whom this book is ascribed in the catalogue, can have been the same person as the author who flourished ninety years earlier. It is also obvious, from the nature of the contents, that of whatever work the edition of 1740 is the second it could hardly be that of 1652.

61. In the fifth part of these notes, read in 1888,† a short notice was devoted to Hugh Plat's *Delights for Ladies*, with an accompanying work, *A Closet for Ladies*, and the paper concluded with these words:

* *Index-Catalogue*, Washington, 1892, XIII., p 347.

† *Transactions of the Glasgow Archæological Society*, N.S. 1889. I., p. 460 (Reprint, Part V., 1889, p. 44).

"There is an edition of 1656 of the *Delights for Ladies* and the *Closet for Ladies*, but I have not seen a copy." Three and twenty years have elapsed since these words were written, and it was only on the third day of August last [1910] that a copy of this edition in its wanderings came to rest along with some of its predecessors in my possession. Some books have taken a longer time to reach me than that, but it is curious how books drift sooner or later to those who require them for a specific purpose.

At the best the book is nothing to boast of, and the copy I have acquired has had considerable usage, though, fortunately, without loss of title-page or any part of the text; it is a poor little volume in its original sheep-skin cover, rebacked. In it the *Closet* comes first, and it was printed in 1656 by R. H., who was the printer of the earlier editions. The second tract is the *Delights*, and it was printed by William Dugard in 1656. The author's name is omitted.

The following is a description of the two sections:

A CLOSET | FOR | *Ladies* and *Gentlewomen*, | OR | The Art of Preſerving, | Conſerving, and | Candying. | *With the manner how to make* | *divers kindes of* Syrups, *and* | *all kindes of* Banqueting-ſtufs: | *Alſo divers Soveraign*-Me- | dicines *and* Salvs. |
Corrected, Amended, and much | Enlarged, by adding a very | uſefull-*Table* thereunto. |
London, Printed by *R. H.* 1656. |

12mo. A to H in twelves, but A1 (blank, or with a half-title ?) wanting, or, ff. [11] 84.

Delights | FOR | LADIES. | To adorn their Persons, | Tables, Cloſets and | Diſtillatories: | WITH | *Beuties, Banquets, Perfumes* | *and Waters.* |
Read, Practice, and Cenſure. |
LONDON. | Printed by *William Dugard.* | 1656. |

12mo. A to H in twelves, but H12, probably blank, is awanting. No leaf or page numbers. Hazlitt gives the reading "adorne." *

---

* *Supplements to the Third and Final Series of Bibliographical Collections and Notes*, London, 1889, p. 82. In a note, he adds :—

"This, like the other editions, contains a good deal of matter common to Plat's *Jewel-House*, 1594: it is the last appearance of the *Delights*." But I have not been able to see any similarity in the contents of the two books.

62. In the sixth supplement, §§ 43, 44, there is a notice of the book, *A Discourse of Artificial Beauty*, published by Royston in 1656 and 1662, and in 1692 by Meredith. It contains a discussion between two ladies on aids to beauty, such as powder and patches. In the 1662 and 1692 editions there is a frontispiece representing the ladies, one in the fashion with a moderately low-cut dress, hair in numerous small curls, her face dotted with patches, and holding a fan in her left hand. The other is in more sober, almost puritanical, attire with a plain deep collar, and her hand on a book, possibly a Bible.

In this same year, 1662, Royston published another tract on the same topic, with the same picture of the ladies as a vignette, and, beneath it, a motto from Euripides: Νοῦν χρὴ θεάσασθαι.*

Its title runs thus:

A WONDER of WONDERS:
OR,
A Metamorphofis of Fair FACES voluntarily transformed into foul VISAGES.
Or, an Invective againft Black-fpotted Faces:
By a Well-willer to modeft Matrons and Virgins,
MISO-SPILUS, i. *Qui maculas odit.*
[Vignette]
Publifhed by *R. SMITH*, Gent.
[Motto from Farnefius]
LONDON, Printed by *J. G.* for *Richard Royfton*, Book-feller in *Ivy-lane*. 1662.

Small 4to. A to E in fours; or, pp [8] 31 [1 blank].

This is an attempt to stop a fashion by argument and the weight of authority. As the authoritative passages are quoted, they are interesting as illustrating the prevalence of the practice of face-painting through the centuries; but the custom was not stopped by any criticism of it. The tract is a rare one.

63. Though the next book does not come within the category of receipt books, and hardly within that of secrets, yet as it is a list of natural curiosities and marvels, and as it is not one of the least of these itself,

* Euripides, *Œdip.* Frag. 6, ed. Dindorf, 1833.

it may very well be taken in here; all the more that I have not found this edition in the British Museum, though it is quoted by Watt.*

Of the author, "R. H. aliàs Forges, Gentleman," I have found no particulars, and do not know whether or not he was identical with the person he styles "the Master of Rarities." If he were so, then, according to his own account, he must have had much correspondence and many dealings with kings and emperors, dukes and landgraves, ambassadors, physicians, and others, at home and abroad. By what means he persuaded these persons to part with the rarities which he says they esteemed so highly, so as to enrich his collection, he does not inform us, but the items appear in this list with the donors' names attached, and there is no reason to doubt the truth of Forges' statements. This, then, may be looked upon as the catalogue of a Natural History Museum of the middle of the seventeenth century, and as such it is instructive.

The title runs as follows:

A | CATALOGUE | of part of thofe | RARITIES | Collected in thirty years | time with a great deal | of Pains and Induftry, | by one of his Maje- | fties fworn Servants. |

R. H. aliàs Forges, | Gentleman. |

They are to be feen at the place | formerly called *the Mufique-* | *Houfe* at the Weft end of *Pauls.* |

Small 12mo. 1 leaf, A twelve, B six; or, pp. [2] 33 [3 blank]. Title within a narrow border of small separate typographical pieces. No date—about 1660?

It would be useless to attempt a description of this small work. The fair thing to do would be to reprint it as it stands, but as that would occupy too much space, a short description of it must suffice.

64. The curiosities are arranged in nine groups: namely, fishes heads; whole fishes; parts of fishes; other rarities; birds; serpents; fruits, and such things; stones, or things turned into stone; corals and sea-plants.

---

* *Bibliotheca Britannica,* Edinburgh, 1824, I., Authors, 378.

Under these general heads the various objects are enumerated and described, and it is on these descriptions that the interest of the book turns, for " R. H. aliàs Forges " was rather vague as to spelling and composition. From that point of view the catalogue is possibly unique, but there can be no question about his enthusiasm for his museum, and his eagerness to enlarge it and to make it attractive and instructive for visitors.

65. Besides those catalogued it is stated that " The Gentleman of these Rarities can shew thousands of other Rarities of Nature to those that are more curious, and will be at some more Charge," and then he mentions the necessary arrangements.

On Mondays and Thursdays, things of the sea, sea-plants, " zoophytes," rare stones, strange corals, etc.

On Tuesdays and Fridays, things of the land, as metals, minerals, transmutations or things turned into stones ; crystals and precious stones, etc., etc.

On Wednesdays and Saturdays, things of the sea, land and air, strange bones and teeth, eggs of fowls and serpents, small creatures of the sea, small creeping venomous creatures of the land, 200 or 300 species of insects, amber with flies or insects involved in the transparent amber.

It is also announced that if any persons of quality with their families and friends would like to see them alone, he would endeavour to give them the opportunity if he had notice the day before, so as to put by the great concourse of people for that day. And he also offers to describe them in three or four languages to any foreign ambassador who would like to pay him a visit at the Musique-house at the west end of St. Paul's. The tract concludes with a list of donors and benefactors.

66. Though this edition is not in the British Museum, there is one of a probably later date, from which it appears that the initials " R. H." stand for Robert Hubert, about whom no particulars are as yet forthcoming. The title-page of this edition is as follows

A
C A T A L O G U E
OF MANY
Natural RARITIES,
WITH
Great Industry, Coſt, and thirty
Years travel in Foreign
Countries,
COLLECTED BY
ROBERT HVBERT, aliàs *FORGES*,
Gent. and ſworn Servant to His Majeſty,
And dayly to be ſeen, at the
place called the *Muſick Houſe*, at
the *MITER*, near the Weſt end of
St. *Pauls* Church.
London, Printed by *Tho. Ratcliffe*, for the
Author, 1664.

Small 8vo. B to E in eights, * four leaves; or, pp. [2] 62, 7 [1 blank]. The title is enclosed within a border of fleurs-de-lys placed close together.

This copy is in Sir Joseph Banks' collection (957. e. 13.).

The contents are differently arranged from those in the previous edition:

I. Rarities Human.
II. Rarities of Beaſts and Parts of them.
III. Birds.
IV. Fishes.
V. Serpents, &c.
VI. Inſects or Flies.

The Second Part of the Catalogue:

I. Vegetables.
II. Sea Plants, &c.
III. Minerals, Stones of ſtrange ſhape, and things turned into Stones, with Chryſtals and precious Stones.
IV. Mettals, Minerals, &c.
V. Things of ſtrange Operation.

At the end is

"A Catalogue of the Names of thoſe Great Princes and perſons of Quality whoſe Love of Vertue, Learning and of the admirable workes of God in Natural Rarities has been ſhewed by their Bountifull adding of ſomething to the increaſe of the forementioned Collection."

"Rarities Added by ſeveral friends, ſince the Printing of the Catalogue."

This catalogue is considerably larger than the other, which has undergone a thorough revision. The spelling has been corrected, the descriptions are more exact, a number of new rarities have been added and old ones omitted. But with these emendations and revisions the catalogue has become deadly commonplace, and the quaint and amusing originality of the other edition, in keeping with the character of the curiosities, has been obliterated.

67. The making of collections of natural rarities and wonders was a common pursuit about this time, and I have seen lists of one or two such gatherings. There was that formed by Palissy, by Pierre Borel, Royal Physician in France, by Tradescant * and the curiosities and rarities to be seen in the Gallery of the Garden of the Academie of Leyden, the English catalogue of which must have been produced in Holland, judging by the mistakes in printing and spelling. I have seen also a MS. list of curiosities which contained such interesting things as a divining rod, the tooth of a hippopotamus, spar, Mallucco beans found in Orkney, King James VII. Irish coyne, petrified water, pins vomited up by a maid bewitched, fulmar oil from St. Kilda, antimony, lapis calaminaris, thunderbolts as they are commonly called, hobgoblins' clause [!], used for medicinal purposes, a petrified wilk, Roman medals, lead ore found in Lorn in Argyleshire, silkworms' eggs, a caul "with which my nephew Ja: Lockwood at Hallhill was born. It covered the head." "An elfe Arrow given by Mrs. Smith daughter of Mr. Wood. It was found in Perth(?)shire. The milk wherein it is boyld given to a beast struck with ane Elf arrow is s^{d} to be an effectuall cure."

68. The objects in these gatherings were selected for their curiosity and rarity, not with any educational idea, or for a scientific purpose. That came later, when collections of many specimens began and comparison became possible. But the contents of these old cabinets were

* Tradescant's *Catalogus*, which is contemporary with Hubert's, cannot vie with it in any respect. There are no such wonders, no such descriptions and no such spelling. But then Tradescant was a naturalist, had almost of course lost the faculty of wonder and surprise and took a matter of fact view of his collections.

to make people stare and wonder, and they bear the same relation to a modern museum as the *Livre des Merveilles du Monde* and other books of a like kind to a modern treatise on Natural History.

69. In paragraph 61 above an instance is given of the appearance, after long waiting, of two books, just when they were wanted to fill a gap in the lists. Another illustration of the same thing has to be recorded in connection with two works which have been often referred to in these notes. Though out of chronological succession, they may be taken together.

One is a copy of the collection known under the name of Queen Henrietta Maria, which has not been recorded before.* It is similar in style and contents to the other editions, but this copy wants the third part: *The Compleat Cook.*

It fills the place left for it in the lists, between that of 1655 and that of 1662, though it is to be regretted that the editions were not numbered as they followed one another, for they could then have been traced more easily and a more exact knowledge of the book arrived at.

THE | QUEENS | CLOSET | OPENED. | Incomparable Secrets in | Physick, Chyrurgery, Prefer- | ving and Candying, &c. | Which were prefented unto the | QUEEN, | By the moft Experienced Perfons of the | Times, many whereof were had in e- | fteem, when fhe pleafed to defcend | to private Recreations. | Corrected and Reviewed, with many | Additions: Together with three ex- | act Tables. |

Vivit post funera Virtus. |

LONDON, | Printed for *Peter Dring,* and are to be | fold at the Sun in the Poultry, next | door to the Rofe Tavern, 1661. |

12mo. Six leaves, including the portrait; C to P in twelves, Q six leaves; or, pp. [12] 300 [23, 1 blank]. The portrait of the Queen is a fine impression, and at the left-hand corner has the words C[e] (?) *Faithorne fec*: and at the right: Sold by Nat. Brooke att y[e] Angell in Cornhill.

On L1 *recto* is the title-page of the second tract: | A | QUEENS DELIGHT: | OR | THE ART | OF | *Preferving, Conferving,* | and *Candying*; | as alfo | A right Knowledge of | making PERFUMES, and | Diftilling the moft | Excellent Waters. | Never before Published. | Printed by *R. Wood,* for *Nath. Brooke,* at | the Angel in *Cornhill,* 1660.

Part III. is wanting.

---

* It was just alluded to in 1907 in the Sixth Supplement, § 52.

70. It seems only fitting that the second book of the two just mentioned should be that which competed for popularity and wide circulation with the Queen's treatise, namely the Countess of Kent's *Choice Manual*, this time of the fifteenth edition. It, too, curiously enough, wants the last part: *A True Gentlewoman's Delight*, as well as the portrait, which has been removed by some unscrupulous and irrational Grangerizer.

This edition, dated 1667, fits according to rule between the fourteenth, 1663, and sixteenth, 1670, as shown in the list of editions.* It is in the diminutive size of the second, fourth, and other editions, and is identical with the fourteenth in its form and appearance. There is no difference in the text. The title may be copied.

A | *Choice Manuall,* | OR | RARE and SELECT | SECRETS | IN | PHYSICK | AND | CHYRURGERY : | Collected and practised | by the Right Honourable, the | Counteſſe of *Kent,* late | deceaſed. | Whereto are added ſeve- | ral Experiments of the Virtue | of Gaſcon pouder, and *Lapis con-* | *tra Yarvam,* by a Profeſſor | of Phyſick. | As also moſt Exquiſite waies | of Preſerving, Conſerving, | Candying, *&c.* | The Fifteenth Edition. |

LONDON, | Printed, by *Gartrude Dawſon,* and | are to be ſold by *Margaret Shears* | at the Sign of the Blew Bible | in *Bedford* Street in *Covent* | *Garden,* 1667. |

24mo. [A] to L in twelves; [A1] is blank, and L11 and 12 are wanting; or, pp. [2 blank ?] [15, 1 blank] 233 [1 blank] [5, 1 blank] [4 blank ? wanting]. A narrow border of fleurs-de-lis within two lines runs round the title.

This copy, as has been already said, wants the portrait and the second part: "*A True Gentlewoman's Delight.*" It will be observed that the business at the Blew Bible was now carried on by Margaret Shears, but whether she was the widow, sister, daughter, or other relation of William Shears, is not divulged.

71. The compilers of the old receipt books had a happy knack of conferring high-sounding titles on the meagrest of pamphlets. Instances have not been awanting in these notes, but choice examples are supplied by one W. Lovel, who describes himself as "Gent. and

* *Transactions of the Glasgow Archæological Society,* N.S. 1910, VI. i., p. 30 (Reprint, Supplement VI., p. 30).

Traveller," as well as "Practitioner in Physick and Chyrurgery." He must not be confused with Robert Lovel, who flourished about the same time. Of W. Lovel I have found no notice; he is not included in the *Dictionary of National Biography*, but there are three little chap-books in the British Museum which bear his name and are worth a brief examination. It is not often that the compiler of a receipt book bursts for a moment into rhyme in the title of his book as Lovel does.

B.M. 7953. a. 50.

APPROVED | RECEIPTS, | OR, | The Queens Reprefentation, | To this our Englifh Nation. | AND | To all the reft of the Kings moft Ex- | cellent Majefties Dominions, as *Wales*, | *Scotland*, and *Ireland*, &c. | DISCOVERING | Divers Infirmities in Mans Body. | AND ALSO | Plain and eafie Rules, for making the In- | gredients, and to cure the fame. | By *W. Lovel*, Gent. and Traveller. | Printed and publifhed for the ufe and | benefit of the poor. |

London, Printed for *Charles Tyus* at the | three Bibles on *London-Bridge*. 1663. |

Small 8vo. A in eight; no pagination. The title-page is within a narrow ornamental border.

The receipts are medical for a few common diseases, and the remedies are taken from old-fashioned garden plants, such as marigolds, basilicon, green ivy, pellatory, scurvy-grass, "allicompane" roots, carduus benedictus, water-cresses, camomile, southernwood, sage, rosemary, spermint, "bittany leaves," leeks, bramble leaves, dock roots, burr roots, bay leaves, and such like. Besides these, tar, wood ashes, aquavitæ, oil of bitter almonds, "the stone of a thunderbolt," "moss growing on a freestone wall," hog's grease, sallet-oyl, "black sope," bay-salt, ginger, liquorice, sugar-candy, turpentine, formed items of the popular pharmacopœia.

The troubles dealt with were as follows:

A2 To purge the head, and to keep it free from all manner of pain.
To cure a scurf head.
A receipt for Deafness in Hearing.
A3 Another receipt for deafness.
For the Catherick or philme which groweth over the eye. For this was employed the "stone of a thunderbolt" beaten to powder as fine as wheat flower.
For the pin and web in the eye.
For a burn or scald.

A4 For the Quinsie in the throat. (An unsavoury decoction in milk was prescribed to be drunk night and morning. "It is a most perfect cure, *probatum est.*")

For fowleness or stoppage in the stomach.

For the consumption in the Lungs.

A5 For the Scurvey.

An ointment to ease the Gout.

To staunch the bleeding of any Vein.

For a broken belly.

A plaister for the gout.

A6 For the paines, and aches, in the limbs and Ioynts.

A Salve for a new cut that will not leave bleeding.

A powder for the Cough.

For the dropsie or Timpany of the Belly called Ascitus.

A7 To prevent sickness at Sea, called the Scurvey. ("This disease cometh of laziness and slothfulness.")

To make an oyntment for the syatica or any other ache.

For a knee that is sore, and swelled in the Joynt.

A8 A Receipt for Children that have the Rickets, of some called the German evil.

This is a very curious little rudely-printed pamphlet, and it brings evidence that the troubles which afflicted the just in those days were identical with those in operation now. But the treatment is different.

72. Another tract by Lovel is to be noted.

B.M. 7953. a. 48.

THE | Dukes Desk | NEWLY | BROKEN UP, | Wherein is Difcovered | Divers rare *Receipts* of *Phyfick* and *Chyrur-* | *gery*, good for Men, Women and Children. | Together with feveral *Medicines*, to Pre- | vent and Cure the moft Peftilent Difeafes | in any *CATTEL.* | With a *Remedy* fent to the Lord Mayor of | *London*, by King *Henry* the Eighth, againft the | *PLAGUE.* | *By* W. Lovell, *Gent. and Traveller.* |

LONDON, Printed by *T. Leach*, for *F. Coles*, at | the *Lamb* in the *Old-Baily*, 1665. | *

Small 8vo. A eight leaves; or, pp. [2] 14. Title-page within a single line.

---

* This title curiously resembles that of a popular booklet of about the same time, entitled:

Mother Bunch's Closet newly broke open. Wherein is contained many rare Secrets of Art & Nature.

The earliest known edition is of 1685; there were others subsequently, but

This chap-book is in two parts; pp. 1-10 contain short receipts for curing various common troubles, and the remedies are again of the usual empirical kind. This is a different collection from the *Approved Receipts*. The second part, pp. 11-14, contains "Medicines for Cattel," as for the staggers and yellowes, the scratches, the bots, the farcy, strain in the joints, murrain, etc. The remedies employed are the same in kind as the others, but several plants are mentioned besides those enumerated above: herb-grace, elder leaves, rue, mandragora, "featherfew," sorrel, barnet, maidenhair, cherry-tree gum, sinkfoyle, "Callenbines," flower-deluce, pennyroyal, hysop, turmeric, saffron, comfrey, smallage, plantine, groundsel, centory, wormwood, garlick, aloes, "jallump," besides alum, powdered skull, oyl of mastick, vinegar, frankincense, green Coperas, brimstone, barrowsgrease, mallows, "oyl of petre," tansey. Some of the receipts were printed in the supplement to *The Compleat Servant-maid*, § 85 below.

73. B.M. 7953. a. 49.

THE | Dukes Desk | NEWLY | REVIVED; | Wherein is difcovered | Divers Rare Receipts of | PHYSICK, and CHYRURGERY, good for | Men, Women, and Children. | Together with feveral Medicines, to Pre- | vent, and Cure the moft Peftilent | Difeafes in any | CATTEL. | By *W. Lovel*, Practitioner in Phyfick, | and Chyrurgery. |
London, Printed by *E. C.* 1672. |

Small 8vo. A in eight, no pagination. The title-page is enclosed in a border of fleurs-de-lis.

This is an inferior reprint of the 1663 edition, and is in the British

---

it, along with a continuation or second part, was edited by Sir G. L. Gomme, and published in 1885 as No. III. of a Series of Chap-Books and Folk-Lore Tracts. The editor suggests that the title indicates the existence of another earlier collection of Mother Bunch's good advices to young men and maidens, but, whether that be so or not, it is probable that there were issues of this work prior to 1685. Anyhow, unless the resemblance of the titles is a mere accident, it is more likely that Lovel borrowed the form of his title from the chap-book, just because it was familiar and ensured wider recognition for his work, than that the chap-book title was copied from Lovel's receipts. Notwithstanding Mother Bunch's promise of rare secrets of art and nature, there are none such, and there is no similarity in the contents of the two tracts.

Museum (7953. a. 49.). At the end, on A8 *recto* (*verso* is blank), is the following amulet printed along the page:

This worn in the boſome or anywhere near the body, cures the Ague.

When Jeſus ſaw the Croſs whereon his body ſhould be Crucified, His Body did ſhake. The Jews ſaid unto him; Haſt thou an Ague. He anſwered, and ſaid: "Whoſoever keeps this in mind or writting, ſhall not be troubled with neither Ague, nor Feavour. So Lord help thy Servant, who putteth his truſt in thee.

God bleſs King Charles, *and our Royal Queen* Catherine.

This is surrounded by the same border of fleurs-de-lis. The amulet is not contained in the previous edition.

In connection with these books there is an obvious difficulty: Who was the duke, and what was he thinking about to allow his desk to be so misused? A still greater difficulty is to know how, after being apparently smashed to pieces, a desk can be revived. Lovel gives no treatment for such a comminuted fracture.

74. Works on cosmetics and personal adornment have been mentioned from time to time, and there are several more to come, but that which falls into place here is somewhat of a curiosity.

The author, though he appends to the dedication the initials M. S., was called Thomas Jeamson, of whom it is said in a note to the copy of his book which I have seen that he was first scholar, afterwards fellow of Wadham College in Oxford, and became Bachelor of Physic and subsequently Doctor of Medicine in 1668. He tried to conceal his connection with the book by the initials aforesaid, but, being betrayed by the publisher, he was nicknamed "Artificial Embellishments." In 1664 he delivered the Music Speech,* and was considered a person of good ability. He died at Paris in 1674.

It was a shabby trick of the publisher, for Jeamson may have regretted the publication of the book when it was beyond recall. It

* Anthony à Wood, *Athenæ Oxonienses*, ed. Bliss, London, 1813, I., p. xciii in Wood's Life.

has become rare, the copy, now described, having been formerly in the possession of Heber.

The title is as follows:

> ARTIFICIALL | Embellifhments. | OR | *ARTS BEST DIRECTIONS* | How to Preferve Beauty | or Procure it. |
> Νικᾷ δὲ καὶ σίδηρον,
> Καὶ πῦρ καλή τις οὖσα. *Anacr.*
> OXFORD, | Printed by *William Hall*, | *Ann.* D. 1665. |
>
> Small 8vo. A to N in eights; or, pp. [16] 192.

75. This seems to be the only edition. It is in the British Museum (1174. d. 6.), and is mentioned by Hazlitt * and by Watt.†

The Dedication to the Honourable and Truly Vertuous A. E. is elegantly and dexterously expressed, even if somewhat overdone.

The Epistle to the Ladies, describing "Deformity" and extolling the benefits of all his preparations, is a mine of mixed metaphor and elaborate rigmarole, and both in this epistle and in the text the author is not free from the diction of the days of Charles II.

I cannot refrain from quoting the first sentence:

> "Deformity (fairest Ladies) is a fingle name, yet a complicated misery; for a young Algebraist in this only word, knows how to read a whole Iliad of evills,"

which may prove true, when understood.

The treatise is in four parts and deals with:

I. The whole body and the beautifying of it, and "here is the method to remove superfluous fat and become thin and elegant," a perennial problem as clamant for solution now as it was two hundred and fifty years ago.

II. The head, neck and breast, and it includes hair dyes the skin, the complexion, and skin diseases.

III. The arms, legs and feet.

IV. Scents and perfumes.

---

* *Collections and Notes*, London, 1876, p. 369, without mention of the author's name.

† *Bibliotheca Britannica*, Edinburgh, 1824, Authors, II., 543 *r*.

The materials used are interesting, for the receipts show how long they have engaged the attention of women. His receipts are peculiar, and his secrets are sometimes more wonderful than attractive.

Jeamson's name does not appear in the *Dictionary of National Biography*.

76. Dr. Walter Charleton, or Charlton,* "physician to the late King" (Charles I.), a man of wide and varied learning, did not restrict himself to medical topics, but extended his observations in many directions, and was a voluminous writer.

In 1669 appeared a little volume containing two discourses which exhibit such a queer incongruity, or else such a subtle connection in the collocation of their themes, that the title-page may be quoted as it stands.

B.M. G. 16397.

TWO | DISCOURSES. | I. Concerning the Different | WITS of MEN: | II. OF THE | MYSTERIE | OF | VINTNERS. |

LONDON, | Printed by *R. W.* for *William Whitwood* at | the Sign of the *Golden-Lion* in *Duck-* | *Lane*, near *Smithfield*, 1669. |

---

* Niceron says the name is Charlton and not Charleton, Watt gives them alternatively, Wood calls him Charlton. He was born at Shepton Mallett, 2nd February, 1619, entered Magdalen Hall, Oxford, in 1635, and graduated M.D. in 1642. He took up his abode in London, 1650, published a number of works, and was physician both to Charles I. and Charles II. He was an original F.R.S. and F.R.C.P., delivered the Harveian Oration in 1680, 1702 and 1706, was President of the College in 1689 and Senior Censor from 1698 to 1706. He died in 1707. *Dictionary of National Biography*, 1887, X., p. 166 (article by Dr. Norman Moore). Bromley mentions a portrait *ad vivum* by Loggan, æt. 56, 1678, in the "Anatomiæ Prælectiones," 4to, and one by P. Lombart in the "Treatise on the Soul," 1657, 8vo (*A Catalogue of Engraved British Portraits*, London, 1793, Period VII. Class VII., p. 233); Munk, *The Roll of the Royal College of Physicians of London*, 1878, I., p. 390.

Notices of Charleton and lists of his works are given by various writers: Wood, *Athenæ Oxonienses*, ed. Bliss, London, 1820, IV., cols. 752-756; Mercklin, *Lindenius Renovatus*, Norimbergæ, 1686, 4to, p. 365, five medical works; Niceron, *Memoires . . . des Hommes illustres*, Paris, 1732, XVIII., pp. 110-117; Zedler, *Grosses vollständiges Universal-Lexicon*, Halle und Leipzig, 1733, V., col. 2022; Jöcher, *Allgemeines Gelehrten-Lexicon*, Leipzig, 1750, I., col. 1844; Watt, *Bibliotheca Britannica*, Glasgow, 1819, Part II., or, Authors, Vol. I., 216, *f-l*. Watt's list of Charleton's writings is the fullest that I have seen.

8vo. A to P in eights, Q four, A1 blank ? and Q4 blank ? wanting ; or, pp. [Titles and Contents 14] 230.

A2 is the general title as above.

A3 is the title of the first discourse :—

A Brief | DISCOURSE | Concerning the Different | WITS of MEN : | Written | At the Requeſt of a Gentleman, | Eminent in Virtue, Learn- | ing, Fortune. | In the Year 1664. | And now Publiſhed with Con- | ſent of the Author. |

LONDON, | Printed by *R. W.* for *William Whitwood* at | the Sign of the *Golden-Lion* in *Duck-* | *Lane*, near *Smithfield*, 1669. |

A4 to A8, the Contents.

B to K5 *recto*, pp. 1-137, text of the first Discourse. P. 138 is blank.

K6, p. [139], title of the second tract :—

THE | MYSTERIE | OF | VINTNERS. | OR | A Brief Diſcourſe concern- | ing the various *Sickneſſes* of | WINES, and their reſpective | *Remedies*, at this Day com- | monly uſed. | Delivered to the | ROYAL SOCIETY, | Aſſembled in *Greſham-Colledge* on | the 26 of *November*, | *Anno Dom.* 1662. |

LONDON, | Printed for *William Whitwood* at the | Sign of the *Golden-Lion in Duck-* | *Lane*, near *Smithfield*, 1669. |

The text runs from p. 141 to p. 200. Then follows, p. 201: "Some Observations Concerning the Ordering of Wines," by Dr. Merret, ending p. 230.

The title-pages are surrounded by double lines, and there are two parallel lines along the top of the pages, between which the running title is printed.

77. In 1675 a reprint appeared, but inferior in quality as a book. A description of it follows from the British Museum copy (232. l. 34.).

TWO | DISCOURSES | I. Concerning the Different | WITS of MEN : | II. OF THE | MYSTERIE | OF | VINTNERS. | By *Walter Charleton* D.M. and Phy- | ſitian in ordinary to his Majeſty. | *The ſecond Edition enlarged.* |

LONDON, | Printed by *F. L.* for *William Whitwood* | at the Sign of the *Golden-Bell* in *Duck-* | *lane* near *Smithfield*, 1675. |

8vo. A to Q in eights, A1 and Q8 are blank ; or, pp. [2 blank, 14 Titles and Contents] 1-144 [145-146, title-leaf of the second tract] 147-235 [1 Errata, 2 blank].

A3 contains the title of the first tract :—

A Brief | DISCOURSE | Concerning the Different | WITS of MEN | Written | At the Requeft of a Gentleman, | Eminent in Virtue, Learn- | ing, Fortune. | In the Year 1664. | And now Publifhed with Con- | fent of the Author. |

LONDON, | Printed by *F. L.* for *William Whitwood* at | the Sign of the *Golden-Bell* in *Duck-* | *Lane*, near *Smithfield*, 1675. |

Both titles are enclosed in a double line. The running title is also between two lines.

A4, the Contents, misprinted Cotnents, to A8.

The text of the first tract begins on B, p. 1, and ends K8, p. 144.

L1, p. [145] contains the title of the second tract :—

THE | MYSTERIE | OF | VINTNERS. | OR A | BRIEF DISCOURSE | CONCERNING | The various *Sickneffes* of | WINES, and their refpective | *Remedies*, at this Day com- | monly ufed. | Delivered to the | ROYAL SOCIETY, | Affembled in *Grefham-Colledge*. |

LONDON, | Printed for *William Whitwood* at the | Sign of the *Golden-Bell in Duck Lane*, | near *Smithfield*, 1675. |

Small 8vo. L1 to Q8 in eights; Q8 is blank; or, pp [2] 147-235 [1 Errata].

Title within double lines. The running title of 'The Mysterie of Vintners' is between lines, but there is no running title to the second treatise, 'The Ordering of Wines,' though the double lines are continued. I have a copy of this second discourse which has been bound separately. We have had Whitwood before, as the publisher of some of John White's books.

This is practically a reprint of the 1669 edition in the same form and style, but it has been enlarged by the insertion of a few Latin quotations and one or two paragraphs, but there is no real difference between them.

78. Then came the third edition enlarged, but in a poor form:

TWO | DISCOURSES | The firft, Concerning the different | WITS of MEN. | The fecond, A brief Difcourfe con- | cerning the various Sickneffes of | Wines, and their refpective reme- | dies; at this day commonly ufed: | Delivered to the | Royal Society | By *Walt. Charleton*, M.D. | To which is added in this Third Edition, the | Art and Mystery of Vintners, and Wine-Coop- | ers: Containing approved Directions for Con- | ferving and curing all manner and forts of | Wines; whether *Spanish*, *Greek*, *Italian*, *Por-* | *tugall*, or *French*: As it is now practifed in | the City of *London*. |

London, Printed for *Will. Whitwood*, at the | *Angel* and *Bible* in *Little Britain*, 1692. |

It is a small 12mo. Sigs. A4, B to I in twelves, K2; or, pp. [8] 183, 1 blank [10, 2]. The preliminary pages contain title, advertisement and contents of the first discourse, the text of which is contained in pp. 1-78. Then the title-leaf of the second discourse "The Mystery of Vintners" and the text of Charleton's discourse which ends p. 112, followed (p. 113) by "Some Observations Concerning the Ordering of Wines" by Dr. Merret. And here it may be remarked that Dr. Merret does not give his readers any advice as to the purchase or "ordering" of wines from, say, a seventeenth century Gilbey, but how to keep wines "in order." This discourse ends p. 126.

Next follows the title-leaf of the tract on the sophistication of wine, added to this third edition:

THE | Art and Myſtery | OF | VINTNERS | AND | Wine-Coopers: | CONTAINING | Approved Directions for the | Conſerving and Curing all man | ner and ſorts | OF | WINES, | whether *Spaniſh, Greek, Ita-* | *lian*, or *French*, very neceſſary | for all ſorts of People. |

London, Printed for *Will. Whitwood* at the Sign | of the *Angel* and *Bible* in *Little Britain*, 1692. |

It ends on p. 183, and the remaining pages are occupied with an alphabetical table of contents, the last leaf of all containing a few book advertisements. This tract is sufficiently replete with trade and technical secrets to give it a place in the present supplement.*

These various treatises about wine are of interest from a good many points of view.

1°. The wines used: Hippocras, Sack, Claret, Rhenish, Xeres, vulgarly Sherry, Malmsey, Muskadel, Canary.

2°. Thoughts about fermentation.

3°. The treatment of wines to restore their colour and flavour, to remove acidity, ropiness and other disorders to which they are liable.

4°. The substances used: alum, sulphur, limestone, salt, white of egg, milk, all sorts of spices, seeds, and extracts and decoctions of fruits for flavour and colour.

The most interesting result is that over two hundred years ago the vintners were well skilled "in the Transmutation or Sophistication of

* It seems to have been published separately. See *Notes*, Part VI., p. 24, among the books brought out by Conyers.

Wines, which they call Trickings or Compassings," of which Dr. Charleton gives "a taste of the more disingenuous practices," which is about as disagreeable in the description as it must have been in reality.

79. John Smith, the author of the following work, was an apprentice of Matthew Cradock, a London merchant who was interested in fisheries. While still an apprentice he was sent in 1633 to examine the trade and industries of Orkney and Shetland. He lived there for over a year, and drew up a report on these subjects and on the fishing trade there. It was afterwards published at London, 1661, in 4to.

This report was incorporated in the present book. Donaldson, in his notice * of Smith, does not give "England's Improvement" much praise; he says the matter is miscellaneous and the subject very passingly treated. Sheep and cattle have a short notice, but the author's attention is chiefly occupied with forestry, which was an absorbing subject at that time.

It is more favourably noticed by E. I. Carlyle,† who adds that it is very practical and is not concerned with economic theory.

80. The book was first issued with the following title-page:

England's | IMPROVEMENT | REVIV'D: | Digested into | SIX BOOKS. | BY | Captain *JOHN SMITH.* |
In the *SAVOY.* | Printed by *Tho. Newcomb* for the *Author,* | *An. Dom.* 1670. |

4to. A four leaves, (a) two leaves, A to Z, Aa to Ll in fours, Ll 4 blank; or, pp. [12], 270, 2 blank.

It was re-issued in 1673. The title-page above was cut away and a new one inserted. The 1673 edition is not a reprint, but consists of the remainder copies with the new title-page. The work seems to have been privately printed, and in the 1673 edition the bookseller states how he had come to the rescue and put it abroad for the common good. There is a copy of this first issue, which seems to be rare, in the Grenville Library in the British Museum (G. 19287.).

---

* *Agricultural Biography,* London, 1854, p. 34, No. XLVI.

† *Dictionary of National Biography,* London, 1898, LIII., p. 75.

81. The 1673 title-page is elaborated in the following fashion :

ENGLAND'S | Improvement | REVIV'D : |
In a Treatife of all manner of | HUSBANDRY & TRADE | By Land and Sea. |
Plainly difcovering the feveral ways of Improveing all forts of Wafte and Barren Grounds, and Enriching all Earths; with the Natural Quality of all Lands, and the feveral Seeds and Plants which moft naturally thrive therein.
Together with the manner of Planting all forts of Timber-trees, and Under-woods, with two feveral Chains to Plant Seeds or Sets by; with feveral Directions to make Walks, Groves, Orchards, Gardens, Planting of Hops and good Fences; with the Vertue of Trees, Plants, and Herbs, and their Phyfical Ufe; With an Alphabet of all Herbs growing in the Kitchin, and Phyfick-gardens; and Phyfical Directions.
ALSO
The way of Ordering Cattel, with feveral Obfervations about Sheep, and choice of Cows for the Dairy, all forts of Dear, Tame Conies, Variety of Fowles, Bees, Silk-worms, Pigeons, Fish-ponds, Decoys: with Directions to make an Aviary. And with accounts of Digging, Delving, and all Charges and Profits arifing in all fore-mentioned: and a particular view of every part of the pleafant Land: With many other Remarks never before extant.
Experienced in thirty years Practife, and digefted into fix Books,
By *JOHN SMITH*, Gent.
Published for the Common good.
London, Printed by *Tho. Newcomb*, for *Benjamin Southwood*, at the *Star* next to *Sergeants-Inn* in *Chancery-lane*; and *Ifrael Harrifon* near *Lincolns-Inn*. 1673.

4to. 1 leaf, A4, a2, A to Z, Aa to Ll, in fours (Ll4 is blank); or, pp. [14] 270 [2 blank].

This title-page is ample enough in its detail to show the extent of the subjects embraced, and exhibits what earnest thought was given in the seventeenth century to the improvement of England, and to the development of agriculture, of trade, and of thrift in every direction. The present work belongs, perhaps, in some measure to Political Economy, but there is a strong practical side to it also, and the discussions upon forestry, gardening, poultry, bees, silkworms, game, rabbits, fish ponds, decoy ducks, not to speak of the practical and medicinal receipts connected with the flowers and herbs, bring it into immediate connection with the work just following, *The Good Housewife*, and with those of Hill, Mascall, Gabriel Plattes, Markham, and others already included in these notes. Apart from that, however, it is worth while considering the view which was taken of all these topics at that time, for what was good then cannot be otherwise now, even though things have changed considerably in two and a half centuries, and not always for the better.

The 1673 edition is quoted by W. C. Hazlitt,* who adds in a note:

> This book was ready about three years prior to its publication, but was delayed by the author's poverty. Dedicated to Lord Brouncker.

Apparently he was not then aware that the book had been already printed.

Subsequently,† however, he gave a complete account of the 1670 edition of the work like that above.

82. The next little volume gives us an insight into English home life in the seventeenth century, by the directions therein imparted by F. B.—whoever that was—to a Good Housewife in her recurring daily avocations. There are none of Mrs. Walley's accomplishments, for this book was intended for a much more active and practical person than came within her horizon. The work of the Good Housewife lay with the cows, pigs, poultry, including hens, ducks, geese, peacocks, turkeys and Indian hens, the Dovehouse, the Garden, both for pot-herbs and flowers, spinning, and, as special pursuits, the tending of silk-worms and bees. This, therefore, is a brief abstract of topics which had been discussed at much greater length by the writers referred to so often.

THE | OFFICE | OF THE | Good Houſe-wife. | With | Necessary Directions for the | Ordering of her *Family* and *Dairy*; | and the keeping of all ſuch Cattle, | as to her particular Charge the over-ſight | belongs. | *ALSO,* | The manner of Keeping and Govern- | ing of *SILK-WORMES,* and | *HONEY-BEES*; both very delight- | ſome and Profitable. | By *F. B.* |

LONDON, | Printed by *T. Ratcliffe,* and *N. Thompſon,* | for *Richard Mills* at the *Pestle* and *Morter* | without *Temple Bar,* Anno Dom. 1672. |

Small 8vo. A to I in eights, K four leaves; or, pp. [6] 143 [1]. Title enclosed in a double line.

This book is quoted by Hazlitt.‡

---

* *Collections and Notes,* London, 1876, pp. 390-391.

† *Second Series of Bibliographical Collections and Notes,* London, 1882, p. 564.

‡ *Bibliographical Collections and Notes,* London, 1903, p. 14.

83. There is another book which may be compared with Shirley's *Servant-Maid's Delightful Companion*,* the *Office of the Good House-Wife*, and others.

Copies of the fifth and eighth editions have come into my hands, and there are two or three others in the British Museum, so that what follows is the fullest survey of the book that can be supplied at the moment.

The Compleat | SERVANT-MAID ; | OR, THE | Young Maidens Tutor. | Directing them how they may fit, and | qualifie themſelves for any of theſe Employ- | ments. | Viz. |

Waiting-Woman,	*Nurſery-Maid,*
Houſe-keeper,	*Dairy-Maid,*
Chamber-Maid,	*Laundry-Maid,*
Cook-Maid,	*Houſe-Maid,*
Vnder Cook-Maid,	*Scullery-Maid.*

Compoſed for the great benefit and advan- | tage of all young Maidens.
LONDON,
Printed for *T. Paſſinger*, at the Three | Bibles on *London Bridge*, 1677.

12mo. A six leaves, B to H in twelves ; or, pp. [11, 1 blank] 167 [3, advertisements of books sold by Thomas Passinger]. Two small engravings of the alphabet and certain lines to serve as a copy for learning to write by. The title is enclosed in a couple of lines. A1—Licenſed. R. L'Eſtrange. January 20, 167$\frac{6}{7}$.

This is the first edition, and it was reprinted practically as it stands in the third, fifth, and sixth editions. To these, however, was added the supplement of medical receipts. This first edition is superior to all the others ; it is on better paper, it is better printed, and it is far more liberally spaced. This copy is in the British Museum (1037. d. 10.).

Among the books sold by Passinger are enumerated Cotgrave's *Dictionary* ; Crollius' *Chymistry*, folio, 10s. ; Reynolds' *God's Revenge against Murther*, folio, 10s. ; *Chymical essays*, by John Beguminus (*sic* !), price 1s. 8*d.* ; *The Pathway to Health*, 18*d.* ; *The Book of Knowledge*, 10*d.* ; *The Book of Palmistry*, 18*d.* ; *The History of Friar Bacon and Friar Bungy ; The Jovial Garland ; Scarborough Span*, by Doctor Wittie ; and others, all very interesting.

* *Transactions of the Glasgow Archæological Society*, N.S., 1890, I., p. 321 (Reprint, 1888, Part IV., p. 23).

84. The second edition appeared probably about 1679. The book was a new one, appealed to a wide *clientèle*, and, as it was probably eagerly bought up, the second edition came quickly. I have not had the fortune to meet with a copy, though doubtless it will appear some day, perhaps to-morrow, perhaps five and twenty years hence.

85. While waiting for it, the editions which I have seen may be put on record.

B.M. 1036. b. 14.

The Compleat | SERVANT-MAID ; | OR, THE | Young Maidens Tutor. | Directing them how they may fit, and | qualifie themſelves for any of theſe | Employments.

Viz.

Waiting-Woman,	*Nurſery-Maid,*
Houſe-keeper,	*Dairy-Maid,*
Chamber-Maid,	*Laundry-Maid,*
Cook-Maid,	*Houſe-Maid,*
Under Cook-Maid,	*Scullery-Maid,*

Whereunto is added a Supplement con- | taining the choiceſt Receipts, and rareſt | Secrets in Phyſick and Chirurgery.

Compoſed for the great benefit and ad- | vantage of all young Maidens.

The Third Edition Corrected and amended. |

LONDON,

Printed for *T. Paſſinger*, at The Three Bibles | on *London-bridge*, 1683. |

12mo. A to G in twelves, H six leaves; or, pp. [1-8] 9-144 [145, title-page of the Supplement, dated 1682], 146-178 [179-180, book advertisements]. The title is enclosed in two lines. Two engraved plates, copies for learning to write by. A1, the frontispiece, contains in the centre another transformation of the portrait of Hannah Wolley or Sarah Gilly. At the top the maid is shown in the still-room, and in the nursery, and below in the kitchen, and in the dairy. But the execution is as wretched as may be. The Supplement is reprinted word for word in the fifth edition.

The title of the Supplement is as follows:

A | SUPPLEMENT | TO THE | Compleat | SERVANT-MAID, | CONTAINING | The Choiceſt Receipts and rareſt Se- |crets in Phyſick and Chirurgery. |

Fitted to the Capacity of the meaneſt Houſe- | keeper, and if put into Practice may pre- | vent the frequent Expences, and Charges | which moſt Families are at upon Docto- | ry and Chyrurgery. |

LONDON,

Printed for Thomas Paſſinger *at the three Bi-* | *bles on* London-bridge, 1682.

The diseases which are included—some of them at least—are of such importance that they could hardly be entrusted to the care of servant-maids or of young maidens at all. They would require all the skill that could be brought to bear upon them, even at the present time, if any good result was to follow.

The two receipts for the plague, one by Henry VIII. and the other for Henry VII., taken from *The Dukes Desk newly broken up, etc.* (§ 72 above), are the first two in that collection, and are followed by that *For the Plague in the Guts* and *Mr.* Baker's *Medicine, for the Cough of the Lungs.* Others also throughout the supplement are taken from *The Dukes Desk,* but in some cases are much enlarged.

86. The fourth edition has hitherto escaped me. It was published probably in 1687. The fifth came in 1691. The title is the same as that of 1683, but Passinger had altered his sign.

The Compleat
SERVANT-MAID:
OR, THE
Young Maidens Tutor.
Directing them how they may fit,
and qualifie themselves for any of
these Employments,
viz. | . . .
The Fifth Edition Corrected and Amended.
LONDON,
Printed for *Tho. Paſſenger* at the Three
Bibles and *Star* on *London-Bridge* 1691.

Small 12mo. A to G in twelves, H six leaves; or, pp. [8] 9-178 [2 of book advertisements]. A1 is the frontispiece, in the upper compartment of which is an altered copy of that in the fourth edition of *The Accomplish'd Ladies Delight* while in the lower compartment, the servant is represented as Cook-maid and Dairy-maid. There is also a folding copper-plate of the alphabet, to serve as a copy when she desired to learn to write.

The Supplement of Receipts and Secrets in Physick and Chirurgery has a title-page included in the pagination of the book, as in the third edition.

There is a copy of this edition in the British Museum (8415. aa. 41.).

It wants the frontispiece and has several leaves mended. It has been used to some purpose.

As may be supposed from its design, this book is full of multifarious receipts, as well as of advice and hints to the different kinds of servants, "tips" as they might be called, at least in one sense of that term. No hint is given of the author.*

87. In the next edition the title underwent a change, the book having passed into new hands. Moreover, the interval between the two editions was longer than usual, which delay may have been due to the same cause. The title is from the copy in the British Museum, 1037. a. 42.

The Compleat | SERVANT-MAID : | OR, THE | Young Maidens Tutor. | Directing them how they may fit, and qua- | lifie themſelves for any of theſe Employ- | ments.

viz.

Waiting-Woman,	*Nurſery-Maid,*
Houſe-Keeper,	*Dairy-Maid,*
Chamber-Maid,	*Laundry-Maid,*
Cook-Maid,	*Houſe-Maid,*
Under-Cook-Maid,	*Scullery-Maid.*

Whereunto is added a Suppliment contain- | ing the choiceſt Receipts and rareſt Se- | crets in Phyſick and Chyrurgery ; Alſo | for Salting and Drying Engliſh Hams | equal to Weſtphalia.

The Compleat Market-Man and Market- | Woman, in Buying Fowl, Fiſh, Fleſh, *&c.* | and to know their Goodneſs or Badneſs | in every reſpect, to prevent being Cheat- | ed. Never before Printed.

The Sixth Edition with Additions.

LONDON.

Printed for *Eben Tracy*, at the *Three Bibles* | on *London-Bridge*. 1700.

12mo. A to H in twelves, I six leaves ; or, pp. [1-8] 9-144 [145, title to the Supplement] 146-174 ; 175-185, Instructions for salting and drying Neat's Tongues and other receipts ; 185-200, rules for buying fowls, fish, meat, &c., &c. [201-02] contain an advertisement of a wonderful balsam from "Chily," endowed with curative virtues of very varied character, to be had only from Eben Tracy at 1s. 6d. the ounce.

This edition contains one of the engraved copies for writing. The title is enclosed within two lines.

* Another work with a similar title is quoted by Westwood and Satchell, *Bibliotheca Piscatoria*, London, 1883, p. 192, as a reprint of *The Accomplish'd Ladies Delight*. (See *Supplement* VI., § 90.)

A1, the frontispiece, is worse than even its predecessors. There is an altered portrait of Sarah Gilly, and below are two very crude cuts of the cook in the kitchen and the dairymaid churning in the dairy. The two at the top, in the third edition, have been left out.

The book itself is a thoroughgoing chap-book in unlined skiver covers, the paper soft and spongy and now brown at the edges and spotted. Even as such, however, it throws light on life in England two centuries ago, in regard to social and family matters, that can be obtained from no other source.

88. The seventh edition may have been published in 1704 or 1705.

89. The latest edition I have seen, that of 1711, shows a slight falling off in the demand for the book. Whether this was the last edition of all or not, I am unable at present to say. Possibly there were one or two more before the book fell into disuse. The title-page is similar to that of 1700, and the book is again an out and out chap-book.

The Compleat | *Servant-Maid* ! | OR, THE | Young Maidens Tutor. |
Directing them how the (*sic*) may fit, and qualifie themfelves for any of thefe Employments. Viz.

Waiting-Woman,	Nurferry-Maid,
Houfe-keeper,	Dairy-Maid,
Chamber-Maid,	Landry-Maid,
Cook-Maid,	Houfe-Maid,
Under Cook-Maid,	Sculery-Maid.

Whereunto is added

A Supplement containing the choiceft Receipts and rareft Secrets in Phyfick and Chyrury (*sic*); Alfo for Salting and Drying Englifh Hams equal to Weftphalia.

The Compleat Market-Man and Market-woman, in Buying Foul, Fifh, Flefh, &c. and to know their Goodnefs or Badnefs in every refpect, to prevent being Cheated. Never before Printed.

Eigbth (*sic*) Editjon with Additions.

London, Printed for *E. Tracy*, at the Three Bibles on *London-Bridge*. 1711.

Small 12mo. A to G in twelves, H six leaves; or, pp. [2 blank, 4] 7-179 [1]. Title within a double line. There is no portrait.

90. The books on natural marvels were not nearly so numerous as the receipt books, but the wonders were not altogether neglected. The following are examples:

MIRACLES | OF | ART and NATURE : | OR, | A Brief Defcription of the | feveral varieties of BIRDS, | BEASTS, FISHES, PLANTS, | and FRUITS of other Countreys. | Together with feveral other | Remarkable Things in the World. | By *R. B.* Gent. |

LONDON, | Printed for *William Bowtel* at the | Sign of the *Golden Key* near *Miter-* | *Court* in *Fleet-Street,* 1678. |

Small 8vo. A to H7 in eights (H8, blank? wanting); or, pp. [6] 120 [2 wanting].

(British Museum 975. a. 13.)

This little volume contains short chapters describing different countries and their products, inhabitants, birds and beasts, etc. Some stories are correct enough, but others are taken from the old naturalists, or from more recent sources, like that one of the lady who had 365 children, 182 Johns, 182 Elizabeths, and a hermaphrodite not named.

This volume seems to be one of the earliest productions of the author, both from the neat printing and its general get up, as well as from what occurs in the remarks to the reader, where he states that the collection was made by him at vacant hours, and if it be well received he might make others of a like kind. That he did so, the following numbers demonstrate:

91. THE SURPRIZING
MIRACLES
OF
Nature and Art.
In TWO PARTS. Containing

I. The Miracles of Nature, or the *Strange Signs* and *Prodigious Afpects and Appearances* in the *Heavens,* the *Earth,* and the *Waters* for many Hundred years paft, with an account of the moft famous *Comets,* and other *Prodigies,* fince the Birth of our Bleffed Saviour, and the dreadful Effects of many of them: Alfo a particular Defcription of the five Blazing Stars feen in *England,* within Eighteen years laft paft, and abundance of other unaccountable *Accidents and* Productions of all kinds, till 1682.

II. The Miracles of Art, defcribing the moft *Magnificent Buildings,* and other *Curious In-*

ventions in all Ages, as *Solomons* Temple. The *Seven Wonders of the World,* and many more Excellent Structures and Rarities throughout the whole Earth.
Beautified with divers Sculptures of many Curiofities therein.
By *R. B.* Author of the Hift. of the Wars of *England*; Remarks of *London*; Wonderful Prodigies; Admirable Curiofities in *England*; and Extraordinary Adventures of feveral famous Men.
London, Printed for *Nath. Crouch,* at the Bell next *Kemp's* Coffee-Houfe in *Exchange-Alley,* over againft the *Royal Exchange* in *Cornhil.* 1683.

12mo. A to I in twelves, K 10; or, pp. [4] 222. 6 "sculptures."

92. One of the books advertised in the preceding volume happens to have appeared in time to be registered.

ADMIRABLE
Curiofities
Rarities and Wonders in
ENGLAND,
Scotland and Ireland.
Being an Account of many Remarkable Perfons and Places; and likewife of the Battles, Sieges, Earthquakes, Inundations, Thunders, Lightnings, Fires, Murders, and other confiderable Occurrences, and Accidents for feveral hundred Years paft. With the Natural and Artificial Rarities of every County, and many other obfervable Paffages.
As they are recorded by credible Hiftoans (*sic*) of former and latter Ages.
With Pictures of divers memorable Paffages.
By *R. B.*
The sixth Edition.
London, Printed for Nath. Crouch at the *Bell* in the *Poultry,* near *Cheafide* (*sic*), 1702.

12mo. A to H in twelves; or, pp. [2] 178 [12 of book advertisements]. Frontispiece not included in the foliation; woodcuts in the text. The frontispiece is in two parts: the upper part represents "K. Canutus commanding the Sea"; the lower "K. Henry 3. preaching to the Monks." This copy wants the first two leaves of the advertisements. The last twelve leaves of the text are in smaller type than the rest of the volume.

93. The initials R. B. in these books stand for Richard, or, perhaps, Robert Burton, but there is some discussion as to who he was. He was not the author of the *Anatomy of Melancholy*; dates preclude that. It has been suggested that the author was Richard Braithwaite, but the commonly accepted view is that he was no other than Nath. Crouch, the publisher, himself. Crouch was born in 1632, became a freeman of the Stationers' Company, and began publishing. Under the initials R. B., or the name R. Burton, he is said to have compiled and printed his popular manuals of history, adventure, natural wonders, and so on, to something like the number of fifty. Considering the time when they were written, and that they were mere chap-books, priced one shilling, they were really of no small merit. The author was, at any rate, a diligent writer, and his books were appreciated, for many of them went through several editions. They are not without interest even now, and a complete collection of them would be of commercial as well as of literary and historical value. Lists of them are given by Lowndes,* Watt,† Allibone,‡ Hazlitt,§ and by Axon,|| who has written his life; but his bibliography is still to do.

A considerable number are in the British Museum, and copies are to be met with now and then in booksellers' lists, but the prices asked for them are in excess of their intrinsic value, and would rather amaze Crouch if he could see them.

Burton's books, however, are scarce, and difficult to get in good condition, which may explain their high price. They must have been eagerly read and roughly used, and many copies of them must have been worn out and at last destroyed. That, too, may account for their scarcity.

One of his books is not without interest from recent events. It is entitled *The General History of Earthquakes*, London, printed for Nath.

---

* *The Bibliographer's Manual*, ed. Bohn, 1865, I., col. 328.

† *Bibliotheca Britannica*, Glasgow, 1819, Authors, I., 176*e*.

‡ *A Critical Dictionary of English Literature*, London, 1859, I., p. 307.

§ They are scattered through the different parts of Hazlitt's bibliographical collections.

|| *Dictionary of National Biography*, 1886, VIII., pp. 14-16.

Crouch at the Bell in the Poultrey near Cheapside, 1694, 12mo, pp. [4] 176 [12, containing advertisements of books printed for N. Crouch], with a folding plate of "The Dreadful Earthquake in Jamaica, June 7, 1692." Making allowance for the time, this is an interesting chronology of notable earthquakes. The account of that in Sicily in 1669 with the eruption of Ætna, and of that in Jamaica, read in details like those of Mont Pelée, of San Francisco in April 1906, and of Messina in December 1908.

94. The next volume inaugurates a set of books on a topic which has not come under notice before, except indirectly. It is interesting for several reasons. It shows what were considered "vermin" at the time, and how little opinion has changed on the subject during two hundred and forty years, in spite of the study of Natural History, of deeper insight into the balance of life and the error of disturbing it. It also shows that after all his efforts during that time man has not been able to get rid of them, for some, at least, and these the worst, are as familiar to him and as abundant and pertinacious as ever. It may be inferred, therefore, that, as they have survived, they are pre-eminently fitted to survive, and that they serve some important purpose, if it be no more than to vindicate their right to attack man in turn. It may be observed, too, that in later editions the list of so-called vermin is much extended, and animals are named which cannot, on any pretext whatever, be included in the category.

Most of the copies which I have seen are in the British Museum.

95. The earliest of them has the following title-page, which, with modifications, recurs in succeeding editions:

B.M. 450. a. 11.

THE | Vermin-killer | BEING | A very neceſſary *Family-Book*, con- | taining *Exact Rules* and *Directi-* | *ons* for the Artificial killing and | deſtroying of all manner of *Ver-* | *min*, &c. |

Rats and *Mice* *Moles* *Piſmires* *Flyes* *Fleas* & *Lice*	*Adders*	*Snakes* *Weaſles* *Catterpillars* *Buggs* *Froggs*, &c.

Whereunto is added the Art of taking | of all forts of *Fiſh* and *Foul*, with many o- | ther Obſervations never before extant. | The Second Edition. | By *W. W.* |

LONDON, | Printed for *Samuel Lee* at the *Feathers* near | the *Poſt-Office* in *Lombard-ſtreet,* 1680. |

12mo. A to C in twelves (C11 is blank, and C12, blank, is wanting); or, pp. [6] 57. Book Advertisements [5; 2 blank, 2 wanting].

The receipts are very brief, but the compiler has given his authorities, among whom are Paxamus, Cor. Agrippa, Didymus, Anatiolus, Alberto, Pliny, Aphrian, Mizaldus, Alexis, Bayrus, Cardanus, Tarentinus, Democritus, Beritus; the most frequently quoted being the first two. The later tracts have copied a good many of the receipts, but they cannot in any sense be called new or enlarged editions. I have not found out who W. W. was, or the date of the first edition. Nor do I know on what ground Cornelius Agrippa is included in this list, for in the long catalogue of his works there is nothing whatever on this subject. It is to be observed, too, as bearing upon the life of the time, that this is called "a very necessary Family-book." The people in 1680 seem to have been sorely plagued by vermin; but then London was a dirty place: quite as dirty as Edinburgh.

96. In this edition the title has been considerably modified:

B.M. 7640. d. 33 (1).

THE | VERMIN-KILLER: | Being a Compleat and neceſſary | FAMILY-BOOK. | Shewing a ready Way to Deſtroy | Adders, Badgers, Birds of all Sorts, Ear- | wigs, Caterpillars, Flies, Fiſh, Foxes, Frogs, | Gnats, Mice, Otters, Piſmires, Pole-cats, | Rabbets, Rats, Snakes, Scorpions, Snails, Spi- | ders, Toads, Waſps, Weaſles, Wants or | Moles, Worms in Houſes and Gardens, Bugs, | Lice Fleas, *&c.* | ALSO | Several excellent RECEIPTS for the Cure of | moſt DISORDERS. | AND | Some uſeful Directions for GARDENING and | HUSBANDRY; and likewiſe for Travellers in re- | gard to the Management of a HORSE on a Jour- | ney, &c. | With many curious SECRETS in ART and NATURE. |

LONDON: | Printed and Sold by W. OWEN, at *Homar's Head,* | *Temple-Bar.* Price 6d. |

Large 12mo. A to F in sixes; or, pp. 72. No date; about 1750?

It is notable that in this and the following examples "birds of all sorts" were reckoned as "vermin," and, therefore, to be destroyed! So, too, were frogs, spiders, wasps, without the slightest consideration of the mischief and annoyance that might ensue from their absence.

97. The title of the next edition I have seen, which is in the British Museum (787. b. 55.), is, with one or two slight rearrangements, identical with that just quoted. The imprint, however, is different:

LONDON. | Printed for W. Bristow, in St. *Paul's* Church-yard; | and Sold by C. Etherington. J. Todd. and | H. Sootheran. and W. Tesseyman. in *York*; | and by all other Bookſellers in *Great Britain* and *Ireland.* |

Large 12mo. A to F in sixes; or, pp. 72. No date; but within a year or two of Owens' edition.

These two editions are line for line and page for page reprints of each other. The first part about vermin, etc., ends at the top of p. 43. Then follow Directions about Bees to p. 46; Physical Receipts for some common disorders to p. 50. P. 51, The Traveller's Pocket Farrier; p. 61, Observations on Gardening; p. 65, Rules for the Months; p. 69. Curiosities in Art and Nature, such as a Receipt for Black Ink, for Red Ink, Secret Writing with juice of lemon and other liquids, and at the end a table showing the amount of certain materials in a definite quantity, as Ream of Paper, Dicker of Hides, a Hogshead of Wine, Barrel of Beer, a Last of Corn, a Truss of Hay, and so on.

98.

The Compleat | English and French | Vermin-Killer: | Being a Necessary | Family-Book. | Shewing | A Readie Way to Destroy Bugs, | Lice, Fleas, Moles, Adders, Badgers, | Birds of all sorts, Ducks, Earwigs, | Flies, Fish, Foxes, Frogs, Gnats, | Mice, Otters, Pismires, Pole-cats, | Rabbets, Rats, Snakes, Scorpions, | Snails, Spiders, Toads, Wants or | Moles, Wasps, Weasles, Worms in | Houses, Gardens, &c. | Also | Directions for Gardeners, | with many Modern Curiosities of | Art and Nature. | London: | Printed and Sold by G. Conyers. at the | Ring in Little-Britain. Price 1.s.

12mo. Pp. 72. No date; probably early in the eighteenth century. At this time ducks apparently were not reckoned among birds.

99. B.M. 11621. b. 29 (22).

THE | VERMIN KILLER; | BEING | A COMPLETE AND NECESSARY | FAMILY BOOK: | *Shewing a Ready Way to Destroy* | Adders, Badgers, Bugs, Birds of all sorts, Caterpillars, | Earwigs, Fish, Fleas, Flies, Foxes, Frogs, Gnats, Lice, | Mice, Moths, Moles, Otters, Pismires, Polecats, Rats, | Scorpions, Snails, Snakes, Spiders, Toads, Wasps, | Weasels, Worms in Gardens, &c. | [Vignette].

Glasgow: | Printed for the Booksellers. |

Small 8vo. No signatures, pp. 24. Ornamental border round the title. The Vignette is a man driving a donkey with panier baskets, a dog follows, and there is a wooded bank and a cottage in the distance. It seems to be a Bewick woodcut. There is no date, but it was printed about 1820-40.

This is a contracted reprint of the first part of *The Vermin Killer.* A great deal has been omitted; what has been retained is nearly in the same words as in the earlier editions. One would like to know, however, when scorpions infested this country.

100. George Hartman, who was Digby's operator, and who edited some of his books, published also one or two on his own account. The following volume, bearing as it does the name of Secrets, belongs apparently to our subject, but in reality the sub-title "Chymical Secrets" denotes the contents more accurately, and the term "Philosophy," here employed, signifies really the deep and recondite study of Nature. The book, therefore, is a treatise on Chemistry, and, as such, should not be included, but on account of its title and partly of its contents, and because other works of its author have been referred to, it too may be described. The title is as follows:

A CHOICE | COLLECTION | OF RARE | SECRETS | AND | EXPERIMENTS | IN | Philosophy. | AS ALSO | Rare and unheard-of *Medicines,* | *Menſtruums,* and *Alkaheſts*; with the | True Secret of Volatilizing the fixt | Salt of TARTAR, | *COLLECTED* | And Experimented by the Honourable and truly | Learned Sir *Kenelm Digby,* Kt. Chancellour | to her Majeſty the Queen-Mother. | Hitherto kept Secret ſince his Deceaſe, but now | Publiſhed for the good and benefit of the | Publick, | By *GEORGE HARTMAN.*

London, Printed for the Author, and are to be Sold by | *William Cooper,* at the *Pelican* in *Little Britain*; and | *Henry Faithorne* and *John Kerſey,* at the *Roſe* in *St.* | *Paul's* Church-yard. 1682. |

Small 8vo. A to S in eights; or, pp. [16] 272. Four engraved plates of chemical apparatus. This book is dedicated to Robert, Baron of Paston, Viscount and Earl of Yarmouth.

101. In the British Museum (1038. d. 25.) is another book by Hartman, which belongs more strictly to the category of books of receipts.

THE TRUE | PRESERVER and RESTORER | OF | HEALTH: | BEING A | CHOICE COLLECTION | OF | Select and Experienced REMEDIES for | all Diſtempers incident to Men, Women and Children. | Selected from, and Experienced by the moſt Famous | Phyſicians and Chyrurgeons of *Europe.* |

TOGETHER WITH | Excellent DIRECTIONS for | COOKERY; | AS ALSO FOR | *PRESERVING*, and *CONSERVING*, | and making all Sorts of *METHEG-* | *LIN*, *SIDER*, *CHERRY-WINE*, &c. | WITH THE | Deſcription of an Ingenious and Uſeful *ENGIN* | for Dreſſing of Meat and for Diſtilling the Choiceſt | Cordial Waters, without Wood, Coals, Candle or Oyl.

Publiſhed for the Publick Good, by G. Hartman, *Chymiſt.* |

London, Printed by *T. B.* for the Author, and are to be ſold | at his Houſe in *Hewes-Court* in *Black-Friers.* 1682. |

Small 8vo. Title-leaf; Epistle dedicatory to Anna, Countess of Sunderland; Signed George Hartman, A2-3. To the Reader, signed G. Hartman, A2 (for A4); Index (A5-8); Engraved folding plate of the furnace and distilling apparatus, without mark; description of the apparatus, 2 leaves without mark; Text of the first or Medicinal Part, B to Z, in eights; Text of the second part, Aa to Ee, in eights; A, B, in eights.

Pp. [15, 1 blank]. [Folding engraved plate, 2 leaves; description, pp. 4]; Text of first part, pp. 351 [1].

The second part has a title-page [1-2]; Receipts for Cookery, etc., 3-80; Receipts for making of Metheglin, etc., 1-32.

In the British Museum copy leaves Aa2 to Aa5 in the second part have been cut out and Aa6 has been pasted to the title-leaf, so that pp. 2-11 are wanting.

The title of the second part is as follows:

EXCELLENT | DIRECTIONS | FOR | COOKERY; | TOGETHER WITH | The Deſcription of an Uſeful *ENGIN* | ſerving for the ſame: and likewiſe | for Diſtilling the Choiceſt and Beſt | *Cordial Waters.* | AS ALSO | SELECT RECEIPTS FOR | *PRESERVING*, *CONSERVING*, | and *CANDYING*, &c. | WITH A | COLLECTION | of the Choiceſt Receipts for making of | *METHEGLIN*, *SIDER*, | *CHERRY-WINE*, &c. | The SECOND PART. |

LONDON, | Printed by *T. B.* for *G. Hartman* Chymiſt. 1682. |

102. Another edition was soon called for. There is a copy in the British Museum (7383. aa. 34.):

> THE TRUE | PRESERVER and RESTORER | OF | HEALTH : | BEING A | CHOICE COLLECTION | OF | Select and Experienced REMEDIES for | all Diſtempers incident to Men, Women and Children. | Selected from, and Experienced by the moſt Famous | Phyſicians and Chyrurgions of *Europe.* |
>
> TOGETHER WITH | Excellent DIRECTIONS for | COOKERY ; | AS ALSO FOR | *PRESERVING*, and *CONSERVING*, | and making all ſorts of *METHEG-* | *LIN, SIDER, CHERRY-VVINE*, &c. | WITH THE | Deſcription of an Ingenious and Uſeful *ENGIN* | for Dreſſing of Meat, and for Diſtilling the Choiceſt | Cordial Waters without Wood, Coals, Candle or Oyl. |
>
> The Second Edition with Additions. |
>
> *Publiſhed for the Publick Good by* G. Hartman, *Chymiſt.* |
>
> *London*, Printed by *T.* B. and are to be ſold by *Randol Taylor* | near *Stationers* Hall. 1684.
>
> Small 8vo. A to Z in eights ; Aa to Ee in eights ; A, B, in eights ; or, pp. [15, 1 blank] 352, 80, 32. In the first part a folding engraved plate of a furnace and distilling apparatus separate. In the second part, Aa4 *recto* (p. 7), a woodcut of the *Engine* for Cookery and Distilling.

The title-page of the second part is a copy of that in the previous edition, and is even dated 1682, but it is not identical with it, for there are various typographical differences to be noted ; in particular the use of *VV* for *W*.

This edition of 1684-82 is quite distinct from the previous one of 1682, though it is almost a literal reprint of it.

103. Hartman was the author of still another book of Medical Receipts :

> THE | FAMILY PHYSITIAN, | OR A | COLLECTION | OF | Choice, Approv'd and Experienc'd Remedies, | for the Cure of Almoſt all Diſeaſes incident | to Humain Bodies, whether Internal or Ex- | ternal, uſeful in Families, and very ſervice- | able to Country People. | The First Volume. | Containing ſome hundreds of conſiderable Receipts | and Secrets of great Vallue, with Obſervations of | great Cures. Together with the true Engliſh | WINE-CELLER, or the right method of | making *Engliſh-Wines*, or *Metheglin* : With a Col- | lection of the Choiceſt and Safeſt *Coſmetick* Reme- | dies for preſerving the Beauty and Complection | of Ladies, never before Publiſh'd. |

By GEO. HARTMAN, Phylo Chymift, who | liv'd and Travell'd with the Honourable and Re- | noun'd Sir KENELM DIGBY in feveral | parts of Europe, the fpace of Seven Years till he | died. |

LONDON, | Printed by *H. Hills* for the Author, at the lower-end of *Cher-* | *ry-Garden-ftreet* near the *Jamaica-houfe* in *Rotherhith*, | 1696. |

8vo. A to Z, Aa to Ll in eights; or, pp. [16] 528. Frontispiece of a still and other appliances. Another plate wanting.

There was apparently no more printed than this first volume. The above title-page was subsequently cancelled and another, printed on stiffer paper of a slightly different colour, was prefixed. The statement on p. 524, "End of the first volume, . . . " remains, of course, unchanged.

104. The British Museum copy (1039. b. 19.) has this cancel title-page as follows:

THE | FAMILY PHYSITIAN, | OR A | COLLECTION | OF | Choice, Approv'd and Experienc'd Remedies, | for the Cure of Almoft all Difeafes incident | to Humane Bodies, whether Internal or Ex- | ternal; ufeful in Families, and very fervice- | able to Country People. | CONTAINING | Some hundreds of confiderable Receipts and Secrets | of great Vallue, with Obfervations of great Cures. | Together with the true Englifh WINE- | CELLER, and the right Method of making | *English-Wines*, or *Metheglin*: With a Collection of | the Choiceft and Safeft *Cofmetick* Remedies for pre- | ferving the Beauty and Complection of Ladies, ne- | ver before Publifh'd. | By GEO. HARTMAN, Phylo Chymift, Author | of the *Preferver and Reftorer of Health*, who liv'd and | Travell'd with the Honourable Sir KENELM | DIGBY in feveral parts of Europe, the fpace of | Seven Years till he died. |

LONDON, | Printed for *Richard Wellington*, at the *Lute* in St. *Pauls-* | *Church-Yard*. 1696. | Where is likewife fold variety of Phyfical Medicines, incerted | at the end of this Book. |

In the British Museum copy, before the title, are the two copper plates of chemical apparatus: a furnace, with its different parts, and a still with its different parts, both rather nicely executed.*

105. This book is dedicated to William, Baron of Paston, Viscount and Earl of Yarmouth, and Hartman refers to the dedication of the

* Only this edition appears in *The Term Catalogues*, ed. Arber, 1905, III., p. 602, and in the British Museum; not the preceding one with Hill's imprint.

previous book (§ 100) to his father as a reason for dedicating this one to him. According to Hartman, the elder man was a friend of the famous physician, Sir Theodore Mayern, from whom he had got manuscripts and secrets. It was the fashion of the day, for every one, from the King downwards, seems to have been dabbling in chemistry, in pharmacy and medicine, in practical arts, in making collections of receipts and processes, and even in specimen-hunting to stock a museum of natural wonders, as we have seen Mr. Forges did (§ 63 *et sqq.*).

106. In the first issue of *The Family Physician*, the imprint is not very clear, and leaves one to doubt whether or not the book was actually printed at Rotherhithe. Had it been so, it would have been most interesting as an example of a press existing there at that time. Examination, however, seems to show that it was the author who lived in Rotherhithe and not the printer. Henry Hils, or Hills, had his office in Southwark at the corner of the George Inn. From St. Margaret Hill to St. Thomas's Street was a printing and publishing centre, and Hills in 1648 lived in St. Thomas's, over against the Hospital. His press and books were destroyed by a mob on December 12, 1688, but he must have recovered from that damage, for the present work was printed eight years later.* Could he possibly have gone to Rotherhithe after the sacking of his house?

Cherry-Garden Street at the present day stretches from the Stairs on the Thames across Jamaica Road. On the west side and parallel with it, as far as Jamaica Road, is Marigold Street, and on the other side, running in an easterly direction, is Paradise Street. These names are remainders and reminders of the time when the banks of the Thames were decked with gardens and orchards, and when it may have been approximately

> "The clear Thames bordered by its gardens green."

Though the fruit trees and flowers have been replaced long since by dingy warehouses and small dwellings, one can possibly see in the narrow

* Wm. Rendle and Philip Norman, *The Inns of Old Southwark and their Associations*, London, 1888, p. 161.

lanes lying between Marigold and Cherry-Garden Streets, the footpaths which crossed the gardens, or, it may be, were boundaries between them.

After it crosses Jamaica Road, Cherry-Garden Street runs down towards Tranton Street, but before reaching it turns off at a right angle to the left and ends in Southwark Park Road. The way, however, is continued to Tranton Street by a short paved foot-passage called New Place. On the right of this passage, going to Tranton Street, and a few steps from it, is a small modern three-storey building with the title "Jamaica House" on the lintel over the door. If it does not actually occupy the site of the old house, it stands presumably on part of the garden, sufficiently near the site to entitle it to the name it bears.

The Cherry Garden was a place of resort in Pepys's time, and he mentions visits he paid there on June 13th and 15th, 1664, by water, and a purchase of cherries which he made.*

> June 13, 1664.—Thence [from the Tower] having a galley down to Greenwich, and there saw the King's works, which are great, a-doing there, and so to the Cherry Garden, and so carried some cherries home.†
>
> June 15, 1664.—And so to the Cherry Garden, and then by water singing finely to the Bridge and there landed.

Jamaica House with a tea garden was also a place of public entertainment at this time, situate at the foot of Cherry-Garden Street. It was still standing so recently as 1854, but in a dilapidated condition. In a picture it is represented as a two-storey building with an ornamental gallery, or balcony, on a level with the first storey, supported on six pillars, access to it being gained by a flight of steps at one end.‡

---

* *The Diary of Samuel Pepys*, ed. Henry B. Wheatley, London, 1894, IV., pp. 159, 161, and notes.

† "The Cherry Garden was a place of public entertainment at Rotherhithe. The site is marked by Cherry Garden Stairs, a landing-pier for Thames steamers and small boats."—Wheatley's note.

‡ The picture is contained in Rendle and Norman's book, quoted above, p. 400, There is no reference to the place under Tea Gardens, in Wm. B. Boulton's book, *The Amusements of Old London*, 1901, 2 vols. 8vo.

This place also is mentioned by Pepys. On April 14th, 1667, he says:

> Took out my wife and the two Mercers and two of our maids, Barker and Jane, and over the water to the Jamaica House, where I never was before, and there the girls did run for wagers over the bowling-green: and then with much pleasure spent little, and so home.*

There is a report to the effect that Cromwell lived in it. At what particular place Hartman resided, it is probably impossible now to tell.

107. A book on medicine for home use was drawn up by a certain John Yarwood, and it bears the following queer title, in keeping with the eccentric character of the middle of the seventeenth century, when very many singular beliefs were current, even as there are now. The little volume has one element at least of interest: it is a London-Bridge book, and was sold there by Thomas Passinger, whose name appears on more books than one in the present paper. As cannot be overlooked, the author excels in the variety and use of adjectives and adverbs. It is a profitable exercise in English to read his writings and observe the effect of omitting his decorations.

The title follows:

תרופה צרוף:

Phyſick Refin'd: | OR, A | LITTLE STREAM | OF | Medicinal Marrow, | Flowing from the | BONES of NATURE. | WHEREIN | Several Signs, particular Rules, | and diſtinct Symptoms; where- | by the moſt ordinary Diſeaſes may be | diſtinctly known, and truly judged, are | perſpicuouſly Delineated: | AND | The moſt proper Way, ſafe Me- | thod, and ſimpathetical Cure, | whereby Nature may be helped, the Sick | eaſed, and Languiſhing Patients relieved | (without the Uſe of Poyſonous Purging | Potions, and Venomous Medicaments) is | ſuccinctly Demonſtrated: By | *John Yarwood*, Profeſſor of Phyſick. |

LONDON, Printed for *Tho. Paſſinger*, at the | *Three Bibles* on *London-Bridge*, 1683. |

12mo. A to E in twelves; or, pp. [1-6] 7-120.

In the British Museum Catalogue (549. a. 34.) the date given is 1682, instead of 1683 which is contained in the copy of the book there, as I ascertained by inspecting it. The entry is a misprint, but a misleading one.

---

* *The Diary of Samuel Pepys*, 1895, VI., p. 272, and note. Wheatley's notes are repeated in his *Pepysiana*, London, 1899, pp. 218, 219.

The book is dedicated to Mrs. Mary Allsop in the author's sumptuous manner, but from it we gain more proof that, at that date, there were few physicians in country places, and that it was considered almost a duty on the part of ladies and gentlewomen to attend to the wants of the sick in their neighbourhood, or, if it was not a duty, many of them made it one.

108. The author deals with forty-five major diseases, such as scurvy, dropsy, gout, headache, consumption, epilepsy, deafness, blindness, apoplexy, palsy, melancholy, phrensy, madness, quinsy, pleurisy, spotted fever, plague, and such like. The arrangement of the book is simple: the disease is described, and the cure is then given. No definite order is followed, and there is no index.

The treatment in some cases is singular, but the paragraphs afford us a fresh insight into the pharmacy and materia medica of England about two and a half centuries ago.

Such substances were employed as the tinctures of tartar, of antimony and of corals, crocus martis and oleum martis, spirit of salt, oil of sulphur, ens Veneris, Bezoar mineral, alkermes, salt of lead, oil of vitriol with spirit of wine, laudanum, native cinnabar, spirit of hartshorn, potable silver, white vitriol, crocus metallorum, tutia, oil of amber, vinegar, sal prunella, spirit of nitre, potable gold, alum, tincture of emeralds (the mode of preparation is not given), ash of hartshorn, spirit of sal ammoniac, bay salt. These are for the most part the mineral drugs, but great use is made besides of plants prepared in various ways. There were prescribed also complex mixtures, like *elixir proprietatis*, and others. After perusal of his catalogue of drugs, one asks, with a feeling of awe, what the author considered a "poisonous purging potion" or "venomous medicament."

Yarwood's tractate is instructive as a popular handbook of medical receipts.

109. The next paper will deal with books from this date onwards to the eighteenth century.

ADDENDUM.

SUPPLEMENT VI. TRANSACTIONS, Vol. VI. p. 33.

In the transcript of the 1684 title of *The Queens Closet Opened* two blanks were left on account of defects in the copy examined. Since then I have seen another copy of the book and can supply the defects. The first word is "Fixed," and the second is "Esteem," both with capitals.

ERRATUM.

SUPPLEMENT V. TRANSACTIONS, VOL V.

P. 185 (Reprint, p. 61), § 60, *for* 1684 *read* 1654.

BOOKS OF SECRETS.

BOOKS OF SECRETS.

A PAPER READ BEFORE THE BIBLIOGRAPHICAL SOCIETY,
APRIL 21, 1913.

BY

PROFESSOR FERGUSON.

LONDON:
REPRINTED BY BLADES, EAST & BLADES, FROM
THE SOCIETY'S *TRANSACTIONS.*

1914.

BOOKS OF SECRETS.

By PROFESSOR FERGUSON.

Read April 21st, 1913.

IF this theme had not, in a manner, been prescribed to me by our Secretary, I myself should hardly have selected it for presentation to the Society. It is true that I have been rummaging among these books for a good many years past, from time to time have printed some results, and ought to be able to say something about them, but I am so far from having exhausted the material which has gathered, that I am hardly prepared as yet to pass a judgment on this branch—historical, I may remind you—of scientific, or rather technological, literature. I repeat technological, for the aim of the greater part of it is not to speculate, or discuss, or describe, but to give directions how to do something, how to produce something tangible, a practical result for human use or convenience. Hence "secrets," as they are called, have, from this point of view, no reference to religious, or philosophical, or masonic, or other mysteries, but simply denote "receipts," which are used even now for effecting certain purposes.

It is unnecessary to dwell on the name. We may, however, deduce from it, that, at one time or other, the artist or craftsman had special knowledge gained by practice and experience which was personal and which he, intentionally or unintentionally, kept to himself. This accumulated skill constituted the "secrets" of his art or craft, and whoever wished to acquire these "secrets" had to serve an apprenticeship to the master, and be initiated by him. That it was personal one can comprehend, for when the productions of different men, with the same material and with the same processes for the same ends are compared, the difference in the results is sufficiently marked. The work of Peter Vischer, for example, or of Benvenuto Cellini, or of Bernard Palissy, stands out conspicuously among that of their contemporaries, to go no farther, through their "secret," or personal skill and knowledge.

Even after the processes had been published and the "secrets" divulged, it did not follow by any means that anyone, still less everyone, could carry out the directions accurately, not to speak of the results achieved by their authors. Something more is wanted than the written or spoken word, for as Opie showed long ago the brains and conscience of the artist are not negligible components of success.

By degrees, presumably, the "secrets" oozed out and were disseminated among the craftsmen and were finally collected and published. They retained their distinctive name, however, for though now displayed to common view, they had been "secrets" once and might bear that title still.

My acquaintance with these books began when I was engaged with a different subject, and I was disposed at first to resent their intrusion, but as their numbers increased and the use made of them in the past became more obvious by the variety of topics dealt with and the number of editions, they acquired a practical importance and a bibliographical interest, not to say significance, and I drifted into the examination of them, without anticipating in what it might ultimately involve me. That, however, is the usual course of research and one of its charms, and I have yielded to it. So starting with a score or two of treatises and editions, I have been led on to an examination and recording of books of which previously I had no knowledge, and which, so far as I know, have never been brought together and catalogued before. After all, the number of them which have passed through my hands is but a small one, a couple of thousand at the outside, and even that number would have been reduced, had I adhered to my original intention. I certainly began the survey, over thirty years ago, with an attempted restriction of the lists to books specifically labelled "secrets." But it was almost immediately apparent that the limitation was artificial and impossible to enforce, that consideration of the mere title must be discarded and the inquiry based on the broader foundation of the theme. So from the very first were included books of receipts in general, even though not designated "secrets."

From these, however, have been deliberately omitted books on gardening, cookery, occult science, and old chemistry.[1] These truly are all books of receipts and are replete with secrets, but they have been often dealt with and I have had no desire to compete with the lists already published, when there are extensive subjects for investigation to which no attention has been paid.

Again, while there are many books of receipts which are not styled "secrets," there are also books of "secrets" which contain no receipts. These have been included in virtue of their name rather than of their contents, it may be thought unduly. They profess to reveal some obscure or unknown matter which does not necessarily entail any practical application or result. The term is especially applied to natural phenomena, Natural History as it might be called now, but then, the secrets, wonders or marvels of Nature. The contemplation of these marvels was very like what it is with the ordinary man now. The daily routine of Nature passes without remark; no one, for instance, is conscious of what Father Beccaria called "the mild and slow electricity which prevails in the atmosphere during serene weather," but when anything out of the common happens, a tempest, a flood, a snowstorm, an earthquake, thunder and lightning,

(1) Nevertheless circumstances have again been too strong for me, and have compelled me to admit some books on these topics, because they were in the company of others which fell within my prescribed area, or were written by authors of whose works it was an object of mine to make as complete an enumeration as possible.

that is at once talked about. So in the earlier times the unusual events were noticed, but defective observation and partial ignorance of their causes coupled with superstitious fears, led to their being credited with sinister significance, and being assigned to malign agencies.

Books of Secrets, therefore, are of diverse sorts, and I have not hesitated about including all the varieties attainable. But just on that account the theme is so extensive and can be contemplated from so many points of view that it is impossible to include them all within a single paper. Attention, consequently, may be restricted to one or two aspects. One, for example, which concerns the Society more particularly, deals with the characteristics of the books themselves; another with their contents, but I shall not attempt to keep these separate. As books they are in all languages, they are of all dates, sizes, and qualities of paper, printing and binding, and they are in every kind of condition, good, bad and indifferent. There are the books which were in everybody's hands, and those which can never have been used at all. In such cases it is not always easy to see what was the merit which carried a book through many editions, when others, seemingly as good, if not better, were quite unsuccessful. That, however, is not now a topic for discussion.

Prior to the sixteenth century there is little in printed form that can be brought into the category of Books of Secrets. The encyclopædias which remain, notably those of Isidorus, of Bartholomæus de Glanvilla and Vincentius Bellovacensis, even though they glance at practical matters, are too comprehensive to be included under such a limited designation. But that collections of technical receipts were in use is shown by the work of Theophilus, which was known only in manuscript till it was printed in French in 1843, and afterwards in English in 1847. This work contains receipts for colours, glass, enamel and metal work. Other manuscripts on colours, painting, glass and other arts of the Middle Ages were published by Mrs. Merrifield in 1849, and the work on colours by Heraclius was edited in 1873 by Albert Ilg, who also began an edition of Theophilus the next year.

But the outstanding book of Secrets of the earlier time is that ascribed to Albertus Magnus and, curiously enough, it was the first book of the kind which came into my possession. That copy is a quaint little volume, printed at Lyons in 1566, and it contains the three tracts, *Secreta Mulierum*, *Liber Aggregationis*, and *De Mirabilibus Mundi*. The popularity of these tracts down to the eighteenth century is extraordinary. Hain enumerates forty-six editions before 1500, and there are some which have escaped his notice, such as those printed in London by Machlinia, and an Italian translation of the *Liber Aggregationis* at Milan in 1495. Many followed in the sixteenth and seventeenth centuries, so that as dealing with the secrets of generation, the properties of stones, plants and animals, and the marvels of the universe, no book had such a circulation in Europe. Whoever has given any attention to medieval literature must have encountered it at some time or other.

Intimately connected with it by subject, and often by publication, was the analogous work *De Physiognomia* of the senior contemporary of Albertus, Michael Scotus, round whom so much romance has been woven, but a consideration of that book would involve a monograph so as to do it and its author justice.

After 1500 books on the present subject made their appearance in greater abundance. Notable rarities are among them and many printed in the earlier part of the sixteenth century are not devoid of bibliographical attractions. Among those in English are many on medicine, such as the editions of "The Treasure of pore men," "The Myrrour or Glasse of helth," "The treasury of healthe," written by Petrus Hispanus, afterwards Pope John XXI, and translated by Humfre Lloyde. There were the works edited or translated by John Hester, "The Joyfull Jewell," "Secrets of Chirurgery," "The Excellence of Physick and Chirurgery," "Three curious Pieces of Secrets," "The Order to distill oyles," "The Key of Philosophie"; Baker's "The Newe Jewell of Health"; Vicary's "Englishman's Treasure"; "An Hospital for the Diseased"; "A Rich Store-house or Treasury for the Diseased";

"Præpositas his Practice" by Leonard Mascall, and others. There was no lack of guides to health, if it could be attained or preserved by secrets and receipts.

Mascall wrote other books on practical affairs such as the planting and grafting of trees, "The Booke of Cattell" concerning the management of live stock, horses, oxen, sheep, goats, hogs, "The Governmente of Poultrie," probably the earliest treatise on the subject in English, "A profitable Boke to take out spottes and staines," which is one of a set of books to which reference is made below. To these may be added the works on gardening by Thomas Hill, his "Parfite orderinge of Bees," his "Physiognomy," and other works. Another little tract similar to some of these is "The Booke of Thrift, containing a perfite order, . . . to profite lands, and other things belonging to Husbandry," printed at London by John Wolfe, in 1589.

In the seventeenth century the output of these books was greater than ever. Occasionally they were respectable small quarto volumes, while those in small octavo were for the most part chap-books. But though cheaply got up they were sometimes decorated, if one may say so, with a woodcut portrait or title-page, or frontispiece. Among the quartos may be mentioned the late reprints of Hill's and Mascall's books on gardening, on cattle and arboriculture, and Gervase Markham's works on farming, and household economy. In this century too appeared Hill's "Legerdemain," which ran through so many editions that one might suppose the art of conjuring was a much cultivated profession. Works on the secrets of medicine were not wanting; Brugis' "Marrow of Physic," Levens' "Pathway to Health," Bonham's "Chyrurgian's Closet," the "Dispensatory" of the two quacks, Salvator Winter and Francisco Dickinson, and reprints of earlier treatises may serve as examples.

Of the little octavos and duodecimos which flowed from the press in a copious stream, mention can be made of only a few that are more or less typical. There were the works of John White, "lover of artificial conclusions," as he styled himself, "Arts Masterpiece," "Arts Treasury of Rarities and curious Inventions," which went through six editions at least, "A rich Cabinet with variety of Inventions," which was in vogue from 1651 to 1689, and "Hocus-Pocus," another book of tricks. There were: "A choice Manual of rare and select Secrets in Physick and Chyrurgery" by Elizabeth Talbot, Countess of Kent, which, between 1653 and 1708 went through twenty-one editions; the parallel collection of Queen Henrietta Maria, entitled "The Queen's Closet opened," with about sixteen editions between 1655 and 1713, and a certain T. P. who compiled "The Accomplished Lady's Delight in Preserving, Physick, Beautifying and Cookery" which ran from 1672 or 1673 to 1719, and had ten editions. Salmon's "Polygraphice," a collection of receipts for a number of Arts, was issued eight times between 1672 and 1701. Besides these, reference may be made to Mrs. Hannah Wolley, or Woolley, an entertaining person, and an adept in all feminine accomplishments, who was the authoress of "The Cooks Guide," "The Gentlewomans Companion," "The Ladies Delight, or a Rich Closet of Choice Experiments and Curiosities," "The Ladies Directory," "The Queen-like Closet," of which there were certainly five editions, all these books appearing between 1664 and 1684. Nor should Sir Kenelm Digby be forgotten, and his operator George Hartman, who between them published "Chymical Secrets," "The Closet opened," "Choice and experimented Receipts in Physick and Chyrurgery." There was also the revelation of a thorough-going secret in his famous "Discourse on the Powder of Sympathy," delivered at Montpellier and first printed in English in 1658, a book said to be of the greatest rarity. The second edition appeared in the same year, others came later and, with other books of Digby's, it was translated into Dutch, French and German. Hartman compiled "A Choice Collection of rare Secrets and Experiments," "The true Preserver and Restorer of Health," "The Family Physitian, . . . containing some hundreds of considerable Receipts and Secrets of great Value" which was printed by Henry Hills in 1696, and from which we gather that the author lived at Rotherhithe. There was Owen Wood's "An Alphabetical Book of Physicall Secrets," of which five editions appeared between 1632 and 1656, when it was issued under the name of the Duchess of Lenox. This book has caused me trouble, for Owen Wood has never got credit for his labour, such as it was.

A representative book of Secrets is that by Thomas Lupton, "A thousand notable things of sundry sorts," which was published in London in 1596 and was often reprinted, the last issue I know of being dated 1815! What was the secret, or receipt, to which this book owed its longevity, does not appear, but one would like to know it. It cannot be owing to its intrinsic merit, for the book is a rifacimento of the extravagancies of the old marvel-mongers.

Towards the close of the century a good many of the books of Secrets were published by G. Conyers at the Ring in Little Britain, but with the unpardonable omission of the date. Other publishers were John Starkey, Edward Brewster, N. Boddington, T. Salusbury, T. Sowle, Andrew Sowle, T. Passenger, W. Whitwood, N. Crouch, Gartrude Dawson, E. Tracy, Charles Tyus, J. Blare, and many others. Among the chap-books were "The way to save Wealth," attributed to Thomas Tryon, the author of several curious essays, who in some of his ideas was far ahead of his time, "The Complete Husbandman," "A New Book of Knowledge" full of curious information and actually with a date, 1697. There were also more receipt books for ladies, such as John Shirley's "Accomplished Ladies Rich Closet of Rarities," which enjoyed some popularity, "The Ladies Companion, or Modern Secrets and Curiosities, never before made Publick" concerned mainly with toilet secrets, "The Accomplished Female Instructor," dated 1704, and "Arts Compleat Master-Piece," containing receipts for all sorts of purposes, and sold by James Hodges at the Looking-glass on London Bridge.

On passing into the eighteenth century one encounters books different in style and to some extent different in theme. Collections of medical secrets almost disappear, but there are more on strictly practical subjects. The term "secrets" is dropped, though it may occur in the course of a title. But the titles now become of prodigious length and take the form of a table of contents. They may have been handy for the readers of a hundred and fifty or two hundred years ago, but they try the patience of the bibliographer of the present time, who has to copy them. As might be expected too the books are not attractive. There is a pretence at ornament which is not only unnecessary, but is ugly.

There was a second flood of manuals of legerdemain, with the name of H. Dean. At first the books were well enough, but as time went on the issues became so bad that some were illegible.

Directions for the ladies and the good wives were supplied in such works as "The whole Duty of a Woman," "The complete Housewife," which, more by merit than looks, managed to reach the fifteenth edition, "The Complete Family Piece," which, in a duodecimo of six hundred pages, with an overwhelming title-page for length and detail, gives a history of the active side of country life in England a hundred and seventy years ago, and "The Lady's Companion, or an infallible Guide to the Female Sex." "The Young Ladies School of Arts" dealt with the refinements rather than the necessities of daily life and is the converse of "The Farmers Wife or the complete Country Housewife." This last contains directions about poultry, about the preparing of pork, bacon and sausages, making wine, cyder, perry, mum, mead, and so on, directions for the dairy, etc., etc., and then the author blossoms into rhyme, like Mascall in his book on cattle:

Instructions full and plain we give
To teach the Farmer's wife
With satisfaction how to live
The happy country life.

There were other books on the fine arts: "The Handmaid to the Arts" is mostly about painting; "Valuable Secrets concerning Arts and Trades" is also largely about painting though other topics are included. There are at least half a dozen editions of this book. "The School of Wisdom" contains a survey of the arts in general. All these books have ponderous title-pages and are disagreeable examples of typography. There are more of the same kind. "The Fountain of Knowledge or Complete Family Guide," which begins with the Indian way of marking silk, linen or woollen, tells how to breed game-cocks and singing birds, to restore an apparently drowned person to life, gives rules for nursing, a cure for the small pox and a receipt to remove freckles. This pamphlet is by "Sarah Saunders, Mother of seventeen Children, and brought them through all Diseases incident to Children"; an excellent mother but weak in syntax. I have seen the first edition, besides the sixth, which is in the British Museum. Amongst other things it contains a description of the "expeditious or fountain pen," which is, therefore, an old invention.

There were also "The New Handmaid to the Arts," "The School of Arts," and "The Laboratory or School of Arts," this last adapted from the German.

In the first quarter or so of the last century books after the old fashion still appeared, as, for example, "The Painters and Varnishers Guide," 1804, from the French; the seventh edition in 1810 of "The Laboratory," just mentioned; "The Female Instructor," on manners, medicine, cookery and domestic economy; "The Family Receipt-book," undated, a quarto of six hundred pages in double columns, with an appalling title-page; "500 Useful and amusing Experiments in the Arts and Manufactures," by George G. Cary. Books on fireworks, on conjuring, on the toilet, are among the receipt books of the time.

All these and many more are of such a quality that interest in them falls to a minimum, whether as regards their contents or their execution.

When, leaving chronology, we examine the contents of the books, they appear to be about as varied as human wants and desires themselves, and to be ready to provide efficient practical guidance for most of the contingencies of every day life. A brief notice of a few of the subjects may be given.

The section of the literature which deals with the Secrets of Nature is a remarkable one, for in it we see the method of descriptive Natural History in its widest aspect, previous to the introduction of the more recent system of observation and experiment. To the ordinary man the world is full of marvels and secrets which he does not understand and of forces which he cannot always control. These phenomena in the course of time were observed, sometimes accurately, and interpreted in many ways, not always conclusively. The familiar authorities from classical times are, of course, Aristotle, Theophrastus, Dioscorides, Pliny, Seneca and others. For the medieval writers, Pliny was the chief guide, but all sources were drawn upon by those who compiled, for popular delectation, narratives of strange appearances and events. Such were the "Liber Aggregationis" of Albertus Magnus, already alluded to, and in the sixteenth century the "Prodigies" of Polydore Vergil and of Julius Obsequens, the "Histoires Prodigieuses" of Boaistuau, the "Occulta Naturæ Miracula" of Levinus Lemnius of Zierikzee, Johnson's "Cornucopiæ," and the "Secrets and Wonders of the World abstracted out of Pliny" in English in 1587.

At the same time Arthur Golding translated Solinus and Pomponius Mela, both of whose histories are storehouses of marvels. There were the "Contemplation of Mysteries" compiled by Thomas Hill and printed by Denham about 1571, and the natural history of John Maplet of Cambridge, called "A greene Forest," published in 1567. As early as 1563, Dr. William Fulke, the controversialist and divine, wrote a book about meteors. What is called the second edition appeared at London long after in 1634 in a little black letter volume entitled "A most pleasant Prospect into the Garden of Naturall Contemplation." It is divided into five sections and describes meteors, fiery meteors, ayry impressions, watry impressions, such as clouds, rain, snow, springs, lakes, rivers, the sea, and earthly meteors, such as metals and stones. Hill's book resembles Fulke's sufficiently to make it worth while comparing them.

A century later, about 1670, came the popular manuals with the initials R.B., or the name R. Burton, published by Nath. Crouch and indeed assigned to Crouch himself as the author.

The self-assigned task of all the writers was to record astonishing and unusual events, signs and wonders in the heavens above and in the earth and sea below, portents, the birth and prophesyings of monsters, and such like. The writers did not seek to verify their reports, or, if they ran counter to ordinary experience and observation, to criticize them, but set them down for what they were worth, to be accepted or rejected. Such collections, if we may judge by their number and variety, were popular and must have supplied what was required, strange and startling narratives, true or not did not matter, just as one accepts an impossible romance now for the passive excitement it affords.

One of the most singular of these collections is in French and it appeared in 1504, entitled "Le livre des Merveilles du Monde," the book of the wonders of the world. It went through some six or seven editions which search for many years has shown me are of rare occurrence. It is divided into two parts; the first is arranged by the countries of the old world, and the wonders of each are enumerated, the second deals with the wonders of beasts, birds, plants, stones, and man, and is arranged by subject. The compiler has withheld his name unfortunately, so he cannot get credit and commendation for his production, but he has been wise and considerate enough to mention his authorities for the marvels he has amassed, so that we can judge of their probability and authenticity and confirm them if so disposed. Solinus's "Polyhistor" is one of his main sources, but, on the whole, the most attractive stories are derived from the collection which Gervasius, called of Tilbury, made for the recreation and edification of the Emperor Otto IV, a book of which there is no proper edition. One can imagine how the Emperor, when he was off duty (the work is entitled "Otia Imperialia," "an Emperor's half holiday") would recline luxuriously and purr while Gervasius tickled the imperial ears with his amazing narratives. It is no exaggeration to call them so, for even the compiler is forced to say about some of them that they "are hard for a man like myself of uncultivated understanding to believe, for in truth without the honour of the deeply learned doctor Gervasius the marvels seem to me impossible to believe according to nature, for Aristotle relates nothing like them. But it may well be that those to whom these marvellous events happened, were cheated by some diabolical illusions or otherwise."

One feels disposed to agree with the compiler and to say of Gervasius what King James VI said of the witches of Tranent, that "they were all extream lyars," though the editor here rather spoils his Higher Criticism by quoting Aristotle and the devil. But "otherwise" is good; it is such a roomy alternative.

In this book the stories are told at some length, but in others they are reduced to the smallest dimensions and only the essential part is preserved. Such is the treatment of the subject in Thomas Johnson's "Cornucopiæ," a thin black letter pamphlet of 24 leaves, printed at London in 1596, and in the translation of Pierre de Changy's abstract of Pliny, printed at London in 1587. As it contains only thirty-six leaves in small 4to, the barest outline of Pliny's thirty-seven books in folio is the result, but the wonders have been culled with due attention to their exceptional character.

Of equal interest and curiosity is the collection of Pierre Boaistuau, which was put out in English as an original work by Edward Fenton in 1569. It too contains remarkable stories and it possesses an unusual feature in the form of woodcuts illustrating different narratives. These do ample justice to the text, and are highly creditable to the imagination of the artist. The author describes monstrous men and animals, but what he says would not be intelligible without the pictures. People were much exercised in those times about the origin of monsters, and small wonder considering the appearance of those depicted and described. There was always a doubt too about the share devils had in producing them, and whether in general devils could have families. Then there are accounts of natural wonders: great floods, burning mountains, including the eruption in which Pliny perished, with a woodcut of the catastrophe;

marvellous appearances in the heavens, comets, dragons, flames, blazing stars, torches, fire forks, pillours, lances, bucklers, two suns and two moons at one instant, and many more, "which whosoever would recompte by order, those which onely have appeared sithens the nativitie of Jesus Christ, together, searching the causes of their beginnings and births, the life of a man would not performe the same," and then the author gives details of some of these portents with their dates and their significations. Some of them must have been frightsome enough, if they at all corresponded to their description and picture. The secrets and wonders of Nature in the sixteenth century were calculated to inspire far more widespread interest and alarm than they do now, for by following the prescription of Albertus Magnus, we have lost the capacity for wonder.

Mizaldus, in the sixteenth century, was a successful collector of natural secrets. He read extensively and noted the wonderful things he met with about the world, about beasts, birds, fishes and reptiles, about men and women, children born and unborn, about the virtues of plants and minerals and the cure of diseases. His collections he arranged in nine centuries, with a rough sort of classification. His book is not common and the Cologne editions of 1572, 1573, 1574, are of rarity. A more curious and uncommon book by him is entitled "Secrets de la Lune," in which he points out its connection with and influence upon the sun, women, certain beasts, birds, fishes, stones, trees, plants, diseases, sick people, etc., etc.

Founded upon Mizaldus's book and taken in part from it, is Lupton's book already alluded to: "A Thousand notable things of Sundry Sorts."

Bodin and Cardan made contributions to this literature, and there may be quoted two much less known works than theirs.

One is "Idea del Giardino del Mondo" by Thomaso Thomai of Ravenna. It was published at Venice in 1593 and often afterwards. Notwithstanding its small bulk it is comprehensive; begins with the earth, deals with all its products, animal, vegetable and mineral; with men, women, monsters; with the preservation of health; with the air, the winds, and aerial demons, comets and other alarming signs in the heavens, thunder and lightning, the rainbow; with fire, animals that live in it, and fiery demons; the seven planets, the signs of the zodiac, and the crystalline sky. His authorities are classical for the most part, but he has got interesting material from some later writers, mostly Italian.

The other is entitled "Jardin de Flores Curiosas," by Antonio de Torquemada. The earliest edition seems to be that published at Salamanca in 1570, and it was printed several times thereafter. It was turned into French by the well known Gabriel de Chappuys under the title *Hexameron*, at Paris, 1582; into Italian by Celio Malespina, *Giardino di Fiori Curiosi*, Vinegia, 1591; into English under the curious title *The Spanish Mandevile of Miracles*, London, 1600, and again in 1618, and into German, *Hexameron, oder Sechs Tage-Zeiten*, Cassel, 1652.

It gets the name Hexameron from its being the report of a six days' conversation on the wonders and secrets of Nature and the topics discussed are recondite and almost incredible.

On the first day the discourses are appropriately inaugurated by stories about strange and abnormal births, monsters, satyrs, pigmies, giants, amazons, tailed men, dog-headed men, tall men, strong men, mermaids, mermen, long livers, people who have renewed their youth, others who have changed their sex, centaurs and such like.

The second conversation turns upon rivers, fountains and lakes, and the unexpected properties which some of these display, and the discussion naturally includes the four rivers of Paradise and its locality, the flood, and winds up with some account of idolatry and heresies of various sorts.

The third discourse is about visions, enchanters, witchcraft, and contains matter which the author himself calls delightful and necessary to be known. It is, quite.

The fourth conversation is somewhat discursive, treats of good and bad fortune, of the influences of the heavenly bodies and of many singular occurrences.

The fifth conversation is about the lands near and around the North Pole, the variation in the length of the day and night, the inhabitants, the climate, and similar details.

The last discussion is about the animals that live in the Northern regions, wolves, hares which become white when snow begins, trees that remain green all the year and other curiosities.

The whole book is an excellent summary. I hope here be truths.

While these writers are mere compilers of strange stories about Nature, others endeavoured to treat Natural History from a common sense point of view. They could not free themselves altogether from the marvellous, but they tried to keep that element within bounds and to give rational descriptions.

One of them was Joannes Jonstonus, who wrote a review of the subject under these heads : the heavens, the four elements, meteors, fossils (mineral substances), plants, birds, quadrupeds, bloodless animals (insects), fishes, man. The classification is a cross one, lizards, for instance, are put among quadrupeds, but the attempt at a classification at all, based upon observation of characters, represents a marked advance, and indicates that the old kind of Natural History was passing away.

Other books that may be noticed in this connection are Robert Lovel's "Panbotanologia," and "Panzooryctologia," the former a treatise on British plants, the latter on animals and minerals. They contain much information, not very critically sifted, with a view specially to pharmacy and therapeutics. Though of little service in that way now, they are not to be despised in connection with the history of medicine and more particularly of medical folk-lore, a branch of the subject of which the significance and value are now fully recognised by historians in Germany.

Lovel's aggregations may serve as a transition to those on medicine itself, with its related and ancillary sciences. They are in all languages, of all degrees of merit, and treat of different sections of the subject. Leaving out the medical authorities of the Middle Ages, Galen and Avicenna, and the protagonists of the revolution of the sixteenth century, Paracelsus and Erastus, with their respective followers and disciples, attention must be confined to the books written for popular use, many of which went by the name of secrets. They were not systematic but practical; their aim was to cure, not to theorize or expound; and the remedies given were as straightforward and simple as possible. They therefore took the form of receipts, which dictated to the patient, or nurse, the drugs to be used, or the course of treatment to be followed, or the diet most suitable, without troubling him about the origin or nature of the disorder. The author probably thought that the patient had trouble enough with the disease itself, and his drugs, without adding knowledge of the cause of it, always supposing that he knew it himself, which was doubtful. In these times, so far as one can gather, the practice of medicine by licensed persons with such training as could be had, was restricted to populous centres, but in country districts few physicians or surgeons were available, and, from what is stated here and there, when they were, they charged large, if not prohibitive fees. So it came about that the treatment of disease had to be managed by each household for itself. As the men were occupied otherwise, it fell almost inevitably into the hands of the women to attend to the health of the children, as well as their own, and to that of the men folks who might be ill or have met with an accident. The necessary knowledge they had to acquire as best they could by experience and trial and from the skill of their elder neighbours. It was for their use largely that the books of medical receipts were compiled and they were often dedicated to ladies and gentlewomen distinguished by their care of the sick, and especially of the sick poor round about them, who could obtain no medical advice or attention. A survival of these times was the manual of our grandmothers and great-grandmothers, Buchan's "Domestic Medicine."

The ladies and housewives had not only the nursing to do, but they had to prepare the medicines also, and as by tradition these were almost exclusively vegetable, it followed that in every garden were cultivated familiar medicinal herbs, which were supplemented by others that grew naturally in the woods and fields, by the roadsides and ditches.

This gave rise to another branch of medical literature, the pharmaceutical, and one of its earliest forms was the "herbal," of which there were many varieties, and which in recent years have afforded much sport to collectors, and especially to booksellers.

As a consequence, part of the equipment of a house of any size was a "still" room, in which were prepared the spirits, oils, extracts, cordials, perfumes, essences and medicines in common use. The products were more or less secret and were kept in a special room, the "Closet" as it seems to have been called, access to which the lady reserved to herself, perhaps because some of her preparations were *too* active—shall we say?—to fall into either ignorant or *too* skilful hands, or because some of the preparations were scarce, or costly, or of her own invention, and she had no mind to disclose them. They were her "Secrets," in fact.

These developments were not without their influence upon the literature, not merely of medicine, but of connected subjects. As illustrations of the herbal for popular use may be cited that ascribed to Aemilius Macer, though it is not the work of Virgil's predecessor or elder contemporary, but is of a later period. It was translated into English and published by Banckes in 1525, then in 1526, and by several others immediately afterwards. There were Jerome Brunswick's "Vertuose boke of Distyllacyon of the waters of all maner of Herbes," printed by Lawrence Andrewe at London in 1527; the "Treasure" of Evonymus, published in 1552 in Latin, translated into English and published by Day about 1559, with plates of the plants and distilling apparatus. It was one of the most popular of the books and appeared in various editions in Latin and in translations in French, German and Italian. The bibliography which I have investigated is interesting.

After the herbals came books on gardening, and, in at least some of them, attention is bestowed on the cultivation of medicinal plants. There is space for only one example. Thomas Hill, Londoner, of whose life nothing definite is known, not even the dates of his birth and death, in addition to a number of other works already referred to, wrote one entitled "The Profitable Arte of Gardening," 1568. In it he not only lays down rules for planting and tending the flowers, vegetables and herbes, but to each appends a paragraph headed "The physicke helpes," that is, its use in medicine. In "The Gardeners Labyrinth" by Didymus Mountain, commonly identified with Thomas Hill, though the accuracy of so doing may be doubted, there are similar sections in the second part of the book where the individual plants are described. The point to be noticed, however, is that in these and similar books there is an overlapping of the subjects of gardening, pharmacy and medicine.

There is still another connection which displayed itself in a marked way in the manuals written by ladies, or under their name. In them, as in the Countess of Kent's "Manual," or in "The Queen's Closet opened," or in "The Queen-like Closet" of Mrs. Wolley, Physicke and Chyrurgerie came to be associated with Conserving and Preserving, and ultimately with Cookery. Such collections appeared frequently, and their popularity is exhibited by those just mentioned having passed through many editions, as has been exhibited above.

Of more interest, perhaps, than some of these books, may be those about the mysteries of reproduction. Here again we meet with works which have been long current, such as the *Secreta Mulierum* of Albertus, and the *Physionomia* of Michael Scotus. But a book much better known now than either of these passes under the name of *The Complete Masterpiece* and is ascribed to Aristotle. Such books were the instructors of people in all that pertained to the subject, theoretical and practical, and one has heard occasionally views expressed the source of which could be found in them. That is not surprising, for the "Masterpiece" has been "Printed for the Booksellers" and in popular circulation for over two hundred years. In fact, the book along with other tracts of similar tenour can be purchased in the year of grace 1913 in shops which deal in a certain class of literature. One of the other tracts, also ascribed to Aristotle, is known as the "Problemata." It was printed as early as 1471, went through several editions before 1500 and ever so many after. It is a sort of catechism of the secrets of the human body and questions are asked to which no reasonable answer could be given, nor would it matter whether they could be answered or not. A supplement, or sort of commentary upon it, was written by Alexander Aphrodisiensis and another by Zimara, and as affording insight into the most recondite secrets of nature, *secreta secretorum*, all these works must have been much esteemed.

Later writers seem to have adopted its method and copied its matter, for similar productions of equal imbecility appeared in the seventeenth century, such as that by Scipion Dupleix and his translator Robert Bassett.

Here then is the natural history of at least four hundred and fifty years ago—but it must be far more since the compilation was made—current in the year 1913, with hardly a change of word or notion, as if our knowledge of the Universe had been stagnant all that time; and the rest of the book, though it has undergone considerable alteration, is equally unsatisfactory. The ramifications of the bibliography of this book are not without interest, but this is not the time for such divagations.

It is difficult to get away from books of medical secrets and it must suffice to mention those of Falloppio, the discoverer of the organs which bear his name, Valescus of Taranta, Varignana, author of the "Philonium," Fioravanti, a quack it may be, but a man of strong individuality, several of whose works were translated into English by John Hester, Riverius, Zapata, Bairo, Venturini, Auda, and ever so many more. There are two authors, however, that may be noticed in passing.

One is an undiscoverable person called Raoul du Mont Verd, whose name is attached to a little book called "Les Fleurs et Secrets de

Médecine," dating from early in the sixteenth century. Brunet enumerates a number of editions; there are some—not in Brunet—in the British Museum; I have one or two in neither, so we may infer that there were a good many issues, that it must have been in demand, and that, as the surviving copies demonstrate, the book got hard usage.

It gives directions how and when to bleed, an art almost forgotten; how to manage sick people in the months of the year; how to treat various diseases; it describes the virtues of certain plants and how to make waters from them, and gives numerous medical receipts. This is followed by an astrology for shepherds, a treatise on comets and their significance according to the signs of the zodiac in which they appear, a section on the magnitude of the Universe, and the book concludes with a description of hell, deep down in the earth, dark, cloudy, redolent of sulphur and other bad smells and enveloped in nine kinds of fire.

This is a real book of secrets, but its origin is obscure. It is said to be compiled by Master Raoul du Mont Verd; translated from Latin into French, the which book Hippocrates sent to Jalius, who was ill of various maladies external and internal. In another place, however, it is said that the book was sent by "Hippocrates to Galen, . . . and that the contained receipts were afterwards approved by Galen." These statements are hard to reconcile, and one would gladly know how Hippocrates bridged the gap of some five hundred years between himself and Galen.

The other book is also of the sixteeenth century and is "The Myrrour or Glasse of helth," usually ascribed to Thomas Moulton. It is a little black letter octavo and deals especially with the plague, the causes of it, how to avoid it, and how to treat one who is attacked by it. Then follow the influences of the planets in the days of the week, and the signs of the zodiac in the months of the year from March to February, and remedies "for dyvers Infyrmytes and dyseases, that hurteth the bodye of man." It has some analogy therefore to the preceding volume.

The history of this book is curious, for it appeared in two distinct forms and with different title-pages. Examination of the two raises doubts about Thomas Moulton, of whom nothing is recorded, so far as I have ascertained, except that he was a Dominican friar, and was urged to compose this tract. In the other issue part of Moulton's work is omitted and some of the remainder is claimed by one who speaks of having been in medical practice for many years. This subject, however, must be discussed elsewhere, along with the other contemporary medical works quoted above, such as the "Treasure of Pore Men" and the "Treasury of Health."

The books of receipts for arts and manufactures are in great numbers. Some of them are general and contain directions for all kinds of practical purposes, while others take up one topic only but with fuller detail.

Amongst the general collections, there is one of the sixteenth century which is too conspicuous to be passed over. It is that which goes by the name of Alexis, or Alessio, of Piedmont.[1] It presents problems as to its origin and history, it went through quite a number of editions and translations and in its final form contained a great deal of matter. It was the model which subsequent compilers followed and was a reservoir of information which could be drawn on if required, so that in a way the book is one of the classics of the subject. It is a remarkable gathering, for though it is not devoid of many singular notions, disagreeable remedies, superstitious beliefs, all the more does it throw light on the state of knowledge and skill at the time, the demand there was for information and guidance, and the practical methods which must have been current, and of which, it is likely, only samples found a place in the book.

It first appeared in 1555, in Italian, and consisted only of the first part or division in six books. A second division followed, possibly by the

(1) Some years ago I discussed the author and the bibliography of the book in a paper read to the Society of Antiquaries, but as the printing of it was postponed and has never been carried out, I am unable to refer to the lists I then drew up.

same author, and then divisions three and four which, almost certainly, are not by Alexis. They were translated into English at different times and published, the earlier ones separately, and latterly all four together. The English version of the first part appeared in 1558, and is rare. The last collected edition in English is of 1615. The authorship is usually assigned with great confidence to Girolamo Ruscelli, and the name Don Alessio is considered a pseudonym. This view, however much it may seem to be confirmed by Ruscelli himself, Sansovino and Muzio, introduces so much contradiction and confusion that one can arrive at no satisfactory result. It is an obscure subject and would take too much time to discuss now, but this question of authorship does not affect the book and its contents.

The First division contains a variety of receipts. The first book furnishes preparations for cure of diseases and healing of sores; the second how to make fragrant and essential oils, perfumes, powders, soaps; next a few receipts for toilet purposes, beautifying the skin, dyeing the hair, preserving the teeth, etc., etc.; the fifth book shows how to make inks and colours, and the sixth book deals with salts, metals, gilding, precious stones, true and false. The Second division contains such a miscellaneous accumulation of receipts, that it is impossible to classify them. To the Third division are assigned medical, sanitary, and cosmetic secrets, but at the end there is the collection of receipts from the German to which reference is made later on. The Fourth division is occupied entirely with medical receipts and preparations.

As can be easily understood it is difficult either to describe, or to criticize, or to appraise the value of the hundreds of secrets, so called, in such an aggregation as this famous book is. On looking over it one can see that some directions are empirically sound, that drugs employed for certain diseases then, are or could be employed still, and that methods of preparation then, attained the same ends as now, but with an expenditure of time and labour which has been reduced to smaller compass by a fuller understanding both of the materials and of the necessary processes.

A noticeable feature of these books is the constancy with which the receipts both for medicine and for everyday wants are repeated without alteration in one edition after another. Apparently, the secrets gained in authority by repetition, and if they had stood the test of time and had yielded the expected result, there was no reason why they should be changed or superseded.

As instances of receipt books concerned with one topic only, examples may be selected again from the Italian.

One is upon dyeing, and is entitled: "Plictho, or the art of the dyer, how to dye wool, silk, linnen." It was written by Gioanventura Rosetti and published at Venice in 1540. It forms a thin small quarto of forty-four leaves, and contains one or two woodcuts depicting parts of the process. This is a collection of dyers' receipts and methods; it describes the quantity of the substances required, the length of time the fabric is to be immersed in the colour bath and the whole treatment. The art of dyeing is of great antiquity and the use of mordants to make colours fast, for example, must have been known at an early period. Of this book there were at least five editions, the first of 1540 already mentioned, the fourth at Venice in 1611, and the fifth in 1672, with some changes and additions, but even with these, one hundred and thirty years is a good life for such a book. Modern technical books are much more short-lived.

When engaged with dyeing one must not overlook receipts for making inks and colours, for taking out spots and stains and for colouring wood, bone, and feathers. These are chemical applications and as such I have described them in a series of papers still in progress.[1] The earliest collections of these receipts which I have seen are in German and of date 1531, in following years tracts of similar character were associated with them and translations of them were made. I have seen some fifty editions of these books, but from internal evidence it is probable that that number

(1) Proceedings of the Royal Philosophical Society of Glasgow, from 1888 to 1913.

does not include all the editions which exist, but which have not yet come to light. Some of the ink and colour receipts taken from these books long survived the originals, for they are to be found repeated verbatim late in the eighteenth century.

Another subject which was the origin of some literature at a later time was miniature painting, specimens of which are now collected with much zeal. One of the most popular was the work ascribed to C. Ballard, "Traité de Mignature," which consists of receipts about colours and their use. It had a wide circulation and was translated into English and Italian.

Another Italian book is the "Pirotechnia" or "Art of Fire," by Vannuccio Biringuccio, a contemporary of Rosetti, for his book was also published at Venice in 1540, and several times thereafter, one edition appearing in 1678, another instance of a life of over 130 years. It was translated into French and a small part relating to gold and silver was included in Eden's version of Peter Martyr's "Decades of the New World," 1555. This is a comprehensive treatise upon all the arts in which fire, or, as we should say now, high temperatures are necessary. It therefore includes metallurgy, the distillation of sulphur, the making of salts of various kinds, the construction of furnaces, the manufacture of glass, crucibles, and other vessels and apparatus, assaying, casting of cannon, preparation of gunpowder, alloys, metal working, wire-drawing and similar arts. Illustrations of apparatus and processes are included, and altogether it is a book of importance for the history of metallurgy and chemistry.

Along with this may be taken another, even more specialised. In 1612 Antonio Neri published his book "L'Arte Vitraria." He was in the works at Murano and his book embodied his experience. It was translated into Latin, into French by De Blancourt, into English by Christopher Merrett, who added notes, and into German by Kunckel, who included Merrett's notes as well. Illustrations were added, and it became a sort of text-book. An encyclopædic edition in French, under the editorship of Baron d'Holbach, was published at Paris in 1752, and finally the English version was printed privately at the Middlehill press in a ridiculous folio in 1826. There are modern works on glass by Sauzay, Peligot and others.

Among the older writers, Babington, Nye, Hanzelet, Jones, have given receipts for fireworks, and there are two or three on the same subject in Dutch.

On a totally different subject from dyeing, Gioanventura Rosetti, the author of "Plictho" above mentioned, published a little book in 1555 and again in 1560. It is entitled "Notandissimi Secreti de l'arte profumatorio," and in it he gives receipts for preparing oils, waters, pastes, muscardines, and what not, and all for the use of ladies.

This art was a favourite one and it would be easy to cite examples of its literature from the days of Mercurialis, Liébaut, Le Fournier, to Hugh Plat and his "Delights for Ladies," "The French Perfumer," Erresalde (De la Serre), Marinello, Jeamson and his "Artificial Embellishments," and others down to Lola Montez and the present time. It is but a few years since I picked up at the bookstall at Dover Station a new pamphlet on the Secrets of the Toilet for one penny sterling. There have been others since then.

While some of the topics touched upon have been secret enough, the most sombre quarter is occupied by books on magic, and especially black magic. Luridly conspicuous among these is the popular French "Dragon Rouge," and the secrets of Albertus Parvus, Le petit Albert, where one can find how to fabricate the hand of glory which endows its possessor with strange powers. There are the extraordinary revelations contained in the "Magick of Kiranus, King of Persia," and the "Trinum Magicum" of Longinus. The volume on Occult Philosophy by Cornelius Agrippa, which is full of mysteries, is less a treatise on magic than an early attempt to construct a speculative Natural Philosophy, and similar efforts were made by Cardan, Baptista Porta, Athanasius Kircher, Hernando Castrillo, Zimara and others, who styled their subject Magia

Naturalis. Caspar Schott advanced a step farther, when he called his books "Magia Universalis" and "Physica Curiosa"; in these he gets rid to a great extent of the "secret" element, and deals with physical phenomena pure and simple. The subject, thereafter, developed along two lines, one the scientific, the other the secret or occult, ending practically in legerdemain and conjuring. An instance of transition of another sort was displayed in the case of the graduation thesis of Martius, *De Magia Naturali*, printed in 1700, which passed through a number of editions and took the form more or less of a conjuring book. Then it worked gradually into a book of experimental Physics, after which a volume containing an account of all the newest discoveries appeared annually for nineteen years.

Legerdemain, like other displays of manual dexterity, merits respect. Its secret is the cultivation assiduously of certain endowments, physical and mental. Conjuring, involving as it does certain appliances unknown to the spectator, is one of the oldest of arts, for it is said that in the Egyptian temples traces of the arrangements required to produce mysterious effects are still to be found.

Modern conjuring literature, so far as it is known to me, began about 1630, or perhaps a little earlier, and has continued at intervals to the present time. The earlier books, as was said above, bore the name of Thomas Hill,[1] later ones in the eighteenth century that of H. Dean. There was another crop early last century, and John Anderson, the famous "Wizard of the North," published a handbook of Magic about 1850. Quite recently came a set of books called "The Secret out" Series, reviving thereby the old name, and at the present moment hand-books of Parlour Magic can be readily procured, which profess to instruct the aspirant in all the secrets of prestidigitation.

---

(1) Perhaps as one of the earliest specimens ought to be reckoned the translation which Thomas Hill published in 1567, under the name of "Natural and Artificial Conclusions," of the fatuous tricks compiled, it is said, by sundry scholars of the University of Padua, much to their discredit.

The books enumerated above have been for the most part in English, but they form only one section of those which exist. They were produced in abundance in Europe generally, in Latin, French, German, Italian, Dutch and other languages, and translations from one language to another were common enough. The same desire for knowledge seems to have prevailed in other countries as in England, and the same methods were adopted to satisfy it. There was no other source free to people anxious for practical instruction and information, there was no technical education, and only in such receipts and directions as could be obtained from the books of "secrets" could people become acquainted with the properties and uses of different substances.

The books are pretty much of the same character everywhere, though they do not cover exactly the same ground. Thus in English medical secrets predominate, while in other languages the collections are more general. Italy, especially, was prolific in books of the kind and excelled in the issue of small pamphlets of from four or eight to twelve or sixteen pages, with florid titles and a little kernel of secrets of varied content. Of these evanescent productions I have seen about thirty, but that must be a fraction of those which have been in circulation. It is noteworthy that the deterioration in the externals of the volumes is as patent in the books produced on the Continent as here, showing that general causes were at work, while in the matter of long-winded title-pages and crude ornament the books of the eighteenth century were the same everywhere.

In a review of these books during four centuries and a half, successive styles of book production cannot escape notice. Those of the sixteenth century are in black letter and are generally pleasing and sometimes even handsome, though some by Wyer and others are not creditable to the printer. For the majority of those produced in the seventeenth century little that is favourable can be said ; the execution of the books gradually deteriorated until, towards the end of that period, printing had fallen to the lowest ebb. The books, even the best of them, are of inferior quality. The paper is coarse, the printing rude, and the cover is a bit of brown sheepskin, the boards that it is stretched over being left unlined. No attempt was made to turn them out in an attractive shape. They were not written by scholars for scholars, but were meant for common use and were got up as economically as possible. With the exception of Albertus Magnus and Aristotle and may be one or two others, there is not a single author known outside the present connection. Thomas Hill, John Hester, John White, Hannah Wolley, John Shirley, Robert Lovel, Edward Fenton, Owen Wood, A. T., G. W., C. B., who were they? Little or nothing is known about them. Their names are attached in Catalogues to one or more books, and some of them may have found their way into the *Dictionary of National Biography*, but their names have not borne their books into fame, any more than the fame of their books has perpetuated their names.

But whatever the exterior may have been the contents seem to have been sufficient, and the books served their day and generation. How hard the service must have been is eloquently, if not pathetically, proclaimed by the few victims that have survived, and appeal to us for sympathy. When it is remembered that the circulation must have been enormous, that there was a market for every book of the kind written, that of the popular books edition after edition followed at three or four years' interval, and when after years on years of waiting not a single copy of some of these editions ever appears, one can realise the consumption and destruction that must have gone on and can understand how these books are now so scarce. Sometimes one has the luck to light on a volume of Secrets as fresh as if it had just come from the Looking-glass on London Bridge, or from the Ring, or the Pelican in Little-Britain, or from Duck-lane, but that is a phenomenal exception. For one cannot go into the market and purchase whatever book one wants, and in the condition one would prefer. Some might be acquired without much delay, but for others one might have to wait a life-time. So there is nothing for it but to watch, and when the desired object comes along to seize it. It may be imperfect, it may be dirty, it may be in tatters, but it will be time enough to think of that when another copy is found which satisfies our fastidious taste. Besides, another copy may never appear, and that which was secured may be the sole survivor of the edition. Moreover, supposing there be another copy, instead of being in better state than that which was despised and allowed to slip, it may be in worse; the lesson whereof is, let no opportunity pass.

When such rarities occur, a copy in unusually fine condition, a hitherto unknown treatise or edition, it is customary to acclaim it as unique, but that is an epithet which should be taken with a lot of salt, for when once the hunt is up it is surprising how other copies come to light and infringe the monopoly. Still, among the books of Secrets there are a few which are provisionally unique. The pursuit and acquisition of such specimens afford some temporary amusement and gratification to the collector.

Research on this subject seemed at first simple, for all that had to be done was to find the books and describe them. But, as has been shown above and as anyone can prove who tries, the finding of the books involves a tedious waiting on protracted time. There is besides the question of cost, which is one of the most significant changes of these latter days. There has been a steady advance in the price of the books as a whole, and a preposterous rise in the case of some, such as gardening and cookery. A raw curiosity is pursuing the old books, and as their number is small and that of their hunters is growing the value of available copies has been correspondingly enhanced. So it happens now that books which were plentiful enough a generation or two ago when nobody cared for them have disappeared from the market and cannot be got at all.

The description of them too is at first easy, and then as usual up spring hydra-headed difficulties which are not despatched at one stroke. There are doubts about authors, doubts about first and other editions. There are different editions in the same year, not impossible with a speedy sale; there is the same edition in different years, or in different size and form in the same year. There are doubts about readings, doubts about meanings, doubts, grave doubts, about the veracity and trustworthiness of editors, doubts about almost everything except the bare facts of paper, printing and binding. These troubles are inherent in the books themselves and in the motives of the publishers. Their aim was to push their wares. It was not likely that the bibliographical minutiæ which give the books a fictitious, or adventitious, value in our eyes, ever entered the head of—let us say—Nicholas Boddington, or of his chapmen, still less of the goodwives with whom they chaffered, or that they bothered about whether the editions were numbered or dated or not, which would have been so useful to us now.

Thus the problems instead of being direct and obvious are often crooked and refractory and from sheer want of data the solution must be relegated to a better informed time. Perhaps the solution could be found at once if it were only seen that the data are lying unrecognised and unheeded at our hand. "All difficulties are but easy when they are known."

The subject is not finished, but it is time to stop. From these random selections it can be seen that it is a big one, even when limited. In one aspect it is of no importance, and yet it attracts attention. "Secrets" of course are always alluring, and the promise of a revelation of the hidden meaning will draw many to listen. But the secrets these books disclose are all done with long ago, and are not used for their purpose any more. They can, however, serve another purpose. In short, the books are historical documents and give insight into the knowledge and skill of the times when they were composed. They throw light on the practical life, as the literature illuminates the mental life of those days. From them we can acquire notions of the beliefs and practice of the times and can compare what the books tell us with what remains of the products. For the study of the history of arts and manufactures these books are indispensable, for however inaccurate or mistaken from our point of view the authors may have been, at least the substances they employed were the same as ours, and it is possible thereby to check their statements and detect where their knowledge was less extensive than ours and their theories proportionally defective. These are the simultaneous stages of error and amending progress in human comprehension of the outer world and its multitudinous possibilities. Such too exist in our own epoch, and it may be that our successors a couple of hundred years hence, or perhaps in less time, will recognise more clearly than we, who are among the eddies, can do, what a transitory stage it was, if ever they think about it at all. "Let time shape, and there an end."

Printed in Dunstable, United Kingdom

91220128R00291